Student Solutions M

Alison Hyslop
St. John's University

With contributions by **Duane Swank**, *Pacific Lutheran University*

to accompany

CHEMISTRY

The Molecular Nature of Matter

Seventh Edition

Neil D. Jespersen
St. John's University

Alison Hyslop
St. John's University

With significant contributions by

James E. Brady *St. John's University*

WILEY

COVER PHOTO © Isak55 / Shutterstock

Printed and bound by Bind-Rite Graphics, Inc.

Founded in 1807, John Wiley & Sons, Inc. has been a valued source of knowledge and understanding for more than 200 years, helping people around the world meet their needs and fulfill their aspirations. Our company is built on a foundation of principles that include responsibility to the communities we serve and where we live and work. In 2008, we launched a Corporate Citizenship Initiative, a global effort to address the environmental, social, economic, and ethical challenges we face in our business. Among the issues we are addressing are carbon impact, paper specifications and procurement, ethical conduct within our business and among our vendors, and community and charitable support. For more information, please visit our website: www.wiley.com/go/citizenship.

ISBN 978-1-118-70494-3

Printed in the United States of America

SKY10056113_092523

Table of Contents

Chapter Zero
A Very Brief History of Chemistry

Practice Exercises

0.1 This answer is student dependent. For example
Chapter 1: Measurements – Careful observations on the laboratory scale
Chapter 2: Elements – Atomic theory
Chapter 3: The Mole – Careful observations on the laboratory scale
Chapter 4: Reactions – Atomic theory
Chapter 5: Reduction Oxidation Reactions – Atomic theory
Chapter 6: Thermochemistry – Energy changes
Chapter 7: Quantum Chemistry – Atomic theory
Chapter 8: Bonding I – Atomic theory
Chapter 9: Bonding II – Geometric shapes of molecules and Atomic theory
Chapter 10: Gases – Careful observations on the laboratory scale
Chapter 11: Solids and liquids - Careful observations on the laboratory scale
Chapter 12: Solutions – Careful observations on the laboratory scale and Energy changes
Chapter 13: Kinetics – Careful observations on the laboratory scale and Energy changes
Chapter 14: Equilibrium – Careful observations on the laboratory scale and Energy changes
Chapter 15: Acids and Bases I – Careful observations on the laboratory scale and Geometric shapes of molecules
Chapter 16: Acids and Bases II – Careful observations on the laboratory scale and Geometric shapes of molecules
Chapter 17: Solubility – Careful observations on the laboratory scale and Energy changes
Chapter 18: Thermodynamics – Energy changes
Chapter 19: Electrochemistry – Energy changes
Chapter 20: Nuclear chemistry – Atomic theory and Energy changes
Chapter 21: Metal complexes – Atomic theory and Geometric shapes of molecules
Chapter 22: Organic Chemistry – Geometric shapes of molecules and Energy changes

0.2 Nucleosynthesis occurs at temperatures around 1 billion degrees and a high density of nucleons.

0.3 Only light elements were synthesized during the big bang because the temperature was too high for the heavier elements to form.

0.4 Core and enriched layers or nuclei are needed for nucleosynthesis in stars because the enriched layers are the lighter layers, which react to form the heavier nuclei. The heavier nuclei are more dense and move to the center of the star.Experimental evidence for the concept of nucleosynthesis lies in the layers of the stars and that stars have different compositions and that the most abundant elements are the lightest ones.Iron is the heaviest element in stars since the nucleosynthesis of iron absorbs heat and causes the collapse of the red giant and a supernova, thus preventing larger nuclei from forming.Supernovas provide for the synthesis of heavier elements because the extremely high temperature of the supernova makes the atomic collisions strong enough for the nuclei to combine. ${}^{240}_{94}Pu$, 94 electronsThe bottom number is the atomic number, found on the periodic table (number of protons). The top number is the mass number (sum of the number of protons and the number of neutrons). Since it is a neutral atom, it has 94 electrons.

0.9 $^{35}_{17}Cl$ contains 17 protons, 17 electrons, and 18 neutrons.We can discard the subscript 17 on the symbol since the 17 tells the number of protons which is information that the symbol "Cl" also provides. In addition, the number of protons equals the number of electrons in a neutral atom, so the symbol "Cl" also indicates the number of electrons. The 35 is necessary to state which isotope of chlorine is in question and therefore the number of neutrons in the atom.

0.10 2.24845 × 12 u = 26.9814 uCopper is 63.546 u ÷12 u = 5.2955 times as heavy as carbonAny other atom could have been used as the standard for atomic mass, for example oxygen used to be the standard, and one atomic mass unit was 1/16 the mass of oxygen, but that was before they took into account the isotopes of oxygen. Scientists needed to use one isotope of an element.(0.199 × 10.0129 u) + (0.801 × 11.0093 u) = 10.8 u (0.90483 × 19.992 u) + (0.00271 × 20.994 u) + (0.09253 × 21.991 u) = 20.18 u

Review Problems

0.31 Using the ratio of the number of atoms of O and X and the atomic mass of O, we can compare that to the ratio of the masses of O and X to calculate the atomic mass of X:

$$\frac{1.125 \text{ g } X}{1.000 \text{ g O}} = \left(\frac{2 \text{ atoms } X}{3 \text{ atoms O}}\right)\left(\frac{\text{u } X}{15.9994 \text{ u O}}\right)$$

u X = 27.00 u The element is aluminum.

0.33 (a) ^{226}Ra has 88 protons, 88 electrons, and 138 neutrons
(b) ^{206}Pb has 82 protons, 82 electrons, and 124 neutrons
(c) ^{14}C has 6 protons, 6 electrons, and 8 neutrons
(d) ^{23}Na has 11 protons, 11 electrons, and 12 neutrons

0.35 ^{131}I has 53 protons, 53 electrons, and 78 neutrons.

0.37 Since we know that the formula is CH_4, we know that one fourth of the total mass due to the hydrogen atom constitutes the mass that may be compared to the carbon. Hence we have 0.33597 g H ÷ 4 = 0.083993 g H and 1.00 g assigned to the amount of C-12 in the compound. Then it is necessary to realize that the ratio 1.00 g C ÷ 12 for carbon is equal to the ratio 0.083993 g H ÷ x, where x equals the relative atomic mass of hydrogen.

$$\left(\frac{1.000 \text{ g C}}{12 \text{ u C}}\right) = \left(\frac{0.083993 \text{ g H}}{x}\right) = 1.008 \text{ u}$$

0.39 Regardless of the definition, the ratio of the mass of hydrogen to that of carbon would be the same. If C–12 were assigned a mass of 24 (twice its accepted value), then hydrogen would also have a mass twice its current value, or 2.01588 u.

0.41 (0.6917 × 62.9296 u) + (0.3083 × 64.9278 u) = 63.55 u

Chapter One
Scientific Measurements

Practice Exercises

1.1 The scientific method is an iterative process of gathering information through making observations and collecting data and then formulating explanations that lead to a conclusion.

1.2 (a) element (b) mixture, homogeneous (c) compound
(d) mixture, heterogeneous (e) element

1.3 (a) chemical change (b) physical change
(c) physical change (d) physical change

1.4 (a) intensive (b) extensive
(c) intensive (d) extensive

1.5 $V = \frac{4}{3}\pi r^3$, the SI unit for radius, r, is meters, the numbers $\frac{4}{3}$ and π do not have units. Therefore, the SI unit for volume is meter^3 or m^3.

1.6 Force equals mass × acceleration ($F = ma$), and acceleration equals change in velocity divided by change in time ($a = \frac{\text{change in } v}{\text{change in } t}$), and velocity equals distance divided by time ($v = \frac{d}{t}$). Put the equations together:

$$F = m\left(\frac{\text{change in } v}{\text{change in } t}\right)$$

$$F = m\left(\frac{\text{change in } \frac{d}{t}}{\text{change in } t}\right) = m\left(\frac{\text{change in } d}{\text{change in } t^2}\right)$$

The unit for mass is kilogram (kg); the unit for distance is meter (m) and the unit for time is second (s). Substitute the units into the equation above:

Unit for force in SI base units = $\text{kg}\left(\frac{\text{m}}{\text{s}^2}\right)$ or kg m s^{-2}

1.7 $$t_F = \left(\frac{9\ °\text{F}}{5\ °\text{C}}\right)t_C + 32\ °\text{F} = \left(\frac{9\ °\text{F}}{5\ °\text{C}}\right)(355\ °\text{C}) + 32\ °\text{F} = 671\ °\text{F}$$

1.8 $$t_C = (t_F - 32\ °\text{F})\left(\frac{5\ °\text{C}}{9\ °\text{F}}\right) = (55\ °\text{F} - 32\ °\text{F})\left(\frac{5\ °\text{C}}{9\ °\text{F}}\right) = 13\ °\text{C}$$

To convert from °F to K we first convert to °C.

$$t_C = (t_F - 32\ °\text{F})\left(\frac{5\ °\text{C}}{9\ °\text{F}}\right) = (68\ °\text{F} - 32\ °\text{F})\left(\frac{5\ °\text{C}}{9\ °\text{F}}\right) = 20\ °\text{C}$$

$$T_K = (273\ °\text{C} + t_C)\left(\frac{1\ \text{K}}{1\ °\text{C}}\right) = (273\ °\text{C} + 20\ °\text{C})\left(\frac{1\ \text{K}}{1\ °\text{C}}\right) = 293\ \text{K}$$

1.9 (a) 21.0233 g + 21.0 g = 42.0233 g: rounded correctly to 42.0 g
(b) 10.0324 g / 11.7 mL = 0.8574 g / mL: rounded correctly to 0.857 g/mL

(c) $\dfrac{14.25 \text{ cm} \times 12.334 \text{ cm}}{(2.223 \text{ cm} - 1.04 \text{ cm})} = 148.57$ cm: rounded correctly to 149 cm

1.10 (a) 54.183 g – 0.0278 g = 54.155 g

(b) 10.0 g + 1.03 g + 0.243 g = 11.3 g (rounded after adding)

(c) $43.4 \text{ in} \times \left(\dfrac{1 \text{ ft}}{12 \text{ in.}}\right) = 3.62$ ft. (1 and 12 are exact numbers)

(d) $\dfrac{1.03 \text{ m} \times 2.074 \text{ m} \times 2.9 \text{ m}}{12.46 \text{ m} + 4.778 \text{ m}} = 0.36 \text{ m}^2$

1.11 $\text{m}^2 = (124 \text{ ft}^2)\left(\dfrac{30.48 \text{ cm}}{1 \text{ ft}}\right)^2\left(\dfrac{1 \text{ m}}{100 \text{ cm}}\right)^2 = 11.5 \text{ m}^2$

1.12 (a) $\text{in.}^3 = (3.00 \text{ yd}^3)\left(\dfrac{3 \text{ ft}}{1 \text{ yd}}\right)^3\left(\dfrac{12 \text{ in.}}{1 \text{ ft}}\right)^3 = 140{,}000 \text{ in.}^3 = 1.40 \times 10^5 \text{ in}^3$

(b) $\text{cm} = (1.25 \text{ km})\left(\dfrac{1000 \text{ m}}{1 \text{ km}}\right)\left(\dfrac{100 \text{ cm}}{1 \text{ m}}\right) = 1.25 \times 10^5 \text{ cm}$

(c) $\text{g} = (3.27 \text{ oz})\left(\dfrac{28.35 \text{ g}}{1 \text{ oz}}\right) = 92.7 \text{ g}$

(d) $\dfrac{\text{km}}{\text{L}} = \left(\dfrac{20.2 \text{ mile}}{1 \text{ gal}}\right)\left(\dfrac{1.609 \text{ km}}{1 \text{ mile}}\right)\left(\dfrac{1 \text{ gal}}{3.785 \text{ L}}\right) = 8.59 \text{ km/L}$

1.13 $\text{Density} = \dfrac{\text{mass}}{\text{volume}} = \dfrac{0.244 \text{ g}}{15.0 \text{ mL}} = 0.0163 \text{ g/mL} = 0.0163 \text{ g cm}^{-3}$

1.14 $\text{Density} = \dfrac{\text{mass}}{\text{volume}} = \dfrac{\text{mass}}{V_f - V_i} = \dfrac{0.547 \text{ g}}{(5.95 \text{ mL} - 5.70 \text{ mL})} = 2.2 \text{ g mL}^{-1} = 2.2 \text{ g cm}^{-3}$

1.15 $\text{Density} = \dfrac{\text{mass}}{\text{volume}}$

$\text{Density of the object} = \dfrac{365 \text{ g}}{22.12 \text{ cm}^3} = 16.5 \text{ g/cm}^3$

The object is not composed of pure gold since the density of gold is 19.3 g/cm^3.

1.16 The density of the alloy is 12.6 g/cm^3. To determine the mass of the 0.822 ft^3 sample of the alloy, first convert the density from g/cm^3 to lb./ft^3, then find the weight.

$\text{Density in lb./ft}^3 = \dfrac{12.6 \text{ g}}{\text{cm}^3}\left(\dfrac{1 \text{ lb}}{453.6 \text{ g}}\right)\left(\dfrac{30.48 \text{ cm}}{1 \text{ ft}}\right)^3 = 787 \text{ lb./ft}^3$

Mass of sample alloy = (0.822 ft^3) (787 lb./ft^3) = 647 lb.

1.17 $\text{specific gravity} = \dfrac{\text{density of substance}}{\text{density of water}}$

$1.090 = \dfrac{\text{density of wine}}{62.4 \text{ lb ft}^3}$

density of wine = 1.090 × 62.4 lb. ft^3 = 68.02 lb. ft^3

$$\text{mass of wine} = 79 \text{ gallons} \times \text{density of wine} \times \frac{\text{ft}^3}{7.481 \text{ gallons}}$$

$$= 79 \text{ gallons} \times \frac{68.02 \text{ lb}}{1 \text{ ft}^3} \times \frac{\text{ft}^3}{7.481 \text{ gallons}} = 720 \text{ lb.}$$

1.18 $\text{specific gravity} = \frac{\text{density of substance}}{\text{density of water}}$

$$\text{density of water} = \frac{1.00 \text{ g}}{1 \text{ mL}}\left(\frac{1 \text{ oz}}{28.3495 \text{ g}}\right)\left(\frac{29.574 \text{ mL}}{1 \text{ liquid oz}}\right) = 1.043 \text{ oz./liquid oz.}$$

$$\text{specific gravity of urine} = \frac{1.008 \text{ oz/liquid oz}}{1.043 \text{ oz/liquid oz}} = 0.966$$

The specific gravity of urine is below the normal range.

Review Problems

1.39 (a) Physical change. Copper does not change chemically when electricity flows through it: It remains copper.

(b) Physical change. Gallium is changes its state, not its chemical composition when it melts.

(c) Chemical change. This is an example of the Maillard reaction describing the chemical reaction of sugar molecules and amino acids.

(d) Chemical change. Wine contains ethanol which can be converted to acetic acid.

(e) Chemical change. Concrete is composed of many different substances that undergo a chemical process called hydration when water is added to it.

1.41 (a) Hydrogen is a gas at room temperature.

(b) Aluminum is a solid at room temperature.

(c) Nitrogen is a gas at room temperature.

(d) Mercury is a liquid at room temperature.

1.43 (a) 0.01 (b) 1000 (c) 10^{12}

(d) 0.1 (e) 0.001 (f) 0.01

1.45 (a) $t_F = \left(\frac{9 \text{ °F}}{5 \text{ °C}}\right)(t_C) + 32 \text{ °F} = \left(\frac{9 \text{ °F}}{5 \text{ °C}}\right)(57 \text{ °C}) + 32 \text{ °F} = 135 \text{ °F}$ when rounded to the proper number of significant figures.

(b) $t_C = \left(\frac{5 \text{ °C}}{9 \text{ °F}}\right)(t_F - 32 \text{ °F}) = \left(\frac{5 \text{ °C}}{9 \text{ °F}}\right)(-25.5 \text{ °F} - 32 \text{ °F}) = -31.9 \text{ °C}$

(c) $t_C = (T_K - 273 \text{ K})\left(\frac{1 \text{ °C}}{1 \text{ K}}\right) = (378 \text{ K} - 273 \text{ K})\left(\frac{1 \text{ °C}}{1 \text{ K}}\right) = 105 \text{ °C}$

(d) $T_K = (t_C + 273 \text{ °C})\left(\frac{1 \text{ K}}{1 \text{ °C}}\right) = (-31 + 273)\left(\frac{1 \text{ K}}{1 \text{ °C}}\right) = 242 \text{ K}$

1.47 Temperature in °C:

$$t_C = (T_K - 273 \text{ K})\left(\frac{1 \text{ °C}}{1 \text{ K}}\right) = (15.7 \times 10^6 \text{ K} - 273 \text{ K})\left(\frac{1 \text{ °C}}{1 \text{ K}}\right) = 15.7 \times 10^6 \text{ °C}$$

Temperature in °F:

$$t_F = \left(\frac{9\ °F}{5\ °C}\right)(°C) + 32\ °F = \left(\frac{9\ °F}{5\ °C}\right)(15.7 \times 10^6\ °C) + 32\ °F = 2.83 \times 10^7\ °F$$

1.49 $$t_C = (t_F - 32\ °F)\left(\frac{5\ °C}{9\ °F}\right) = (103.5\ °F - 32\ °F)\left(\frac{5\ °C}{9\ °F}\right) = 39.7\ °C$$

This dog has a fever; the temperature above is out of normal canine range.

1.51 9.2 cm, 2 significant figures; 9.15 cm, 3 significant figures

1.53 (a) 4 significant figures (b) 5 significant figures
(c) 4 significant figures (d) 2 significant figures
(e) 2 significant figures

1.55 (a) 0.72 m^2 (b) 84.24 kg
(c) 4.19 g/cm^3 (dividing a number with 4 sig. figs by one with 3 sig. figs)
(d) 19.42 g/mL (e) 858.0 cm^2

1.57 (a) finite number of significant figures
(b) exact number
(c) finite number of significant figures
(d) finite number of significant figures

1.59 (a) $$\text{km/hr} = (32.0\ \text{dm/s})\left(\frac{1\ \text{m}}{10\ \text{dm}}\right)\left(\frac{1\ \text{km}}{1000\ \text{m}}\right)\left(\frac{3600\ \text{s}}{1\ \text{h}}\right) = 11.5\ \text{km/h}$$

(b) $$\mu\text{g/L} = (8.2\ \text{mg/mL})\left(\frac{1\ \text{g}}{1000\ \text{mg}}\right)\left(\frac{1 \times 10^6\ \mu\text{g}}{1\ \text{g}}\right)\left(\frac{1000\ \text{mL}}{1\ \text{L}}\right) = 8.2 \times 10^6\ \mu\text{g/L}$$

(c) $$\text{kg} = (75.3\ \text{mg})\left(\frac{1\ \text{g}}{1000\ \text{mg}}\right)\left(\frac{1\ \text{kg}}{1000\ \text{g}}\right) = 7.53 \times 10^{-5}\ \text{kg}$$

(d) $$\text{L} = (137.5\ \text{mL})\left(\frac{1\ \text{L}}{1000\ \text{mL}}\right) = 0.1375\ \text{L}$$

(e) $$\text{mL} = (0.025\ \text{L})\left(\frac{1000\ \text{mL}}{1\ \text{L}}\right) = 25\ \text{mL}$$

(f) $$\text{dm} = (342\ \text{pm}^2)\left(\frac{1\times10^{-12}\ \text{m}}{1\ \text{pm}}\right)^2\left(\frac{10\ \text{dm}}{1\ \text{m}}\right)^2 = 3.42 \times 10^{-20}\ \text{dm}$$

1.61 (a) $$\text{cm} = (36\ \text{in.})\left(\frac{2.54\ \text{cm}}{1\ \text{in.}}\right) = 91\ \text{cm}$$

(b) $$\text{kg} = (5.0\ \text{lb})\left(\frac{1\ \text{kg}}{2.205\ \text{lb}}\right) = 2.3\ \text{kg}$$

(c) $$\text{mL} = (3.0\ \text{qt})\left(\frac{946.4\ \text{mL}}{1\ \text{qt}}\right) = 2800\ \text{mL}$$

(d) $$\text{mL} = (8\ \text{oz})\left(\frac{29.6\ \text{mL}}{1\ \text{oz}}\right) = 200\ \text{mL}$$

(e) $\text{km/hr} = \left(55\text{ mi/hr}\right)\left(\dfrac{1.609\text{ km}}{1\text{ mi}}\right) = 88\text{ km/hr}$

(f) $\text{km} = \left(50.0\text{ mi}\right)\left(\dfrac{1.609\text{ km}}{1\text{ mi}}\right) = 80.4\text{ km}$

1.63 (a) $\text{cm}^2 = \left(8.4\text{ ft}^2\right)\left(\dfrac{30.48\text{ cm}}{1\text{ ft}}\right)^2 = 7{,}800\text{ cm}^2$

(b) $\text{km}^2 = \left(223\text{ mi}^2\right)\left(\dfrac{1.609\text{ km}}{1\text{ mi}}\right)^2 = 577\text{ km}^2$

(c) $\text{cm}^3 = \left(231\text{ ft}^3\right)\left(\dfrac{30.48\text{ cm}}{1\text{ ft}}\right)^3 = 6.54\times10^6\text{ cm}^3$

1.65 $\text{mL} = \left(4.2\text{ qt}\right)\left(\dfrac{946.35\text{ mL}}{1\text{ qt}}\right) = 4.0\times10^3\text{ mL}$ (stomach volume)

4.0×10^3 mL ÷ 0.9 mL = 4,000 pistachios (don't try this at home)

1.67 $\dfrac{\text{m}}{\text{s}} = \left(\dfrac{200\text{ mi}}{1\text{ hr}}\right)\left(\dfrac{5280\text{ ft}}{1\text{ mi}}\right)\left(\dfrac{30.48\text{ cm}}{1\text{ ft}}\right)\left(\dfrac{1\times10^{-2}\text{m}}{1\text{ cm}}\right)\left(\dfrac{1\text{ hr}}{60\text{ min}}\right)\left(\dfrac{1\text{ min}}{60\text{ s}}\right) = 89.4\ \dfrac{\text{m}}{\text{s}}$

1.69 $\dfrac{\text{mi}}{\text{hr}} = \left(\dfrac{2230\text{ ft}}{1\text{ s}}\right)\left(\dfrac{1\text{ mi}}{5280\text{ ft}}\right)\left(\dfrac{60\text{ s}}{1\text{ min}}\right)\left(\dfrac{60\text{ min}}{1\text{ hr}}\right) = 1520\ \dfrac{\text{mi}}{\text{hr}}$

1.71 $1\text{ light year} = 1\text{ y}\left(\dfrac{365.25\text{ d}}{1\text{ y}}\right)\left(\dfrac{24\text{ h}}{1\text{ d}}\right)\left(\dfrac{3600\text{ s}}{1\text{ h}}\right)\left(\dfrac{3.00\times10^8\text{ m}}{1\text{ s}}\right) = 9.47\times10^{15}\text{ m}$

$\text{miles} = 8.7\text{ light years}\left(\dfrac{9.47\times10^{15}\text{m}}{1\text{ light year}}\right)\left(\dfrac{1\text{ km}}{1000\text{ m}}\right)\left(\dfrac{1\text{ mi}}{1.609\text{ km}}\right) = 5.1\times10^{13}\text{ mi}$

1.73 density = mass/ volume = 36.4 g/45.6 mL = 0.798 g/mL

1.75 $\text{mL} = 25.0\text{ g}\left(\dfrac{1\text{ mL}}{0.791\text{ g}}\right) = 31.6\text{ mL}$

1.77 $\text{g} = 185\text{ mL}\left(\dfrac{1.492\text{ g}}{1\text{ mL}}\right) = 276\text{ g}$

1.79 mass of silver = 62.00 g – 27.35 g = 34.65 g
volume of silver = 18.3 mL –15 mL = 3.3 mL or 3.3 cm^3
density of silver = (mass of silver)/(volume of silver) = (34.65 g)/(3.3 cm^3) = 11 g/cm^3

1.81 $\text{density} = \left(\dfrac{227{,}641\text{ lb}}{385{,}265\text{ gal}}\right) = 0.591\text{ lb gal}^{-1}$

$\text{density} = 0.591\text{ lb. gal}^{-1}\left(\dfrac{453.6\text{ g}}{1\text{ lb}}\right)\left(\dfrac{3785\text{ mL}^{-1}}{1\text{ gal}^{-1}}\right) = 0.0708\text{ g mL}^{-1}$

Chapter Two
Elements, Compounds, and the Periodic Table

Practice Exercises

2.1 (a) SF_6 contains 1 S and 6 F atoms per molecule
(b) $(C_2H_5)_2N_2H_2$ contains 4 C, 12 H, and 2 N per molecule
(c) $Ca_3(PO_4)_2$ contains 3 Ca, 2 P, and 8 O atoms per formula unit
(d) $Co(NO_3)_2{\cdot}6H_2O$ contains 1 Co, 2 N, 12 O, and 12 H per formula unit

2.2 (a) NH_4NO_3 contains 2 nitrogen, 4 hydrogen, 3 oxygen atoms per formula unit
(b) $FeNH_4(SO_4)_2$ contains 1 iron, 1 nitrogen, 4 hydrogen, 2 sulfur, 8 oxygen atoms per formula unit
(c) $Mo(NO_3)_2{\cdot}5H_2O$ contains 1 molybdenum, 2 nitrogen, 11 oxygen, and 10 hydrogen atoms per formula unit
(d) $C_6H_4ClNO_2$ contains 6 carbon, 4 hydrogen, 1 chlorine, 1 nitrogen, and 2 oxygen atoms per molecule

2.3 Reactants: 4 N, 12 H, and 6 O; Products: 4 N, 12 H, and 6 O.

2.4 Reactants: 6 N, 42 H, 2 P, 20 O, 3 Ba, 12 C; Products: 3 Ba, 2 P, 20 O, 6 N, 42 H, 12 C; The reaction is balanced.

2.5 The number of protons is equal to the atomic number of an element. The number of electrons is equal to the atomic number for a neutral atom. If the atom has a positive charge, the number of electrons is determined by subtracting the charge from the atomic number. If the atom has a negative charge the number of electrons is determined by adding the charge to the atomic number.
(a) Fe has 26 protons, and 26 electrons
(b) Fe^{3+} has 26 protons, and 23 electrons
(c) N^{3-} has 7 protons, and 10 electrons
(d) N has 7 protons, and 7 electrons.

2.6 The number of protons is equal to the atomic number of an element. The number of electrons is equal to the atomic number for a neutral atom. If the atom has a positive charge, the number of electrons is determined by subtracting the charge from the atomic number. If the atom has a negative charge the number of electrons is determined by adding the charge to the atomic number.
(a) O has 8 protons, and 8 electrons (b) O^{2-} has 8 protons, and 10 electrons
(c) Al^{3+} had 13 protons, and 10 electrons (d) Al has 13 protons, and 13 electrons

2.7 (a) NaF (b) Na_2O (c) MgF_2 (d) Al_4C_3

2.8 (a) Ca_3N_2 (b) $AlBr_3$ (c) K_2S (d) CsCl

2.9 (a) $CrCl_3$ and $CrCl_2$, Cr_2O_3 and CrO
(b) CuCl, $CuCl_2$, Cu_2O and CuO

2.10 (a) Au_2S and Au_2S_3, Au_3N and AuN
(b) TiS, Ti_2S_3 and TiS_2; Ti_3N_2, TiN and Ti_3N_4

2.11 (a) Na_2CO_3 (b) $(NH_4)_2SO_4$

2.12 (a) $KC_2H_3O_2$ (b) $Sr(NO_3)_2$ (c) $Fe(C_2H_3O_2)_3$

2.13 (a) K_2S (b) $BaBr_2$ (c) NaCN (d) $Al(OH)_3$
(e) Ca_3P_2

2.14 (a) aluminum chloride (b) barium hydroxide
(c) sodium bromide (d) calcium fluoride
(e) potassium phosphide

2.15 lithium sulfide, magnesium phosphide, nickel(II) chloride, titanium(II) chloride, iron(III) oxide

2.16 (a) Al_2S_3 (b) SrF_2 (c) TiO_2 (d) CoO
(e) Au_2O_3

2.17 (a) lithium carbonate (b) potassium permanganate
(c) iron(III) hydroxide

2.18 (a) $KClO_3$ (b) NaOCl (c) $Ni_3(PO_4)_2$

2.19 The term "octa' means eight, therefore there are 8 carbon atoms in octane. The formula for an alkane is C_nH_{2n+2}, so octane has 8 carbons and $((2 \times 8) + 2) = 18$ H. The molecular formula is C_8H_{18} and the condensed structural formula is $CH_3CH_2CH_2CH_2CH_2CH_2CH_2CH_3$.

2.20 (a) Propanol: C_3H_8O, $CH_3CH_2CH_2OH$
(b) Butanol: $C_4H_{10}O$, $CH_3CH_2CH_2CH_2OH$

2.21 (a) phosphorus trichloride (b) sulfur dioxide
(c) dichlorine heptoxide (d) hydrogen sulfide

2.22 (a) $AsCl_5$ (b) SCl_6 (c) S_2Cl_2 (d) H_2Te

2.23 diiodine pentoxide

2.24 chromium(III) acetate

Review Problems

2.67 1 Cr, 6 C, 9 H, 6 O

2.69 $MgSO_4$

2.71 CH_3COOH or $C_2H_4O_2$

2.73 NH_3

2.75 (b)

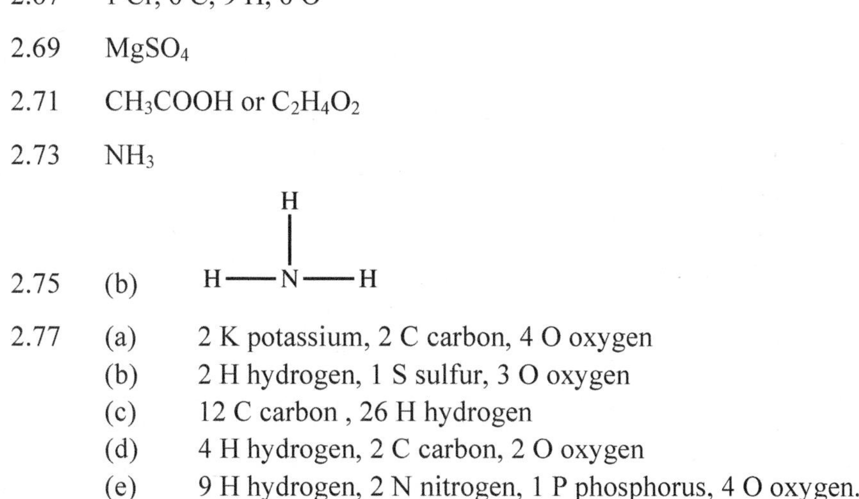

2.77 (a) 2 K potassium, 2 C carbon, 4 O oxygen
(b) 2 H hydrogen, 1 S sulfur, 3 O oxygen
(c) 12 C carbon , 26 H hydrogen
(d) 4 H hydrogen, 2 C carbon, 2 O oxygen
(e) 9 H hydrogen, 2 N nitrogen, 1 P phosphorus, 4 O oxygen.

2.79 (a) 1 Ni nickel, 2 Cl chlorine, 8 O oxygen
(b) 1 C carbon, 1 O oxygen, 2 Cl chlorine
(c) 2 K potassium, 2 Cr chromium, 7 O oxygen
(d) 2 C carbon, 4 H hydrogen, 2 O oxygen
(e) 2 N nitrogen, 9 H hydrogen, 1 P phosphorus, 4 O oxygen.

2.81 (a) 14 C, 28 H, 14 O
(b) 4 N, 8 H, 2 C, 2 O
(c) 15 C, 40 H, 15 O

2.83 (a) 6 Na (b) 3 C (c) 27 O (d) 2 Fe

2.85 No, $S(CH_3)_2$, O_2, SO_2, CO_2, H_2O

2.87 Reaction is not balanced as written.

2.89 (a) K^+ (b) Br^- (c) Mg^{2+}
(d) S^{2-} (e) Al^{3+}

2.91 (a) NaBr (b) KI (c) BaO
(d) $MgBr_2$ (e) Al_2S_3 (f) BaF_2

2.93 (a) $CoCl_2$ and $CoCl_3$
(b) $TiCl_2$, $TiCl_3$ and $TiCl_4$
(c) $MnCl_2$ and $MnCl_3$

2.95 (a) KNO_3 (b) $Ca(C_2H_3O_2)_2$ (c) NH_4Cl
(d) $Fe_2(CO_3)_3$ (e) $Mg_3(PO_4)_2$

2.97 (a) PbO and PbO_2 (b) SnO and SnO_2 (c) MnO and Mn_2O_3
(d) FeO and Fe_2O_3 (e) Cu_2O and CuO

2.99 (a) calcium sulfide (b) aluminum bromide
(c) sodium phosphide (d) barium arsenide
(e) rubidium sulfide

2.101 (a) silicon dioxide (b) xenon tetrafluoride
(c) tetraphosphorus decoxide (d) dichlorine heptoxide

2.103 (a) iron(II) sulfide (b) copper(II) oxide
(c) tin(IV) oxide (d) cobalt(II) chloride hexahydrate

2.105 (a) sodium nitrite (b) potassium permanganate
(c) magnesium sulfate heptahydrate (d) potassium thiocyanate

2.107 (a) ionic chromium(II) chloride
(b) molecular disulfur dichloride
(c) molecular sulfur trioxide
(d) ionic ammonium acetate
(e) ionic potassium iodate
(f) molecular tetraphosphorus hexoxide
(g) ionic calcium sulfite
(h) ionic silver cyanide
(i) ionic zinc(II) bromide

(j) molecular hydrogen selenide

2.109 (a) Na_2HPO_4 (b) Li_2Se (c) $Cr(C_2H_3O_2)_3$ (d) S_2F_{10}
(e) $Ni(CN)_2$ (f) Fe_2O_3 (g) SbF_5

2.111 (a) $(NH_4)_2S$ (b) $Cr_2(SO_4)_3{\cdot}6H_2O$ (c) SiF_4 (d) MoS_2
(e) $SnCl_4$ (f) H_2Se (g) P_4S_7

2.113 (a) Incorrect Hg_2Cl_2
(b) Correct
(c) Correct
(d) Incorrect: $Al_2(CO_3)_3$
(e) Incorrect: $KMnO_4$
(f) Incorrect: CaF_2
(g) Incorrect: Na_2O

2.115 (a) Incorrect: iron(II) sulfide, iron(III) sulfide
(b) Incorrect: sodium chloride
(c) Incorrect: calcium bromide
(d) Correct: $Al_2(CO_3)_3$
(e) Correct: $Ca(OCl)_2$
(f) Incorrect: magnesium phosphate
(g) Incorrect: potassium sulfide

2.117 diselenium hexasulfide and diselenium tetrasulfide

Chapter Three
The Mole and Stoichiometry

Practice Exercise

3.1 mol Al = 3.47 g Al × $\left(\frac{1 \text{ mol Al}}{26.98 \text{ g Al}}\right)$ = 0.129 mol Al

3.2 MM I_2 = 2 × 126.91 g mol^{-1} I_2 = 253.82 g mol^{-1} I_2

mass I_2 = g I_2 = 0.023 mol I_2 × $\left(\frac{253.82 \text{ g } I_2}{1 \text{ mol } I_2}\right)$ = 5.84 g I_2

3.3 Uncertainty in moles = ±0.002 g$\left(\frac{1 \text{ mol } K_2SO_4}{174.25 \text{ g } K_2SO_4}\right)$ = ±1.15 × 10^{-5} mol K_2SO_4

3.4 atoms of Au = 1 oz. Au × $\left(\frac{28.35 \text{ g Au}}{1 \text{ oz Au}}\right) \times \left(\frac{1 \text{ mol Au}}{196.967 \text{ g Au}}\right) \times \left(\frac{6.022 \times 10^{23} \text{ atoms Au}}{1 \text{ mol Au}}\right)$

= 8.67 × 10^{22} atoms Au

3.5 Mass of one molecular of table sugar, $C_{12}H_{22}O_{11}$
MM $C_{12}H_{22}O_{11}$ = 12 × 12.01g/mol + 22 × 1.01 g/mol + 11 × 16.00 g/mol = 342.34 g mol^{-1}
Mass of one molecular of table sugar, $C_{12}H_{22}O_{11}$ = 1 atom $C_{12}H_{22}O_{11}$ ×

$$\left(\frac{1 \text{ mol } C_{12}H_{22}O_{12}}{6.022 \times 10^{23} \text{ atoms } C_{12}H_{22}O_{12}}\right) \times \left(\frac{342.34 \text{ g } C_{12}H_{22}O_{12}}{1 \text{ mol } C_{12}H_{22}O_{12}}\right) = 5.68 \times 10^{-22} \text{ g}$$

3.6 Find the mass of 5.64 × 10^{18} molecules of $C_{18}H_{38}$ (MW = 254.50 g/mol)

$$g = 5.64 \times 10^{18}\left(\frac{1 \text{ mol } C_{18}H_{38}}{6.022 \times 10^{23} \text{ molecules } C_{18}H_{38}}\right)\left(\frac{254.50 \text{ g } C_{18}H_{38}}{1 \text{ mol } C_{18}H_{38}}\right) = 2.38 \times 10^{-3} \text{ g}$$

g = 2.38 × 10^{-3} g = 0.00238 g

The mass is 2 mg, and while a balance can measure 1 mg, the uncertainty is ±1 mg, so the mass cannot be measured accurately.

3.7 Aluminum sulfate: $Al_2(SO_4)_3$, the aluminum is Al^{3+}

$$\text{mole } Al^{3+} = 0.0774 \text{ mol } SO_4^{2-}\left(\frac{2 \text{ mol } Al^{3+}}{3 \text{ mol } SO_4^{2-}}\right) = 0.0516 \text{ mol } Al^{3+}$$

3.8 The formula of dinitrogen pentoxide is N_2O_5

$$\text{mol N} = (8.60 \text{ mol O})\left(\frac{2 \text{ mol N}}{5 \text{ mol O}}\right) = 3.44 \text{ mol N atoms}$$

3.9 The formula for sodium bicarbonate is $NaHCO_3$.

$$\text{mol O} = 1.29 \text{ mol } NaHCO_3\left(\frac{3 \text{ mol O}}{1 \text{ mol } NaHCO_3}\right) = 3.87 \text{ mol O atoms}$$

3.10 $$\text{g Fe} = (25.6 \text{ g O})\left(\frac{1 \text{ mol O}}{16.00 \text{ g O}}\right)\left(\frac{2 \text{ mol Fe}}{3 \text{ mol O}}\right)\left(\frac{55.85 \text{ g Fe}}{1 \text{ mol Fe}}\right) = 59.6 \text{ g Fe}$$

3.11 $$\text{g F} = 0.163 \text{ g Br}\left(\frac{1 \text{ mol Br}}{79.90 \text{ g Br}}\right)\left(\frac{3 \text{ mol F}}{1 \text{ mol Br}}\right)\left(\frac{19.00 \text{ g F}}{1 \text{ mol F}}\right) = 0.116 \text{ g F}$$

3.12 $$\% \text{N} = \frac{0.2012 \text{ g N}}{0.5462 \text{ g sample}} \times 100\% = 36.84\% \text{ N}$$

$$\% \text{O} = \frac{0.3450 \text{ g O}}{0.5462 \text{ g sample}} \times 100\% = 63.16\% \text{ O}$$

Since these two values constitute 100%, there are no other elements present.

3.13 $$\% \text{H} = \left(\frac{\text{mass H}}{\text{total mass}}\right) \times 100\% = \left(\frac{0.0870 \text{ g H}}{0.6672 \text{ g total}}\right) \times 100\% = 13.04\% \text{ H}$$

$$\% \text{C} = \left(\frac{\text{mass C}}{\text{total mass}}\right) \times 100\% = \left(\frac{0.3481 \text{ g C}}{0.6672 \text{ g total}}\right) \times 100\% = 52.17\% \text{ C}$$

It is likely that the compound contains another element since the percentages do not add up to 100%.

3.14 We first determine the number of grams of each element that are present in one mol of N_2O:

2 mol N × 14.01 g/mol = 28.02 g N
1 mol O × 16.00 g/mol = 16.00 g O

The percentages by mass are then obtained using the formula mass of N_2O (44.02 g):

$$\% \text{N} = \frac{28.02 \text{ g N}}{44.02 \text{ g sample}} \times 100\% = 63.65\% \text{ N}$$

$$\% \text{O} = \frac{16.00 \text{ g O}}{44.02 \text{ g sample}} \times 100\% = 36.35\% \text{ O}$$

3.15 NO: Formula mass = 30.01 g/mol
1 mol N × 14.01 g/mol = 14.01 g N % N = (14.01/30.01) × 100% = 46.68% N
1 mol O × 16.00 g/mol = 16.00 g O % O = (16.00/30.01) × 100% = 53.32% O

NO_2: Formula mass = 46.01 g/mol
1 mol N × 14.01 g/mol = 14.01 g N % N = (14.01/46.01) × 100% = 30.45% N
2 mol O × 16.00 g/mol = 32.00 g O % O = (32.00/46.01) × 100% = 69.55% O

N_2O_3: Formula mass = 76.02 g/mol
2 mol N × 14.01 g/mol = 28.02 g N % N = (28.02/76.02) × 100% = 36.86% N
3 mol O × 16.00 g/mol = 48.00 g O % O = (48.00/76.02) × 100% = 63.14% O

N_2O_4: Formula mass = 92.02 g/mol
2 mol N × 14.01 g/mol = 28.02 g N % N = (28.02/92.02) × 100% = 30.45% N
4 mol O × 16.00 g/mol = 64.00 g O % O = (64.00/92.02) × 100% = 69.55% O

The compound N_2O_3 corresponds to the data in Practice Exercise 3.12.

3.16 We first determine the number of mol of each element as follows:

$$\text{mol N} = 1.333 \text{ g N} \times \frac{1 \text{ mol N}}{14.007 \text{ g N}} = 0.095167 \text{ mol N}$$

We need to determine the mass of H in the sample. Since there is a total mass of 1.525 g of compound and 1.333 g of H, the mass of H is:

mass H = 1.525 g sample – 1.333 g N = 0.192 g H

$$\text{mol H} = 0.192 \text{ g H} \times \frac{1 \text{ mol H}}{1.00794 \text{ g H}} = 0.1905 \text{ mol H}$$

Empirical formula: $N_{0.095167}H_{0.190} = N_{\frac{0.095167}{0.095167}}H_{\frac{0.190}{0.095167}} = NH_2$

3.17 First, find the number of moles of each element, then determine the empirical formula by comparing the ratio of the number of moles of each element.
Start with the number of moles of S:

$$\text{mol S} = 0.7625 \text{ g S}\left(\frac{1 \text{ mol S}}{32.066 \text{ g S}}\right) = 0.02378 \text{ mol S}$$

Then find the number of moles of O: since there are only two elements in the compound, S and O, the remaining mass is O

$$\text{g O} = 1.525 \text{ g compound} - 0.7625 \text{ g S} = 0.7625 \text{ g O}$$

$$\text{mol O} = 0.7625 \text{ g O}\left(\frac{1 \text{ mol O}}{15.9994 \text{ g O}}\right) = 0.04766 \text{ mol O}$$

The empirical formula is

$S_{0.02378}O_{0.4766}$

The empirical formula must be in whole numbers, so divide by the smaller subscript:

$S_{\frac{0.02378}{0.02378}}O_{\frac{0.04766}{0.02378}}$ which becomes SO_2.

3.18 $$\text{mol } 0.259 \text{ g N} = 0.259 \text{ g N} \times \frac{1 \text{ mol N}}{14.007 \text{ g N}} = 0.0185 \text{ mol N}$$

$$\text{mol I} = 7.04 \text{ g I} \times \frac{1 \text{ mol I}}{126.9 \text{ g I}} = 0.0555 \text{ mol I}$$

Empirical formula: $N_{0.0185}I_{0.0555} = N_{\frac{0.0185}{0.0185}}I_{\frac{0.0555}{0.0185}} = NI_3$

3.19 $$\text{mol Al} = 5.68 \text{ tons Al}\left(\frac{2000 \text{ lb Al}}{1 \text{ ton Al}}\right)\left(\frac{454 \text{ g Al}}{1 \text{ lb Al}}\right)\left(\frac{1 \text{ mol Al}}{26.98 \text{ g Al}}\right) = 1.91 \times 10^5 \text{ mol Al}$$

$$\text{mol O} = 5.04 \text{ tons O}\left(\frac{2000 \text{ lb O}}{1 \text{ ton O}}\right)\left(\frac{454 \text{ g O}}{1 \text{ lb O}}\right)\left(\frac{1 \text{ mol O}}{16.00 \text{ g O}}\right) = 2.86 \times 10^5 \text{ mol O}$$

Empirical Formula: $Al_{1.91\times10^5}O_{2.86\times10^5}$

In whole numbers: $Al_{\frac{1.91\times10^5}{1.91\times10^5}}O_{\frac{2.86\times10^5}{1.91\times10^5}}$ which is $AlO_{1.5}$ and multiply the subscripts by 2: Al_2O_3

3.20 It is convenient to assume that we have 100 g of the sample, so that the percentage by mass values may be taken directly to represent masses. Thus there is 32.4 g of Na, 22.6 g of S and (100.00 – 32.4 – 22.6) = 45.0 g of O. Now, convert these masses to a number of mol:

$$\text{mol Na} = (32.4 \text{ g Na})\left(\frac{1 \text{ mol Na}}{23.00 \text{ g Na}}\right) = 1.40 \text{ mol Na}$$

$$\text{mol S} = (22.6 \text{ g S})\left(\frac{1 \text{ mol S}}{32.06 \text{ g S}}\right) = 0.705 \text{ mol S}$$

$$\text{mol O} = (45.0 \text{ g O})\left(\frac{1 \text{ mol O}}{16.00 \text{ g O}}\right) = 2.81 \text{ mol O}$$

Next, we divide each of these mole amounts by the smallest in order to deduce the simplest whole number ratio:

For Na: $\frac{1.40 \text{ mol}}{0.705 \text{ mol}} = 1.99$

For S: $\frac{0.705 \text{ mol}}{0.705 \text{ mol}} = 1.00$

For O: $\frac{2.81 \text{ mol}}{0.705 \text{ mol}} = 3.99$

The empirical formula is Na_2SO_4.

3.21 It is convenient to assume that we have 100 g of the sample, so that the percentage by mass values may be taken directly to represent masses. Thus there is 81.79 g of C, 6.10 g of H and (100.00 – 81.79 – 6.10) = 12.11 g of O. Now, convert these masses to a number of mole:

$$\text{mol C} = (81.79 \text{ g C})\left(\frac{1 \text{ mol C}}{12.01 \text{ g C}}\right) = 6.81 \text{ mol C}$$

$$\text{mol H} = (6.10 \text{ g H})\left(\frac{1 \text{ mol H}}{1.008 \text{ g H}}\right) = 6.05 \text{ mol H}$$

$$\text{mol O} = (12.11 \text{ g O})\left(\frac{1 \text{ mol O}}{16.00 \text{ g O}}\right) = 0.757 \text{ mol O}$$

Next, we divide each of these mol amounts by the smallest in order to deduce the simplest whole number ratio:

For C: $\frac{6.81 \text{ mol}}{0.757 \text{ mol}} = 9.00$

For H: $\frac{6.05 \text{ mol}}{0.757 \text{ mol}} = 7.99$

For O: $\frac{0.757 \text{ mol}}{0.757 \text{ mol}} = 1.00$

The empirical formula is C_9H_8O.

3.22 Find the moles of S and C using the stoichiometric ratios, and then find the empirical formula from the ratio of moles of S and C.

Molar mass of $SO_2 = 64.06 \text{ g mol}^{-1}$ Molar mass of $CO_2 = 44.01 \text{ g mol}^{-1}$

$$\text{mol S} = 0.640 \text{ g } SO_2\left(\frac{1 \text{ mol } SO_2}{64.06 \text{ g } SO_2}\right)\left(\frac{1 \text{ mol S}}{1 \text{ mol } SO_2}\right) = 9.99 \times 10^{-3} \text{ mol}$$

$$\text{mol C} = 0.220 \text{ g } CO_2\left(\frac{1 \text{ mol } CO_2}{44.01 \text{ g } CO_2}\right)\left(\frac{1 \text{ mol C}}{1 \text{ mol } CO_2}\right) = 5.00 \times 10^{-3} \text{ mol}$$

Empirical Formula $C_{5.00 \times 10^{-3}}S_{9.99 \times 10^{-3}}$ divide both subscripts by 5.00×10^{-3} to get CS_2.

3.23 Since the entire amount of carbon that was present in the original sample appears among the products only as CO_2, we calculate the amount of carbon in the sample as follows:

$$\text{g C} = (7.406 \text{ g } CO_2)\left(\frac{1 \text{ mol } CO_2}{44.01 \text{ g } CO_2}\right)\left(\frac{1 \text{ mol C}}{1 \text{ mol } CO_2}\right)\left(\frac{12.01 \text{ g C}}{1 \text{ mol C}}\right) = 2.021 \text{ g C}$$

Similarly, the entire mass of hydrogen that was present in the original sample appears among the products only as H_2O. Thus the mass of hydrogen in the sample is:

$$\text{g H} = (3.027 \text{ g } H_2O)\left(\frac{1 \text{ mol } H_2O}{18.02 \text{ g } H_2O}\right)\left(\frac{2 \text{ mol H}}{1 \text{ mol } H_2O}\right)\left(\frac{1.008 \text{ g H}}{1 \text{ mol H}}\right) = 0.3386 \text{ g H}$$

The mass of oxygen in the original sample is determined by difference:

$5.048 \text{ g} - 2.021 \text{ g} - 0.3386 \text{ g} = 2.688 \text{ g O}$

Next, these mass amounts are converted to the corresponding mole amounts:

$$\text{mol C} = (2.021 \text{ g C})\left(\frac{1 \text{ mol C}}{12.01 \text{ g C}}\right) = 0.1683 \text{ mol C}$$

$$\text{mol H} = (0.3386 \text{ g H})\left(\frac{1 \text{ mol H}}{1.008 \text{ g H}}\right) = 0.3359 \text{ mol H}$$

$$\text{mol O} = (2.688 \text{ g O})\left(\frac{1 \text{ mol O}}{16.00 \text{ g O}}\right) = 0.1680 \text{ mol O}$$

The simplest formula is obtained by dividing each of these mole amounts by the smallest:

For C: $\frac{0.1683 \text{ mol}}{0.1680 \text{ mol}} = 1.002$ For H: $\frac{0.3359 \text{ mol}}{0.1680 \text{ mol}} = 1.999$

For O: $\frac{0.1680 \text{ mol}}{0.1680 \text{ mol}} = 1.000$

These values give us the simplest formula directly, namely CH_2O.

3.24 The formula mass of the empirical unit is 1 N + 2 H = 16.03. Since this is half of the molecular mass, the molecular formula is N_2H_4.

$$32.0 \text{ g mol}^{-1} \text{ hydrazine} \times \frac{1 \text{ mol } NH_2}{16.03 \text{ g}} = 2 \text{ mol } NH_2/\text{mol hydrazine}$$

To determine the molecular formula, multiply the subscripts of NH_2 by 2 to obtain N_2H_4.

3.25 To find the molecular formula, divide the molecular mass by the formula mass of the empirical formula, then multiply the subscripts of the empirical formula by that value.

Formula mass of CH_2Cl: 49.48 g mol^{-1}

Formula mass of CHCl: 48.47 g mol^{-1}

For CH_2Cl $\frac{100}{49.48} = 2.02$ and $\frac{289}{49.48} = 5.84$

For CHCl: $\frac{100}{48.47} = 2.06$ and $\frac{289}{48.47} = 5.96$

The CH_2Cl rounds better using the molecular mass of 100, therefore multiply the subscripts by 2 and the formula is $C_2H_4Cl_2$.

For CHCl, the molecular mass of 289 gives a multiple of 6, and therefore the formula is $C_6H_6Cl_6$.

3.26 $3Ba(NO_3)_2(aq) + 2(NH_4)_3PO_4(aq) \longrightarrow Ba_3(PO_4)_2(s) + 6NH_4NO_3(aq)$

3.27 $CaCl_2(aq) + K_2CO_3(aq) \longrightarrow CaCO_3(s) + 2KCl(aq)$

3.28 $\text{mol } O_2 = 6.76 \text{ mol } SO_3 \times \frac{1 \text{ mol } O_2}{2 \text{ mol } SO_3} = 3.38 \text{ mol } O_2$

3.29 $\text{mol } H_2SO_4 = 0.366 \text{ mol NaOH} \times \frac{1 \text{ mol } H_2SO_4}{2 \text{ mol NaOH}} = 0.183 \text{ mol } H_2SO_4$

3.30 $Fe_2O_3(s) + 2Al(s) \longrightarrow 2Fe(l) + Al_2O_3(s)$

$$\text{g } Al_2O_3 = (86.0 \text{ g Fe})\left(\frac{1 \text{ mol Fe}}{55.85 \text{ g Fe}}\right)\left(\frac{1 \text{ mol } Al_2O_3}{2 \text{ mol Fe}}\right)\left(\frac{102.0 \text{ g } Al_2O_3}{1 \text{ mol } Al_2O_3 \text{ mol } Al_2O_3}\right) = 78.5 \text{ g } Al_2O_3$$

3.31 $g\ CO_2 = (1.50 \times 10^2\ g\ CaO)\left(\frac{1\ mol\ CaO}{56.08\ g\ CaO}\right)\left(\frac{1\ mol\ CO_2}{1\ mol\ CaO}\right)\left(\frac{44.01\ g\ CO_2}{1\ mol\ CO_2}\right) = 1.18 \times 10^2\ g\ CO_2$

3.32 First determine the number of grams of $CaCO_3$ that would be required to react completely with the given amount of HCl:

$$g\ CaCO_3 = (125\ g\ HCl)\left(\frac{1\ mol\ HCl}{36.461\ g\ HCl}\right)\left(\frac{1\ mol\ CaCO_3}{2\ mol\ HCl}\right)\left(\frac{100.088\ g\ CaCO_3}{1\ mol\ CaO_3}\right)$$

$$= 171.57\ g\ CaCO_3$$

Since this is more than the amount that is available, we conclude that $CaCO_3$ is the limiting reactant. The rest of the calculation is therefore based on the available amount of $CaCO_3$:

$$g\ CO_2 = (125\ g\ CaCO_3)\left(\frac{1\ mol\ CaCO_3}{100.088\ g\ CaCO_3}\right)\left(\frac{1\ mol\ CO_2}{1\ mol\ CaCO_3}\right)\left(\frac{44.01\ g\ CO_2}{1\ mol\ CO_2}\right) = 55.0\ g\ CO_2$$

For the number of grams of left over HCl, the excess reagent, find the amount of HCl used and then subtract that from the amount of HCl started with, 125 g.

$$g\ HCl\ used = (125\ g\ CaCO_3)\left(\frac{1\ mol\ CaCO_3}{100.088\ g\ CaCO_3}\right)\left(\frac{2\ mol\ HCl}{1\ mol\ CaCO_3}\right)\left(\frac{36.461\ g\ HCl}{1\ mol\ HCl}\right)$$

$$= 91.1\ g\ HCl$$

g HCl remaining = 125 g – 91.1 g = 34 g HCl remaining

3.33 First determine the number of grams of O_2 that would be required to react completely with the given amount of ammonia:

$$g\ O_2 = (30.00\ g\ NH_3)\left(\frac{1\ mol\ NH_3}{17.03\ g\ NH_3}\right)\left(\frac{5\ mol\ O_2}{4\ mol\ NH_3}\right)\left(\frac{32.00\ g\ O_2}{1\ mol\ O_2}\right)$$

$$= 70.46\ g\ O_2$$

Since this is more than the amount that is available, we conclude that oxygen is the limiting reactant. The rest of the calculation is therefore based on the available amount of oxygen:

$$g\ NO = (40.00\ g\ O_2)\left(\frac{1\ mol\ O_2}{32.00\ g\ O_2}\right)\left(\frac{4\ mol\ NO}{5\ mol\ O_2}\right)\left(\frac{30.01\ g\ NO}{1\ mol\ NO}\right)$$

$$= 30.01\ g\ NO$$

3.34 First determine the number of grams of salicylic acid, $HOOCC_6H_4OH$ that would be required to react completely with the given amount of acetic anhydride, $C_4H_6O_3$:

$$g\ HOOCC_6H_4OH = (15.6\ g\ C_4H_6O_3) \times$$

$$\left(\frac{1\ mol\ C_4H_6O_3}{102.09\ g\ C_4H_6O_3}\right)\left(\frac{2\ mol\ HOOCC_6H_4OH}{1\ mol\ C_4H_6O_3}\right)\left(\frac{138.12\ g\ HOOCC_6H_4OH}{1\ mol\ HOOCC_6H_4OH}\right)$$

$$= 42.2\ g\ HOOCC_6H_4OH$$

Since more salicylic acid is required than is available, it is the limiting reagent. Once 28.2 g of salicylic acid is reacted the reaction will stop, even though there are 15.6 g of acetic anhydride present. Therefore the salicylic acid is the limiting reactant. The theoretical yield of aspirin $HOOCC_6H_4O_2C_2H_3$ is therefore based on the amount of salicylic acid added. This is calculated:

$$g\ HOOCC_6H_4O_2C_2H_3 = (28.2\ g\ HOOCC_6H_4OH) \times$$

$$\left(\frac{1\ mol\ HOOCC_6H_4OH}{138.12\ g\ HOOCC_6H_4OH}\right)\left(\frac{2\ mol\ HOOCC_6H_4O_2C_2H_3}{2\ mol\ HOOCC_6H_4OH}\right)\left(\frac{180.16\ g\ HOOCC_6H_4O_2C_2H_3}{1\ mol\ HOOCC_6H_4O_2C_2H_3}\right)$$

$$= 36.78\ g\ HOOCC_6H_4O_2C_2H_3$$

Now the percentage yield can be calculated from the amount of acetyl salicylic acid actually produced, 30.7 g:

$$\text{percentage yield} = \left(\frac{\text{actual yield}}{\text{theoretical yield}}\right)\times 100\% = \left(\frac{30.7\text{ g HOOCC}_6\text{H}_4\text{O}_2\text{C}_2\text{H}_3}{36.78\text{ g HOOCC}_6\text{H}_4\text{O}_2\text{C}_2\text{H}_3}\right)\times 100\% = 83.5\%$$

3.35 First determine the number of grams of C_2H_5OH that would be required to react completely with the given amount of sodium dichromate:

$$\text{g C}_2\text{H}_5\text{OH} = (90.0\text{ g Na}_2\text{Cr}_2\text{O}_7)\left(\frac{1\text{ mol Na}_2\text{Cr}_2\text{O}_7}{262.0\text{ g Na}_2\text{Cr}_2\text{O}_7}\right)\left(\frac{3\text{ mol C}_2\text{H}_5\text{OH}}{2\text{ mol Na}_2\text{Cr}_2\text{O}_7}\right)\left(\frac{46.08\text{ g C}_2\text{H}_5\text{OH}}{1\text{ mol C}_2\text{H}_5\text{OH}}\right)$$

$$= 23.7\text{ g C}_2\text{H}_5\text{OH}$$

Once this amount of C_2H_5OH is reacted the reaction will stop, even though there are 24.0 g C_2H_5OH present, because the $Na_2Cr_2O_7$ will be used up. Therefore $Na_2Cr_2O_7$ is the limiting reactant. The theoretical yield of acetic acid ($HC_2H_3O_2$) is therefore based on the amount of $Na_2Cr_2O_7$ added. This is calculated below:

$$\text{g HC}_2\text{H}_3\text{O}_2 = (90.0\text{ g Na}_2\text{Cr}_2\text{O}_7)\left(\frac{1\text{ mol Na}_2\text{Cr}_2\text{O}_7}{262.0\text{ g Na}_2\text{Cr}_2\text{O}_7}\right)\left(\frac{3\text{ mol HC}_2\text{H}_3\text{O}_2}{2\text{ mol Na}_2\text{Cr}_2\text{O}_7}\right)\left(\frac{60.06\text{ g HC}_2\text{H}_3\text{O}_2}{1\text{ mol HC}_2\text{H}_3\text{O}_2}\right)$$

$$= 30.9\text{ g HC}_2\text{H}_3\text{O}_2$$

Now the percentage yield is calculated from the amount of acetic acid actually produced, 26.6 g:

$$\text{percent yield} = \left(\frac{\text{actual yield}}{\text{theoretical yield}}\right)\times 100\% = \left(\frac{26.6\text{ g HC}_2\text{H}_3\text{O}_2}{30.9\text{ g HC}_2\text{H}_3\text{O}_2}\right)\times 100\% = 86.1\%$$

3.36 Three step synthesis overall yield = $(0.872 \times 0.911 \times 0.863) \times 100\% = 68.6\%$
Two step synthesis overall yield = $(0.855 \times 0.843) \times 100\% = 72.1\%$
Therefore, the two-step process is the preferred process.

Review Problems

3.31 (a) $$\text{g Fe} = (1.35\text{ mol Fe})\left(\frac{55.85\text{ g Fe}}{1\text{ mole Fe}}\right) = 75.4\text{ g Fe}$$

(b) $$\text{g O} = (24.5\text{ mol O})\left(\frac{16.0\text{ g O}}{1\text{ mole O}}\right) = 392\text{ g O}$$

(c) $$\text{g Ca} = (0.876)\left(\frac{40.08\text{ g Ca}}{1\text{ mole Ca}}\right) = 35.1\text{ g Ca}$$

3.33 (a) $NaHCO_3$ = 1Na + 1H + 1C + 3O
= (22.98977) + (1.00794) + (12.0107) + (3 × 15.9994)
= 84.00661 g/mole = 84.0066 g/mol

(b) $(NH_4)_2CO_3$ = 2N + 8H + C + 3O
= (2 × 14.0067) + (8 × 1.00794) + (12.0107) + (3 × 15.9994)
= 96.08582 g/mole = 96.0858 g/mol

(c) $CuSO_4 \cdot 5H_2O$ = 1Cu + 1S +9O + 10H
= 63.546 + 32.065 + (9 × 15.9994) + (10 × 1.00794)
= 249.685 g/mole

(d) Potassium dichromate: $K_2Cr_2O_7 = 2K + 2Cr + 7O$
$= (2 \times 39.0983) + (2 \times 51.9961) + (7 \times 15.9994)$
$= 294.1846$ g/mole

(e) Aluminum sulfate: $Al_2(SO_4)_3 = 2Al + 3S + 12O$
$= (2 \times 26.98154) + (3 \times 32.065) + (12 \times 15.9994)$
$= 342.15088$ g/mole $= 342.151$ g/mol

3.35 (a) $$\text{g } Ca_3(PO_4)_2 = (1.25 \text{ mol } Ca_3(PO_4)_2)\left(\frac{310.18 \text{ g } Ca_3(PO_4)_2}{1 \text{ mol } Ca_3(PO_4)_2}\right) = 388 \text{ g } Ca_3(PO_4)_2$$

(b) $$\text{g } C_4H_{10} = (0.600\ \mu\text{mol } C_4H_{10})\left(\frac{1 \text{ mol } C_4H_{10}}{10^6\ \mu\text{mol } C_4H_{10}}\right)\left(\frac{58.12 \text{ g } C_4H_{10}}{1 \text{ mol } C_4H_{10}}\right) = 3.49 \times 10^{-5} \text{ g } C_4H_{10}$$

(c) $$\text{mg } Fe(NO_3)_3 = (0.625 \text{ mmol } Fe(NO_3)_3)\left(\frac{241.86 \text{ mg } Fe(NO_3)_3}{1 \text{ mmol } Fe(NO_3)_3}\right) = 151 \text{ mg } Fe(NO_3)_3$$

$= 0.151$ g $Fe(NO_3)_3$

(d) $$\text{g } (NH_4)_2CO_3 = (1.45 \text{ mol } (NH_4)_2CO_3)\left(\frac{96.09 \text{ g } (NH_4)_2CO_3}{1 \text{ mol } (NH_4)_2CO_3}\right) = 139 \text{ g } (NH_4)_2CO_3$$

3.37 (a) $$\text{mol } NH_3 = 1.56 \text{ ng } NH_3\left(\frac{1 \text{ g } NH_3}{10^9 \text{ ng } NH_3}\right)\left(\frac{1 \text{ mole } NH_3}{17.03 \text{ g } NH_3}\right) = 9.16 \times 10^{-11} \text{ mol } NH_3$$

(b) $$\text{mol } Na_2CrO_4 = 6.98\ \mu\text{g } Na_2CrO_4\left(\frac{1 \text{ g } Na_2CrO_4}{10^6\ \mu\text{g } Na_2CrO_4}\right)\left(\frac{1 \text{ mole } Na_2CrO_4}{162.0 \text{ g } Na_2CrO_4}\right)$$

$= 4.31 \times 10^{-8}$ mol Na_2CrO_4

(c) $$\text{mol } CaCO_3 = (21.5 \text{ g } CaCO_3)\left(\frac{1 \text{ mol } CaCO_3}{100.09 \text{ g } CaCO_3}\right) = 0.215 \text{ mol } CaCO_3$$

(d) $$\text{mol } Sr(NO_3)_2 = 16.8 \text{ g } Sr(NO_3)_2 \times \left(\frac{1 \text{ mole } Sr(NO_3)_2}{211.6 \text{ g } Sr(NO_3)_2}\right) = 7.94 \times 10^{-2} \text{ mol } Sr(NO_3)_2$$

3.39 $$\text{mol Ni} = 17.7 \text{ g Ni}\left(\frac{1 \text{ mol Ni}}{58.69 \text{ g Ni}}\right) = 0.302 \text{ mol Ni}$$

3.41 $$1.56 \times 10^{21} \text{ atoms Ta}\left(\frac{1 \text{ mol Ta}}{6.022 \times 10^{23} \text{ atoms Ta}}\right) = 2.59 \times 10^{-3} \text{ mole Ta}$$

3.43 $$\text{g K} = 2.00 \times 10^{12} \text{ atoms K}\left(\frac{1 \text{ mol K}}{6.022 \times 10^{23} \text{ atoms K}}\right)\left(\frac{39.10 \text{ g K}}{1 \text{ mol K}}\right) = 1.30 \times 10^{-10} \text{ g K}$$

3.45 $$\text{mol C-12} = 6 \text{ g} \times \left(\frac{1 \text{ mol C-12}}{12.00 \text{ g C-12}}\right) = 0.5 \text{ mol C-12}$$

$$\text{Number of atoms C-12} = 0.5 \text{ mol}\left(\frac{6.022 \times 10^{23} \text{ atoms C-12}}{1 \text{ mol C-12}}\right) = 3.01 \times 10^{23} \text{ atoms C-12}$$

3.47 (a) 6 atom C:11 atom H
(b) 12 mole C:11 mole O
(c) 2 atom H: 1 atom O
(d) 2 mole H: 1 mole O

5.49 $\text{mol Bi} = (1.58 \text{ mol O})\left(\frac{2 \text{ mol Bi}}{3 \text{ mol O}}\right) = 1.05 \text{ mol Bi}$

3.51 $\text{mol Cr} = (2.16 \text{ mol } Cr_2O_3)\left(\frac{2 \text{ mol Cr}}{1 \text{ mol } Cr_2O_3}\right) = 4.32 \text{ mol Cr}$

3.53 (a) $\left(\frac{2 \text{ mol Al}}{3 \text{ mol S}}\right)$ or $\left(\frac{3 \text{ mol S}}{2 \text{ mol Al}}\right)$

(b) $\left(\frac{3 \text{ mol S}}{1 \text{ mol } Al_2(SO_4)_3}\right)$ or $\left(\frac{1 \text{ mol } Al_2(SO_4)_3}{3 \text{ mol S}}\right)$

(c) $\text{mol Al} = (0.900 \text{ mol S})\left(\frac{2 \text{ mol Al}}{3 \text{ mol S}}\right) = 0.600 \text{ mol Al}$

(d) $\text{mol S} = (1.16 \text{ mol } Al_2(SO_4)_3)\left(\frac{3 \text{ mol S}}{1 \text{ mol } Al_2(SO_4)_3}\right) = 3.48 \text{ mol S}$

3.55 Based on the balanced equation:
$2\ NH_3(g) \longrightarrow N_2(g) + 3H_2(g)$
From this equation the conversion factors can be written:
$\left(\frac{1 \text{ mol } N_2}{2 \text{ mol } NH_3}\right)$ and $\left(\frac{3 \text{ mol } H_2}{2 \text{ mol } NH_3}\right)$
To determine the moles produced, simply convert from starting moles to end moles:
$\text{mol } N_2 = 0.145 \text{ mol } NH_3\left(\frac{1 \text{ mol } N_2}{2 \text{ mol } NH_3}\right) = 0.0725 \text{ mol } N_2$
The moles of hydrogen are calculated similarly:
$\text{mol } H_2 = 0.145 \text{ mol } NH_3\left(\frac{3 \text{ mol } H_2}{2 \text{ mol } NH_3}\right) = 0.218 \text{ mol } H_2$

3.57 $\text{mol } UF_6 = 1.25 \text{ mol } CF_4\left(\frac{4 \text{ mol F}}{1 \text{ mol } CF_4}\right)\left(\frac{1 \text{ mol } UF_6}{6 \text{ mol F}}\right) = 0.833 \text{ mol } UF_6$

3.59 $\text{atoms C} = 4.13 \text{ mol H}\left(\frac{1 \text{ mol } C_3H_8}{8 \text{ mol H}}\right)\left(\frac{6.022 \times 10^{23} \text{ molecules } C_3H_8}{1 \text{ mol } C_3H_8}\right) \times$

$$\left(\frac{3 \text{ atoms C}}{1 \text{ molecule } C_3H_8}\right) = 9.33 \times 10^{23} \text{ atoms C}$$

3.61 Number of C, H and O atoms in glucose = 6 atoms C + 12 atoms H + 6 atoms O = 24 atoms
Number of atoms = (0.260 mol glucose) ×

$$\left(\frac{6.022 \times 10^{23} \text{ molecules glucose}}{1 \text{ mol glucose}}\right)\left(\frac{24 \text{ atoms}}{1 \text{ molecule glucose}}\right) = 3.76 \times 10^{24} \text{ atoms}$$

3.63 The formula CaC_2 indicates that there is 1 mole of Ca for every 2 moles of C. Therefore, if there are 0.150 moles of C there must be 0.0750 moles of Ca.

$$\text{g Ca} = (0.075 \text{ mol Ca})\left(\frac{40.078 \text{ g Ca}}{1 \text{ mol Ca}}\right) = 3.01 \text{ g Ca}$$

3.65 $$\text{mol N} = (0.650 \text{ mol } (NH_4)_2CO_3)\left(\frac{2 \text{ moles N}}{1 \text{ mole } (NH_4)_2CO_3}\right) = 1.30 \text{ mol N}$$

$$\text{g } (NH_4)_2CO_3 = (0.650 \text{ mol } (NH_4)_2CO_3)\left(\frac{96.09 \text{ g } (NH_4)_2CO_3}{1 \text{ mole } (NH_4)_2CO_3}\right) = 62.5 \text{ g } (NH_4)_2CO_3$$

3.67 $$\text{kg fertilizer} = (1 \text{ kg N})\left(\frac{1000 \text{ g N}}{1 \text{ kg N}}\right)\left(\frac{1 \text{ mol N}}{14.01 \text{ g N}}\right)\left(\frac{1 \text{ mol } (NH_4)_2CO_3}{2 \text{ mol N}}\right)$$

$$\times\left(\frac{96.09 \text{ g } (NH_4)_2CO_3}{1 \text{ mol } (NH_4)_2CO_3}\right)\left(\frac{1 \text{ kg } (NH_4)_2CO_3}{1000 \text{ g } (NH_4)_2CO_3}\right) = 3.43 \text{ kg fertilizer}$$

3.69 Assume one mole total for each of the following.

(a) The molar mass of $NH_4H_2PO_4$ is 115.05 g/mol.

$$\%\text{ N} = \frac{14.0 \text{ g N}}{115.05 \text{ g } NH_4H_2PO_4} \times 100\% = 12.2\%$$

$$\%\text{ H} = \frac{6.05 \text{ g H}}{115.05 \text{ g } NH_4H_2PO_4} \times 100\% = 5.26\%$$

$$\%\text{ P} = \frac{31.0 \text{ g P}}{115.05 \text{ g } NH_4H_2PO_4} \times 100\% = 26.9\%$$

$$\%\text{ O} = \frac{64.0 \text{ g O}}{115.05 \text{ g } NH_4H_2PO_4} \times 100\ \% = 55.6\ \%$$

(b) The molar mass of $(CH_3)_2CO$ is 58.08 g/mol

$$\%\text{ C} = \frac{36.0 \text{ g C}}{58.08 \text{ g } (CH_3)_2CO} \times 100\% = 62.0\%$$

$$\%\text{ H} = \frac{6.05 \text{ g H}}{58.08 \text{ g } (CH_3)_2CO} \times 100\% = 10.4\%$$

$$\%\text{ O} = \frac{16.0 \text{ g O}}{58.08 \text{ g } (CH_3)_2CO} \times 100\% = 27.6\%$$

(c) The molar mass of NaH_2PO_4 is 119.98 g/mol.

$$\%\text{ Na} = \frac{23.0 \text{ g Na}}{119.98 \text{ g } NaH_2PO_4} \times 100\% = 19.2\%$$

$$\%\text{ H} = \frac{2.02 \text{ g H}}{119.98 \text{ g } NaH_2PO_4} \times 100\% = 1.68\%$$

$$\%\text{ P} = \frac{31.0 \text{ g P}}{119.98 \text{ g } NaH_2PO_4} \times 100\% = 25.8\%$$

$$\%\text{ O} = \frac{64.0 \text{ g O}}{119.98 \text{ g } NaH_2PO_4} \times 100\ \% = 53.3\ \%$$

(d) The molar mass of calcium sulfate dihydrate is 172.2 g/mol.

$$\% \text{ Ca} = \frac{40.1 \text{ g Ca}}{172.2 \text{ g CaSO}_4 \bullet 2\text{H}_2\text{O}} \times 100\% = 23.3\%$$

$$\% \text{ S} = \frac{32.1 \text{ g S}}{172.2 \text{ g CaSO}_4 \bullet 2\text{H}_2\text{O}} \times 100\% = 18.6\%$$

$$\% \text{ O} = \frac{96.0 \text{ g O}}{172.2 \text{ g CaSO}_4 \bullet 2\text{H}_2\text{O}} \times 100\% = 55.7\%$$

$$\% \text{ H} = \frac{4.03 \text{ g H}}{172.2 \text{ g CaSO}_4 \bullet 2\text{H}_2\text{O}} \times 100 \% = 2.34 \%$$

3.71 $\% \text{ O in morphine} = \dfrac{48.00 \text{ g O}}{285.36 \text{ g C}_{17}\text{H}_{19}\text{NO}_3} \times 100\% = 16.82\% \text{ O}$

$$\% \text{ O in heroin} = \frac{80.00 \text{ g O}}{369.44 \text{ g C}_{21}\text{H}_{23}\text{NO}_5} \times 100\% = 21.65\% \text{ O}$$

Therefore heroin has a higher percentage oxygen.

3.73 $\% \text{ Cl in Freon-12} = \dfrac{70.90 \text{ g Cl}}{120.92 \text{ g CCl}_2\text{F}_2} \times 100\% = 58.63\% \text{ Cl}$

$$\% \text{ Cl in Freon 141b} = \frac{70.9 \text{ g Cl}}{116.95 \text{ g C}_2\text{H}_3\text{Cl}_2\text{F}} \times 100\% = 60.62\% \text{ Cl}$$

Therefore Freon 141b has a higher percentage chlorine.

3.75 $\% \text{ P} = \dfrac{0.539 \text{ g P}}{2.35 \text{ g compound}} \times 100\% = 22.9\%$

$$\% \text{ Cl} = 100\% - 22.9\% = 77.1\%$$

3.77 For $C_{17}H_{25}N$, the molar mass (17C + 25H + 1N) equals 243.43 g/mole, and the three theoretical values for % by weight are calculated as follows:

$$\% \text{ C} = \frac{204.2 \text{ g C}}{243.4 \text{ g C}_{17}\text{H}_{25}\text{N}} \times 100\% = 83.89\%$$

$$\% \text{ H} = \frac{25.20 \text{ g H}}{243.4 \text{ g C}_{17}\text{H}_{25}\text{N}} \times 100\% = 10.35\%$$

$$\% \text{ N} = \frac{14.01 \text{ g N}}{243.4 \text{ g C}_{17}\text{H}_{25}\text{N}} \times 100\% = 5.76\%$$

These data are consistent with the experimental values cited in the problem.

3.79 $\text{g O} = 7.14 \times 10^{21} \text{ atoms N}\left(\dfrac{1 \text{ mol N}}{6.022 \times 10^{23} \text{ atoms N}}\right)\left(\dfrac{5 \text{ mol O}}{2 \text{ mol N}}\right)\left(\dfrac{16.00 \text{ g O}}{1 \text{ mol O}}\right) = 0.474 \text{ g O}$

3.81 The molecular formula is some integer multiple of the empirical formula. This means that we can divide the molecular formula by the largest possible whole number that gives an integer ratio among the atoms in the empirical formula.

(a) SCl (b) CH_2O (c) NH_3 (d) AsO_3
(e) HO

3.83 We begin by realizing that the mass of oxygen in the compound may be determined by difference:

0.896 g total – (0.111 g Na + 0.477 g Tc) = 0.308 g O

Next we can convert each mass of an element into the corresponding number of moles of that element as follows:

$$\text{mol Na} = (0.111 \text{ g Na})\left(\frac{1 \text{ mol Na}}{23.00 \text{ g Na}}\right) = 4.83 \times 10^{-3} \text{ mol Na}$$

$$\text{mol Tc} = (0.477 \text{ g Tc})\left(\frac{1 \text{ mol Tc}}{98.9 \text{ g Tc}}\right) = 4.82 \times 10^{-3} \text{ mol Tc}$$

$$\text{mol O} = (0.308 \text{ g O})\left(\frac{1 \text{ mol O}}{16.0 \text{ g O}}\right) = 1.93 \times 10^{-2} \text{ mol O}$$

Now we divide each of these numbers of moles by the smallest of the three numbers, in order to obtain the simplest mole ratio among the three elements in the compound:

$$\text{for Na, } \frac{4.83 \times 10^{-3} \text{ mol}}{4.82 \times 10^{-3} \text{ mol}} = 1.00$$

$$\text{for Tc, } \frac{4.82 \times 10^{-3} \text{ mol}}{4.82 \times 10^{-3} \text{ mol}} = 1.00$$

$$\text{for O, } \frac{1.93 \times 10^{-2} \text{ mol}}{4.82 \times 10^{-3} \text{ mol}} = 4.00$$

These relative mole amounts give us the empirical formula: $NaTcO_4$.

3.85 To solve this problem we will assume that we have a 100 g sample.

To find the number of moles of O, first we have to find the number of grams of O:

100 g total = (72.96 g C) + (5.40 g H) + (x g O)

x g O = 21.64 g O

Then we determine the number of moles of each element in the compound:

$$\text{mol C} = (72.96 \text{ g C})\left(\frac{1 \text{ mol C}}{12.011 \text{ g C}}\right) = 6.074 \text{ mol C}$$

$$\text{mol H} = (5.40 \text{ g H})\left(\frac{1 \text{ mol H}}{1.008 \text{ g H}}\right) = 5.36 \text{ mol H}$$

$$\text{mol O} = (21.64 \text{ g O})\left(\frac{1 \text{ mol O}}{15.999 \text{ g O}}\right) = 1.353 \text{ mol O}$$

Now we divide each of these numbers of moles by the smallest of the three numbers, in order to obtain the simplest mole ratio among the three elements in the compound:

$$\text{for C, } \frac{6.074 \text{ mol}}{1.353 \text{ mol}} = 4.49$$

$$\text{for H, } \frac{5.36 \text{ mol}}{1.353 \text{ mol}} = 3.96$$

$$\text{for O, } \frac{1.353 \text{ mol}}{1.353 \text{ mol}} = 1.00$$

These relative mole amounts give us the empirical formula $C_{4.5}H_4O$

Since we cannot have decimals as subscripts, multiply all of the subscripts by 2 to get the formula: $C_9H_8O_2$

3.87 To solve this problem assume that we have a 100 g sample: 14.5 g C, and 85.5 g Cl:

$$\text{mol C} = (14.5\text{ g C})\left(\frac{1\text{ mol C}}{12.01\text{ g C}}\right) = 1.21\text{ mol C}$$

$$\text{mol Cl} = (85.5\text{ g Cl})\left(\frac{1\text{ mol Cl}}{35.45\text{ g Cl}}\right) = 2.41\text{ mol Cl}$$

Now we divide each of these numbers of moles by the smallest of the three numbers, in order to obtain the simplest mole ratio among the three elements in the compound:

for C, $\frac{1.21\text{ mol}}{1.21\text{ mol}} = 1.00$

for Cl, $\frac{2.42\text{ mol}}{1.21\text{ mol}} = 2.000$

These relative mole amounts give us the empirical formula CCl_2

3.89 All of the carbon is converted to carbon dioxide so,

$$\text{g C} = (1.312\text{ g }CO_2)\left(\frac{1\text{ mol }CO_2}{44.01\text{ g }CO_2}\right)\left(\frac{1\text{ mol C}}{1\text{ mol }CO_2}\right)\left(\frac{12.01\text{ g C}}{1\text{ mol C}}\right) = 0.358\text{ g C}$$

$$\text{mol C} = (0.358\text{ g C})\left(\frac{1\text{ mol C}}{12.01\text{ g C}}\right) = 2.98 \times 10^{-2}\text{ mol C}$$

All of the hydrogen is converted to H_2O, so

$$\text{g H} = (0.805\text{ g }H_2O)\left(\frac{1\text{ mol }H_2O}{18.02\text{ g }H_2O}\right)\left(\frac{2\text{ mol H}}{1\text{ mol }H_2O}\right)\left(\frac{1.008\text{ g H}}{1\text{ mol H}}\right) = 0.0901\text{ g H}$$

$$\text{mol H} = (0.0901\text{ g H})\left(\frac{1\text{ mol H}}{1.008\text{ g H}}\right) = 8.93 \times 10^{-2}\text{ mol H}$$

The amount of O in the compound is determined by subtracting the mass of C and the mass of H from the sample.

$\text{g O} = 0.684\text{ g} - 0.358\text{ g} - 0.0901\text{ g} = 0.236\text{ g O}$

$$\text{mol O} = (0.236\text{ g O})\left(\frac{1\text{ mol O}}{16.00\text{ g O}}\right) = 1.48 \times 10^{-2}\text{ mol O}$$

The relative mole ratios are:

for C, $\frac{0.0298\text{ mol}}{0.0148\text{ mol}} = 2.01$

for H, $\frac{0.0893\text{ mol}}{0.0148\text{ mol}} = 6.03$

for O, $\frac{0.0148\text{ mol}}{0.0148\text{ mol}} = 1.00$

The relative mole amounts give the empirical formula C_2H_6O

3.91 This type of combustion analysis takes advantage of the fact that the entire amount of carbon in the original sample appears as CO_2 among the products. Hence the mass of carbon in the original sample must be equal to the mass of carbon that is found in the CO_2.

$$\text{g C} = 19.73 \times 10^{-3}\text{ g }CO_2\left(\frac{1\text{ mole }CO_2}{44.01\text{ g }CO_2}\right)\left(\frac{1\text{ mole C}}{1\text{ mole }CO_2}\right)\left(\frac{12.011\text{ g C}}{1\text{ mole C}}\right) = 5.385 \times 10^{-3}\text{ g C}$$

Similarly, the entire mass of hydrogen that was present in the original sample ends up in the products as H_2O:

$$g\ H = 6.391 \times 10^{-3}\ g\ H_2O\left(\frac{1\ mole\ H_2O}{18.02\ g\ H_2O}\right)\left(\frac{2\ mole\ H}{1\ mole\ H_2O}\right)\left(\frac{1.008\ g\ H}{1\ mole\ H}\right) = 7.150 \times 10^{-4}\ g\ H$$

The mass of oxygen is determined by subtracting the mass due to C and H from the total mass:
6.853 mg total – (5.385 mg C + 0.7150 mg H) = 0.753 mg O
Now, convert these masses to a number of moles:

$$mol\ C = 5.385 \times 10^{-3}\ g\ C\left(\frac{1\ mol\ C}{12.011\ g\ C}\right) = 4.484 \times 10^{-4}\ mol\ C$$

$$mol\ H = 7.150 \times 10^{-4}\ g\ H\left(\frac{1\ mol\ H}{1.0079\ g\ H}\right) = 7.094 \times 10^{-4}\ mol\ H$$

$$mol\ O = 7.53 \times 10^{-4}\ g\ O\left(\frac{1\ mol\ O}{15.999\ g\ O}\right) = 4.71 \times 10^{-5}\ mol\ H$$

The relative mole amounts are:

for C, $\frac{4.484 \times 10^{-4}\ mol}{4.71 \times 10^{-5}\ mol} = 9.52$

for H, $\frac{7.094 \times 10^{-4}\ mol}{4.71 \times 10^{-5}\ mol} = 15.1$

for O, $\frac{4.71 \times 10^{-5}\ mol}{4.71 \times 10^{-5}\ mol} = 1.00$

The relative mole amounts are not whole numbers as we would like. However, we see that if we double the relative number of moles of each compound, there are approximately 19 moles of C, 30 moles of H and 2 moles of O. If we assume these numbers are correct, the empirical formula is $C_{19}H_{30}O_2$, for which the formula weight is 290 g/mole.
In most problems where we determine an empirical formula, the relative mole amounts should work out to give a "nice" set of values for the formula. Rarely will a problem be designed that gives very odd coefficients. With experience and practice, you will recognize when a set of values is reasonable.

3.93 (a) Empirical formula mass = 135.1 g

$$\frac{270.4\ g/mol}{135.1\ g/mol} = 2.001$$

The molecular formula is $Na_2S_4O_6$

(b) Empirical formula mass = 73.50 g

$$\frac{147.0\ g/mol}{73.50\ g/mol} = 2.000$$

The molecular formula is $C_6H_4Cl_2$

(c) Empirical formula mass = 60.48 g

$$\frac{181.4\ g/mol}{60.48\ g/mol} = 2.999$$

The molecular formula is $C_6H_3Cl_3$

3.95 The formula mass for the compound $C_{19}H_{30}O_2$ is 290 g/mol. Thus, the empirical and molecular formulas are equivalent.

3.97 From the information provided, we can determine the mass of mercury as the difference between the total mass and the mass of bromine:

g Hg = 0.389 g compound – 0.111 g Br = 0.278 g Hg

To determine the empirical formula, first convert the two masses to a number of moles.

$$\text{mol Hg} = (0.278\text{ g Hg})\left(\frac{1\text{ mole Hg}}{200.59\text{ g Hg}}\right) = 1.39 \times 10^{-3}\text{ mol Hg}$$

$$\text{mol Br} = (0.111\text{ g Br})\left(\frac{1\text{ mole Br}}{79.904\text{ g Br}}\right) = 1.39 \times 10^{-3}\text{ mol Br}$$

Now, divide each of these values by the smaller quantity to determine the simplest mole ratio between the two elements. By inspection, though, we can see there are the same number of moles of Hg and Br. Consequently, the simplest mole ratio is 1:1 and the empirical formula is HgBr.

To determine the molecular formula, recall that the ratio of the molecular mass to the empirical mass is equivalent to the ratio of the molecular formula to the empirical formula. Thus, we need to calculate an empirical mass:

$$(1\text{ mole Hg})\frac{200.59\text{ g Hg}}{1\text{ mol Hg}} + (1\text{ mole Br})\frac{79.904\text{ g Br}}{1\text{ mol Br}} = 280.49\text{ g/mole HgBr.}$$

The molecular mass, as reported in the problem is 561 g/mole. The ratio of these is:

$$\frac{561\text{ g/mole}}{280.49\text{ g/mole}} = 2.00$$

So, the molecular formula is two times the empirical formula or Hg_2Br_2.

3.99 To solve this problem we will assume that we have a 100 g sample. This implies that we have 27.91 g C, 2.341 g H, 32.56 g N and 37.19 g O. The amount of oxygen was determined by subtracting the total amounts of the other three elements from the total assumed mass of 100 g. Convert each of these masses into a number of moles:

$$\text{mol C} = (27.91\text{ g C})\left(\frac{1\text{ mole C}}{12.011\text{ g C}}\right) = 2.324\text{ mol C}$$

$$\text{mol H} = (2.341\text{ g H})\left(\frac{1\text{ mole H}}{1.008\text{ g H}}\right) = 2.322\text{ mol H}$$

$$\text{mol N} = (32.56\text{ g N})\left(\frac{1\text{ mole N}}{14.01\text{ g N}}\right) = 2.323\text{ mol N}$$

$$\text{mol O} = (37.19\text{ g O})\left(\frac{1\text{ mole O}}{16.00\text{ g O}}\right) = 2.324\text{ mol O}$$

The relative mole amounts are at a 1:1:1:1 ratio, therefore the empirical formula is CHNO.

The empirical formula weight is 43.02 g/mole. The formula mass is 129 g/mol.

$$\frac{129\text{ g mol}^{-1}}{43.02\text{ g mol}^{-1}} = 3.02$$

Multiply the subscripts of the empirical formula by 3 to get the molecular formula: $C_3H_3N_3O_3$.

3.101 (a) $Mg(OH)_2 + 2HBr \longrightarrow MgBr_2 + 2H_2O$

(b) $2HCl + Ca(OH)_2 \longrightarrow CaCl_2 + 2H_2O$

(c) $Al_2O_3 + 3H_2SO_4 \longrightarrow Al_2(SO_4)_3 + 3H_2O$

(d) $2KHCO_3 + H_3PO_4 \longrightarrow K_2HPO_4 + 2H_2O + 2CO_2$

(e) $C_9H_{20} + 14O_2 \longrightarrow 9CO_2 + 10H_2O$

3.103 (a) $Ca(OH)_2 + 2HCl \longrightarrow CaCl_2 + 2H_2O$

(b) $2AgNO_3 + CaCl_2 \longrightarrow Ca(NO_3)_2 + 2AgCl$

(c) $Pb(NO_3)_2 + Na_2SO_4 \longrightarrow PbSO_4 + 2NaNO_3$

(d) $2Fe_2O_3 + 3C \longrightarrow 4Fe + 3CO_2$

3.105 $4NH_2CHO + 5O_2 \longrightarrow 4CO_2 + 6H_2O + 2N_2$

3.107

3.109 $4Fe(s) + 3O_2(g) \longrightarrow 2Fe_2O_3(s)$

3.111 $2Ba(OH)_2{\bullet}8H_2O$ contains:
2 atoms of Ba, 20 atoms of O, and 36 atoms of H
2 moles of Ba, 20 moles of O, and 36 moles of H

3.113 (a) $\text{mol } Na_2S_2O_3 = (0.12 \text{ mol } Cl_2)\left(\frac{1 \text{ mole } Na_2S_2O_3}{4 \text{ mole } Cl_2}\right) = 0.030 \text{ mol } Na_2S_2O_3$

(b) $\text{mol HCl} = (0.12 \text{ mol } Cl_2)\left(\frac{8 \text{ mole HCl}}{4 \text{ mole } Cl_2}\right) = 0.24 \text{ mol HCl}$

(c) $\text{mol } H_2O = (0.12 \text{ mol } Cl_2)\left(\frac{5 \text{ mole } H_2O}{4 \text{ mole } Cl_2}\right) = 0.15 \text{ mol } H_2O$

(d) $\text{mol } H_2O = (0.24 \text{ mol HCl})\left(\frac{5 \text{ mole } H_2O}{8 \text{ mole HCl}}\right) = 0.15 \text{ mol } H_2O$

3.115 (a) $3.45 \text{ mol } C_3H_8\left(\frac{5 \text{ mol } O_2}{1 \text{ mol } C_3H_8}\right)\left(\frac{32.00 \text{ g } O_2}{1 \text{ mol } O_2}\right) = 552 \text{ g } O_2$

(b) $0.177 \text{ mol } C_3H_8\left(\frac{3 \text{ mol } CO_2}{1 \text{ mol } C_3H_8}\right)\left(\frac{44.01 \text{ g } CO_2}{1 \text{ mol } CO_2}\right) = 23.4 \text{ g } CO_2$

(c) $4.86 \text{ mol } C_3H_8\left(\frac{4 \text{ mol } H_2O}{1 \text{ mol } C_3H_8}\right)\left(\frac{18.01 \text{ g } H_2O}{1 \text{ mol } H_2O}\right) = 350. \text{ g } H_2O$

3.117 (a) $4P + 5O_2 \longrightarrow P_4O_{10}$

(b) $\text{g } O_2 = (6.85 \text{ g P})\left(\frac{1 \text{ mol P}}{30.97 \text{ g P}}\right)\left(\frac{5 \text{ mol } O_2}{4 \text{ mol P}}\right)\left(\frac{32.0 \text{ g } O_2}{1 \text{ mol } O_2}\right) = 8.85 \text{ g } O_2$

(c) $\text{g } P_4O_{10} = (8.00 \text{ g } O_2)\left(\frac{1 \text{ mol } O_2}{32.00 \text{ g } O_2}\right)\left(\frac{1 \text{ mol } P_4O_{10}}{5 \text{ mol } O_2}\right)\left(\frac{283.9 \text{ g } P_4O_{10}}{1 \text{ mol } P_4O_{10}}\right) = 14.2 \text{ g } P_4O_{10}$

(d) $\text{g P} = (7.46 \text{ g } P_4O_{10})\left(\frac{1 \text{ mol } P_4O_{10}}{283.9 \text{ g } P_4O_{10}}\right)\left(\frac{4 \text{ mol P}}{1 \text{ mol } P_4O_{10}}\right)\left(\frac{30.97 \text{ g P}}{1 \text{ mol P}}\right) = 3.26 \text{ g P}$

3.119 $\text{g } HNO_3 = 11.45 \text{ g Cu}\left(\frac{1 \text{ mol Cu}}{63.546 \text{ g Cu}}\right)\left(\frac{8 \text{ mol } HNO_3}{3 \text{ mol Cu}}\right)\left(\frac{63.013 \text{ g } HNO_3}{1 \text{ mol } HNO_3}\right) = 30.28 \text{ g } HNO_3$

3.121 $kg\ O_2 = 1.0\ kg\ H_2O_2 \left(\frac{1000\ g\ H_2O_2}{1\ kg\ H_2O_2}\right)\left(\frac{1\ mol\ H_2O_2}{34.01\ g\ H_2O_2}\right)\left(\frac{1\ mol\ O_2}{2\ mol\ H_2O_2}\right)\left(\frac{32.00\ g\ O_2}{1\ mol\ O_2}\right)\left(\frac{1\ kg\ O_2}{1000\ g\ O_2}\right)$

$= 0.47\ kg\ O_2$

3.123 The picture shows 9 molecules of O_2 and 3 molecules of C_2H_6S. The balanced reaction shows that 2 molecules of C_2H_6S react with 9 moles of O_2. Therefore, O_2 is the limiting reagent.
Molecules of SO_2 = (9 molecules O_2)(2 molecules SO_2/9 molecules O_2) = 2 molecules of SO_2

3.125 (a) Determine the amount of Fe_2O_3 that would be required to react completely with the Al:

$$mol\ Fe_2O_3 = (4.20\ mol\ Al)\left(\frac{1\ mol\ Fe_2O_3}{2\ mol\ Al}\right) = 2.10\ mol\ Fe_2O_3$$

Since only 1.75 mol of Fe_2O_3 are supplied, it is the limiting reactant. This can be confirmed by calculating the amount of Al that would be required to react completely with all of the available Fe_2O_3:

$$mol\ Al = (1.75\ mol\ Fe_2O_3)\left(\frac{2\ mol\ Al}{1\ mol\ Fe_2O_3}\right) = 3.50\ mol\ Al$$

Since an excess (4.20 mol – 3.50 mol = 0.70 mol) of Al is present, Fe_2O_3 must be the limiting reactant, as determined above.

(b) $g\ Fe = (1.75\ mol\ Fe_2O_3)\left(\frac{2\ mol\ Fe}{1\ mol\ Fe_2O_3}\right)\left(\frac{55.847\ g\ Fe}{1\ mol\ Fe}\right) = 195\ g\ Fe$

3.127 $3AgNO_3 + FeCl_3 \longrightarrow 3AgCl + Fe(NO_3)_3$
Calculate the amount of $FeCl_3$ that are required to react completely with all of the silver nitrate:

$$g\ FeCl_3 = 18.0\ g\ AgNO_3\left(\frac{1\ mol\ AgNO_3}{169.87\ g\ AgNO_3}\right)\left(\frac{1\ mol\ FeCl_3}{3\ mol\ AgNO_3}\right)\left(\frac{162.21\ g\ FeCl_3}{1\ mol\ FeCl_3}\right) = 5.73\ g\ FeCl_3$$

Since more than this minimum amount is available, $FeCl_3$ is present in excess, and $AgNO_3$ must be the limiting reactant.
We know that only 5.73 g $FeCl_3$ will be used. Therefore, the amount left unused is:
32.4 g total – 5.73 g used = 26.7 g $FeCl_3$
Notice that we decided in the beginning to calculate the amount of $FeCl_3$ required to react with $AgNO_3$. Instead, we could have calculated the grams of $AgNO_3$ required to react with $FeCl_3$. In that case we would have seen that $AgNO_3$ was in excess.

3.129 First calculate the number of moles of water that are needed to react completely with the given amount of NO_2:

$$g\ H_2O = 0.0010\ g\ NO_2\left(\frac{1\ mol\ NO_2}{46.01\ g\ NO_2}\right)\left(\frac{1\ mol\ H_2O}{3\ mol\ NO_2}\right)\left(\frac{18.02\ g\ H_2O}{1\ mol\ H_2O}\right) = 1.3 \times 10^{-4}\ g\ H_2O$$

Since this is less than the amount of water that is supplied, the limiting reactant must be NO_2.
Therefore, to calculate the amount of HNO_3:

$$g\ HNO_3 = 0.0010\ g\ NO_2\left(\frac{1\ mol\ NO_2}{46.01\ g\ NO_2}\right)\left(\frac{2\ mol\ HNO_3}{3\ mol\ NO_2}\right)\left(\frac{63.02\ g\ HNO_3}{1\ mol\ HNO_3}\right)\left(\frac{1000\ mg\ HNO_3}{1\ g\ HNO_3}\right)$$

$= 0.91\ mg\ HNO_3$

3.131 First determine the theoretical yield:

$$\text{g BaSO}_4 = (75.00\text{ g Ba(NO}_3)_2\left(\frac{1\text{ mol Ba(NO}_3)_2}{261.34\text{ g Ba(NO}_3)_2}\right)\left(\frac{1\text{ mol BaSO}_4}{1\text{ mol Ba(NO}_3)_2}\right)\left(\frac{233.39\text{ g BaSO}_4}{1\text{ mol BaSO}_4}\right)$$

$$= 66.98\text{ g BaSO}_4$$

Then calculate a % yield:

$$\%\text{ yield} = \frac{\text{actual yield}}{\text{theoretical yield}} \times 100\% = \frac{64.45\text{ g}}{66.98\text{ g}} \times 100\% = 96.22\%$$

3.133 First, determine how much H_2SO_4 is needed to completely react with the $AlCl_3$

$$\text{g H}_2\text{SO}_4 = 25.0\text{ g AlCl}_3\left(\frac{1\text{ mol AlCl}_3}{133.34\text{ g AlCl}_3}\right)\left(\frac{3\text{ mol H}_2\text{SO}_4}{2\text{ mol AlCl}_3}\right)\left(\frac{98.08\text{ g H}_2\text{SO}_4}{1\text{ mol H}_2\text{SO}_4}\right) = 27.58\text{ g H}_2\text{SO}_4$$

There is an excess of H_2SO_4 present.

Determine the theoretical yield:

$$\text{g Al}_2(\text{SO}_4)_3 = 25.0\text{ g AlCl}_3\left(\frac{1\text{ mol AlCl}_3}{133.33\text{ g AlCl}_3}\right)\left(\frac{1\text{ mol Al}_2(\text{SO}_4)_3}{2\text{ mol AlCl}_3}\right)\left(\frac{342.17\text{ g Al}_2(\text{SO}_4)_3}{1\text{ mol Al}_2(\text{SO}_4)_3}\right)$$

$$= 32.08\text{ g Al}_2(\text{SO}_4)_3$$

$$\%\text{ yield} = \frac{\text{actual yield}}{\text{theoretical yield}} \times 100\% = \frac{28.46\text{ g}}{32.08\text{ g}} \times 100\% = 88.72\%$$

3.135 First, determine how much MnI_2 is needed to completely react with the F_2

$$\text{g MnI}_2 = 10.0\text{ g F}_2\left(\frac{1\text{ mol F}_2}{38.00\text{ g F}_2}\right)\left(\frac{2\text{ mol MnI}_2}{13\text{ mol F}_2}\right)\left(\frac{308.75\text{ g MnI}_2}{1\text{ mol MnI}_2}\right) = 12.5\text{ g MnI}_2$$

There is an excess of F_2 present.

Note that MnI_2 is the limiting reactant and that MnI_2 and MnF_3 are in a 1:1 ratio, so the number of mole of MnI_2 equals the number of moles of MnF_3. According to the problem statement, we will only prepare 56% of this number of moles of MnF_3.

$$\text{g MnF}_3 = 10.0\text{ g MnI}_2\left(\frac{1\text{ mol MnI}_2}{308.75\text{ g MnI}_2}\right)\left(\frac{2\text{ mol MnF}_3}{2\text{ mol MnI}_2}\right)\left(\frac{111.93\text{ g MnF}_3}{1\text{ mol MnF}_3}\right)(0.56) = 2.03\text{ g MnF}_3$$

Chapter Four
Molecular View of Reactions in Aqueous Solutions

Practice Exercises

4.1 (a) $CaI_2(s) \longrightarrow Ca^{2+}(aq) + 2I^-(aq)$
(b) $K_3PO_4(s) \longrightarrow 3K^+(aq) + PO_4^{3-}(aq)$

4.2 (a) $Al(NO_3)_3(s) \longrightarrow Al^{3+}(aq) + 3NO_3^-(aq)$
(b) $Na_2CO_3(s) \longrightarrow 2Na^+(aq) + CO_3^{2-}(aq)$

4.3 Molecular: $(NH_4)_2SO_4(aq) + Ba(NO_3)_2(aq) \longrightarrow BaSO_4(s) + 2NH_4NO_3(aq)$
Ionic: $2NH_4^+(aq) + SO_4^{2-}(aq) + Ba^{2+}(aq) + 2NO_3^-(aq) \longrightarrow BaSO_4(s) + 2NH_4^+(aq) + 2NO_3^-(aq)$
Net ionic: $Ba^{2+}(aq) + SO_4^{2-}(aq) \longrightarrow BaSO_4(s)$

4.4 Molecular: $CdCl_2(aq) + Na_2S(aq) \longrightarrow CdS(s) + 2NaCl(aq)$
Ionic: $Cd^{2+}(aq) + 2Cl^-(aq) + 2Na^+(aq) + S^{2-}(aq) \longrightarrow CdS(s) + 2Na^+(aq) + 2Cl^-(aq)$
Net ionic: $Cd^{2+}(aq) + S^{2-}(aq) \longrightarrow CdS(s)$

4.5 $HC_3H_7O_2(aq) \rightleftharpoons H_3O^+(aq) + C_3H_7O_2^-(aq)$

4.6 $HNO_3(aq) + H_2O \longrightarrow H_3O^+(aq) + NO_3^-(aq)$
$HNO_2(aq) + H_2O \rightleftharpoons H_3O^+(aq) + NO_2^-(aq)$

4.7 $(C_2H_5)_3N(aq) + H_2O \rightleftharpoons (C_2H_5)_3NH^+(aq) + OH^-(aq)$

4.8 The drawing below is the cation formed, $CH_3CH_2NH_3^+$ and does not include the positive charge of the ion.

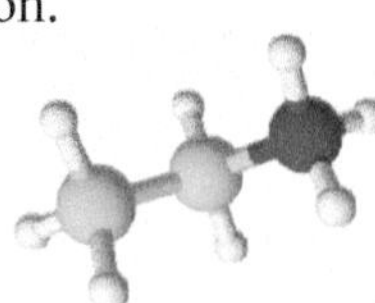

$CH_3CH_2NH_2(aq) + H_2O \rightleftharpoons CH_3CH_2NH_3^+(aq) + OH^-(aq)$

4.9 $H_3C_6H_5O_7(s) + H_2O \rightleftharpoons H_3O^+(aq) + H_2C_6H_5O_7^-(aq)$
$H_2C_6H_5O_7^-(aq) + H_2O \rightleftharpoons H_3O^+(aq) + HC_6H_5O_7^{2-}(aq)$
$HC_6H_5O_7^{2-}(aq) + H_2O \rightleftharpoons H_3O^+(aq) + C_6H_5O_7^{3-}(aq)$

4.10 $H_2S(aq) + H_2O \rightleftharpoons H_3O^+(aq) + HS^-(aq)$
$HS^-(aq) + H_2O \rightleftharpoons H_3O^+(aq) + S^{2-}(aq)$

4.11 HF: Hydrofluoric acid
HI: Hydroiodic acid

4.12 Iodic acid

4.13 Molecular: $Zn(NO_3)_2(aq) + Ca(OH)_2(aq) \longrightarrow Zn(OH)_2(aq) + Ca(NO_3)_2(aq)$
Ionic: $Zn^{2+}(aq) + 2NO_3^-(aq) + Ca^{2+}(aq) + 2OH^-(aq) \longrightarrow$
$Zn^{2+}(aq) + 2OH^-(aq) + Ca^{2+}(aq) + 2NO_3^-(aq)$

Net ionic: No reaction

4.14 (a) Molecular: $AgNO_3(aq) + NH_4Cl(aq) \longrightarrow AgCl(s) + NH_4NO_3(aq)$

Ionic: $Ag^+(aq) + NO_3^-(aq) + NH_4^+(aq) + Cl^-(aq) \longrightarrow AgCl(s) + NH_4^+(aq) + NO_3^-(aq)$

Net ionic: $Ag^+(aq) + Cl^-(aq) \longrightarrow AgCl(s)$

(b) Molecular: $Na_2S(aq) + Pb(C_2H_3O_2)_2(aq) \longrightarrow 2NaC_2H_3O_2(aq) + PbS(s)$

Ionic: $2Na^+(aq) + S^{2-}(aq) + Pb^{2+}(aq) + 2C_2H_3O_2^-(aq) \longrightarrow 2Na^+(aq) + 2C_2H_3O_2^-(aq) + PbS(s)$

Net ionic: $S^{2-}(aq) + Pb^{2+}(aq) \longrightarrow PbS(s)$

4.15 Molecular: $2HNO_3(aq) + Ca(OH)_2(aq) \longrightarrow Ca(NO_3)_2(aq) + 2H_2O$

Ionic: $2H^+(aq) + 2NO_3^-(aq) + Ca^{2+}(aq) + 2OH^-(aq) \longrightarrow Ca^{2+}(aq) + 2NO_3^-(aq) + 2H_2O$

Net ionic: $H^+(aq) + OH^-(aq) \longrightarrow H_2O$

4.16 (a) Molecular: $HCl(aq) + KOH(aq) \longrightarrow H_2O + KCl(aq)$

Ionic: $H^+(aq) + Cl^-(aq) + K^+(aq) + OH^-(aq) \longrightarrow H_2O + K^+(aq) + Cl^-(aq)$

Net ionic: $H^+(aq) + OH^-(aq) \longrightarrow H_2O$

(b) Molecular: $2HCl(aq) + Ba(OH)_2(aq) \longrightarrow 2H_2O + BaCl_2(aq)$

Ionic: $2H^+(aq) + 2Cl^-(aq) + Ba^{2+}(aq) + 2OH^-(aq) \longrightarrow 2H_2O + Ba^{2+}(aq) + 2Cl^-(aq)$

Net ionic: $H^+(aq) + OH^-(aq) \longrightarrow H_2O$

4.17 Molecular: $CH_3NH_2(aq) + HCHO_2(aq) \longrightarrow CH_3NH_3CHO_2(aq)$

Ionic: $CH_3NH_2(aq) + HCHO_2(aq) \longrightarrow CH_3NH_3^+(aq) + CHO_2^-(aq)$

Net ionic: $CH_3NH_2(aq) + HCHO_2(aq) \longrightarrow CH_3NH_3^+(aq) + CHO_2^-(aq)$

4.18 $NaHSO_3$, sodium hydrogen sulfite

4.19 $H_3AsO_4(aq) + NaOH(aq) \longrightarrow NaH_2AsO_4(aq) + H_2O$ sodium dihydrogen arsenate

$NaH_2AsO_4(aq) + NaOH(aq) \longrightarrow Na_2HAsO_4(aq) + H_2O$ sodium hydrogen arsenate

$Na_2HAsO_4(aq) + NaOH(aq) \longrightarrow Na_3AsO_4(aq) + H_2O$sodium arsenate

The acid is completely neutralized in the last step.

4.20 Molecular: $2HCHO_2(aq) + Co(OH)_2(s) \longrightarrow Co(CHO_2)_2(aq) + 2H_2O$

Ionic: $2HCHO_2(aq) + Co(OH)_2(s) \longrightarrow 2CHO_2^-(aq) + Co^{2+}(aq) + 2H_2O$

Net ionic: $2HCHO_2(aq) + Co(OH)_2(s) \longrightarrow 2CHO_2^-(aq) + Co^{2+}(aq) + 2H_2O$

4.21 Molecular: $MgCl_2(aq) + (NH_4)SO_4(aq) \longrightarrow MgSO_4(aq) + 2NH_4Cl(aq)$

Ionic: $Mg^{2+}(aq) + 2Cl^-(aq) + 2NH_4^+(aq) + SO_4^{2-}(aq) \longrightarrow$
$Mg^{2+}(aq) + 2Cl^-(aq) + 2NH_4^+(aq) + SO_4^{2-}(aq)$

Net ionic: No reaction

4.22 $CuO(s) + 2HNO_3(aq) \longrightarrow Cu(NO_3)_2(aq) + H_2O$

Or

$Cu(OH)_2(s) + 2HNO_3(aq) \longrightarrow Cu(NO_3)_2(aq) + 2H_2O$

4.23 You want to use a metathesis reaction that produces CoS, which is insoluble, and a second product that is soluble. You may want the reactants to be soluble.

$CoCl_2(aq) + Na_2S(aq) \longrightarrow CoS(s) + 2NaCl(aq)$

4.24 $M\text{ LiBr} = \frac{\text{mol LiBr}}{\text{L solution}} = \frac{0.175\text{ mol LiBr}}{0.350\text{ L solution}} = 0.500\ M\text{ LiBr}$

4.25 $\text{mol KI} = 2.75\text{ g KI}\left(\frac{1\text{ mol KI}}{166.0\text{ g KI}}\right) = 0.0166\text{ mol KI}$

$M\text{ KI} = \frac{\text{mol KI}}{\text{L solution}} = \left(\frac{0.0.0166\text{ mol LiBr}}{125\text{ mL solution}}\right)\left(\frac{1000\text{ mL solution}}{1\text{ L solution}}\right) = 0.133\ M\text{ KI}$

4.26 $0.142\text{ mol NaNO}_3\left(\frac{1.00\text{ L solution}}{0.500\text{ mol NaNO}_3}\right)\left(\frac{1000\text{ mL solution}}{1.00\text{ L solution}}\right) = 284\text{ mL NaNO}_3\text{ solution}$

4.27 $\text{mol HCl} = 175\text{ mL HCl solution}\left(\frac{1\text{ L solution}}{1000\text{ mL solution}}\right)\left(\frac{0.250\text{ mol HCl}}{1\text{ L solution}}\right) = 0.0438\text{ mol HCl}$

$\text{g HCl} = 0.0438\text{ mol HCl}\left(\frac{36.45\text{ g HCl}}{1\text{ mol HCl}}\right) = 1.60\text{ g HCl}$

4.28 If we were working with a full liter of this solution, it would contain 0.2 mol of $Sr(NO_3)_2$. The molar mass of the salt is 211.62 g mol^{-1}, so 0.2 mol is slightly more than 40 g. However, we are working with just 50 mL, so the amount of $Sr(NO_3)_2$ needed is slightly more than a twentieth of 40 g, or 2 g. The answer, 2.11 g, is close to this, so it makes sense.

$\text{g Sr(NO}_3)_2 = 50\text{ mL}\left(\frac{1\text{ L solution}}{1000\text{ mL solution}}\right)\left(\frac{0.2\text{ mol Sr(NO}_3)_2}{1\text{ L solution}}\right)\left(\frac{211.62\text{ g Sr(NO}_3)_2}{1\text{ mol Sr(NO}_3)_2}\right) = 2.11\text{ g Sr(NO}_3)_2$

4.29 $\text{g AgNO}_3 = (250\text{ mL solution})\left(\frac{1\text{ L solution}}{1000\text{ mL sol'n}}\right)\left(\frac{0.0125\text{ mol AgNO}_3}{1\text{ L solution}}\right)\left(\frac{169.9\text{ g AgNO}_3}{1\text{ mol AgNO}_3}\right) = 0.531\text{ g AgNO}_3$

4.30 $\text{g Na}_2\text{SO}_4 = 500.0\text{ mL solution}\left(\frac{1\text{ L solution}}{1000\text{ mL solution}}\right)\left(\frac{0.0150\text{ mol Na}_2\text{SO}_4}{1\text{ L solution}}\right)\left(\frac{142.0\text{ g Na}_2\text{SO}_4}{1\text{ mol Na}_2\text{SO}_4}\right)$

$= 1.065\text{ g Na}_2\text{SO}_4$

4.31 $\text{mL} = 100.0\text{ mL solution}\left(\frac{1\text{ L solution}}{1000\text{ mL solution}}\right)\left(\frac{0.125\text{ mol H}_2\text{SO}_4}{1\text{ L solution}}\right)\left(\frac{1\text{ L solution}}{0.0500\text{ mol H}_2\text{SO}_4}\right) = 250.0\text{ mL}$

4.32 $\text{mol HCl} = 150\text{ mL solution}\left(\frac{1\text{ L solution}}{1000\text{ mL solution}}\right)\left(\frac{0.50\text{ mol HCl}}{1\text{ L solution}}\right) = 0.075\text{ mol HCl}$

$\text{mL HCl solution} = 0.075\text{ mol HCl}\left(\frac{1\text{ L solution}}{0.10\text{ mol HCl}}\right)\left(\frac{1000\text{ mL solution}}{1\text{ L solution}}\right) = 750\text{ mL}$

To find the number of mL of water to add to the solution subtract the number of mL of the concentrated solution from the total volume:
750 mL solution – 150 mL = 600 mL
Add 600 mL of water.

4.33 $\text{mol H}_3\text{PO}_4 = 45.0\text{ mL KOH}\left(\frac{1\text{ L solution}}{1000\text{ mL solution}}\right)\left(\frac{0.100\text{ mol KOH}}{1\text{ L solution}}\right)\left(\frac{1\text{ mol H}_3\text{PO}_4}{3\text{ mol KOH}}\right)$

$= 1.5 \times 10^{-3}\text{ mol H}_3\text{PO}_4$

$$\text{mL } H_3PO_4 = 1.5 \times 10^{-3} \text{ mol } H_3PO_4\left(\frac{1 \text{ L solution}}{0.0475 \text{ mol } H_3PO_4}\right)\left(\frac{1000 \text{ mL solution}}{1 \text{ L solution}}\right) = 31.6 \text{ mL } H_3PO_4$$

4.34 $$\text{mL NaOH} = (15.4 \text{ mL } H_2SO_4)\left(\frac{1 \text{ L } H_2SO_4}{1000 \text{ mL } H_2SO_4}\right)\left(\frac{0.108 \text{ mol } H_2SO_4}{1 \text{ L } H_2SO_4}\right)$$
$$\times\left(\frac{2 \text{ mol NaOH}}{1 \text{ mol } H_2SO_4}\right)\left(\frac{1 \text{ L NaOH}}{0.124 \text{ mol NaOH}}\right)\left(\frac{1000 \text{ mL NaOH}}{1 \text{ L NaOH}}\right) = 26.8 \text{ mL NaOH}$$

4.35 $FeCl_3 \longrightarrow Fe^{3+} + 3Cl^-$

$$M\,Fe^{3+} = \left(\frac{0.40 \text{ mol } FeCl_3}{1 \text{ L } FeCl_3 \text{ soln}}\right)\left(\frac{1 \text{ mol } Fe^{3+}}{1 \text{ mol } FeCl_3}\right) = 0.40\ M\,Fe^{3+}$$

$$M\,Cl^- = \left(\frac{0.40 \text{ mol } FeCl_3}{1 \text{ L } FeCl_3 \text{ soln}}\right)\left(\frac{3 \text{ mol } Cl^-}{1 \text{ mol } FeCl_3}\right) = 1.2\ M\,Cl^-$$

4.36 $$M\,Na^+ = \left(\frac{0.250 \text{ mol } PO_4^{3-}}{1 \text{ L } Na_3PO_4 \text{ soln}}\right)\left(\frac{3 \text{ mol } Na^+}{1 \text{ mol } PO_4^{3-}}\right) = 0.750\ M\,Na^+$$

4.37 $$\text{mol } CaCl_2 = 18.4 \text{ mL } AgNO_3\left(\frac{1 \text{ L } AgNO_3}{1000 \text{ mL } AgNO_3}\right)\left(\frac{0.100 \text{ mol } AgNO_3}{1 \text{ L } AgNO_3}\right)\left(\frac{1 \text{ mol } Ag^+}{1 \text{ mol } AgNO_3}\right)$$
$$\times\left(\frac{1 \text{ mol } Cl^-}{1 \text{ mol } Ag^+}\right)\left(\frac{1 \text{ mol } CaCl_2}{2 \text{ mol } Cl^-}\right) = 9.20 \times 10^{-4} \text{ mol } CaCl_2$$

$$M\,CaCl_2 = \left(\frac{9.20 \times 10^{-4} \text{ mol } CaCl_2}{20.5 \text{ mL } CaCl_2}\right)\left(\frac{1000 \text{ mL } CaCl_2}{1 \text{ L } CaCl_2}\right) = 0.0449\ M\,CaCl_2$$

4.38 The balanced net ionic equation is: $Fe^{2+}(aq) + 2OH^-(aq) \longrightarrow Fe(OH)_2(s)$.
First determine the number of moles of Fe^{2+} present.

$$\text{mol } Fe^{2+} = 60.0 \text{ mL } Fe^{2+}\left(\frac{0.250 \text{ mol } FeCl_2}{1000 \text{ mL solution}}\right)\left(\frac{1 \text{ mol } Fe^{2+}}{1 \text{ mol } FeCl_2}\right) = 1.50 \times 10^{-2} \text{ mol } Fe^{2+}$$

Now, determine the amount of KOH needed to react with the Fe^{2+}.

$$\text{mL KOH} = 1.50 \times 10^{-2} \text{ mol } Fe^{2+}\left(\frac{2 \text{ mol } OH^-}{1 \text{ mol } Fe^{2+}}\right)\left(\frac{1 \text{ mol KOH}}{1 \text{ mol } OH^-}\right)\left(\frac{1000 \text{ mL solution}}{0.500 \text{ mol KOH}}\right) = 60.0 \text{ mL KOH}$$

4.39 $$\text{g } Na_2SO_4 = 28.40 \text{ mL } BaCl_2\left(\frac{0.150 \text{ mol } BaCl_2}{1000 \text{ mL } BaCl_2}\right)\left(\frac{1 \text{ mol } Ba^{2+}}{1 \text{ mol } BaCl_2}\right)\left(\frac{1 \text{ mol } SO_4^{2-}}{1 \text{ mol } Ba^{2+}}\right)$$
$$\times\left(\frac{1 \text{ mol } Na_2SO_4}{1 \text{ mol } SO_4^{2-}}\right)\left(\frac{142.05 \text{ g } Na_2SO_4}{1 \text{ mol } Na_2SO_4}\right) = 0.605 \text{ g } Na_2SO_4$$

4.40 $$\text{g } Na_2CO_3 = 29.06 \text{ mL } CaCl_2\left(\frac{0.125 \text{ mol } CaCl_2}{1000 \text{ mL } CaCl_2}\right)\left(\frac{1 \text{ mol } NaCO_3}{1 \text{ mol } CaCl_2}\right)\left(\frac{105.99 \text{ g } NaCO_3}{1 \text{ mol } NaCO_3}\right) = 0.385 \text{ g } Na_2CO_3$$

4.41 Balanced equation:

$H_2SO_4(aq) + 2NaOH(aq) \longrightarrow Na_2SO_4(aq) + 2H_2O$

$$\text{mol } H_sSO_4 = 36.42 \text{ mL NaOH}\left(\frac{0.147 \text{ mol NaOH}}{1000 \text{ mL NaOH soln}}\right)\left(\frac{1 \text{ mol } H_2SO_4}{2 \text{ mol NaOH}}\right) = 2.68 \times 10^{-3} \text{ mol } H_2SO_4$$

$$M\, H_2SO_4 = \left(\frac{2.68 \times 10^{-3} \text{ mol } H_2SO_4}{15.00 \text{ mL } H_2SO_4}\right)\left(\frac{1000 \text{ mL } H_2SO_4}{1 \text{ L } H_2SO_4}\right) = 0.178\, M\, H_2SO_4$$

4.42 $$\text{mol HCl} = 11.00 \text{ mL}\left(\frac{0.0100 \text{ mol KOH soln}}{1000 \text{ mL KOH soln}}\right)\left(\frac{1 \text{ mol HCl}}{1 \text{ mol KOH}}\right) = 1.1 \times 10^{-4} \text{ mol HCl}$$

$$M\, \text{HCl} = \left(\frac{1.1 \times 10^{-4} \text{ mol HCl}}{5.00 \text{ mL HCl soln}}\right)\left(\frac{1000 \text{ mL HCl soln}}{1 \text{ L HCl soln}}\right) = 0.0220\, M \text{ HCl}$$

$$\text{g HCl} = 5.00 \text{ mL HCl}\left(\frac{0.0220 \text{ mol HCl}}{1000 \text{ mL HCl}}\right)\left(\frac{36.5 \text{ g HCl}}{1 \text{ mol HCl}}\right) = 4.015 \times 10^{-3} \text{ g HCl}$$

$$\text{weight \%} = \frac{4.015 \times 10^{-3} \text{ g}}{5.00 \text{ g}} \times 100\% = 0.0803\% \text{ HCl}$$

4.43 $$\text{mol } Ca^{2+} = 0.736 \text{ g } CaSO_4\left(\frac{1 \text{ mol } CaSO_4}{136.14 \text{ g } CaSO_4}\right)\left(\frac{1 \text{ mol } Ca^{2+}}{1 \text{ mol } CaSO_4}\right) = 5.41 \times 10^{-3} \text{ mol } Ca^{2+}$$

Since all of the Ca^{2+} is precipitated as $CaSO_4$, there were originally 5.41×10^{-3} moles of Ca^{2+} in the sample.

All of the Ca^{2+} comes from $CaCl_2$, so there were 5.41×10^{-3} moles of $CaCl_2$ in the sample.

$$\text{g } CaCl_2 = 5.41 \times 10^{-3} \text{ mol } CaCl_2\left(\frac{110.98 \text{ g } CaCl_2}{1 \text{ mol } CaCl_2}\right) = 0.600 \text{ g } CaCl_2$$

$$\% \, CaCl_2 = \frac{0.600 \text{ g } CaCl_2}{2.000 \text{ g sample}} \times 100\% = 30.0\% \, CaCl_2$$

$\% \, MgCl_2 = 100\,\% - 30\,\% = 70\% \, MgCl_2$

Review Problems

4.43 (a) strong electrolyte

(b) nonelectrolyte

(c) strong electrolyte

(d) nonelectrolyte

4.45 (a) $LiF(aq) \longrightarrow Li^+(aq) + F^-(aq)$

(b) $KMnO_4(aq) \longrightarrow K^+(aq) + MnO_4^-(aq)$

(c) $K_2CO_3(aq) \longrightarrow 2K^+(aq) + CO_3^{2-}(aq)$

(d) $Co(NO_3)_2(aq) \longrightarrow Co^{2+}(aq) + 2NO_3^-(aq)$

4.47 (a) Ionic: $2NH_4^+(aq) + CO_3^{2-}(aq) + Ba^{2+}(aq) + 2Cl^-(aq) \longrightarrow 2NH_4^+(aq) + 2Cl^-(aq) + BaCO_3(s)$

Net: $Ba^{2+}(aq) + CO_3^{2-}(aq) \longrightarrow BaCO_3(s)$

(b) Ionic: $Cu^{2+}(aq) + 2Cl^-(aq) + 2Na^+(aq) + 2OH^-(aq) \longrightarrow Cu(OH)_2(s) + 2Na^+(aq) + 2Cl^-(aq)$

Net: $Cu^{2+}(aq) + 2OH^-(aq) \longrightarrow Cu(OH)_2(s)$

(c) Ionic: $3Fe^{2+}(aq) + 3SO_4^{2-}(aq) + 6Na^+(aq) + 2PO_4^{3-}(aq) \longrightarrow$
$Fe_3(PO_4)_2(s) + 6Na^+(aq) + 3SO_4^{2-}(aq)$

Net: $3Fe^{2+}(aq) + 2PO_4^{3-}(aq) \longrightarrow Fe_3(PO_4)_2(s)$

(d) Ionic: $2Ag^+(aq) + 2C_2H_3O_2^-(aq) + Ni^{2+}(aq) + 2Cl^-(aq) \longrightarrow$
$2AgCl(s) + Ni^{2+}(aq) + 2C_2H_3O^-(aq)$

Net: $Ag^+(aq) + Cl^-(aq) \longrightarrow AgCl(s)$

4.49 Spectator ions: Na^+ and Cl^-
Net ionic equation: $Co^{2+}(aq) + 2OH^-(aq) \longrightarrow Co(OH)_2(s)$

4.51 This is an ionization reaction: $HClO_4(aq) + H_2O \longrightarrow H_3O^+(aq) + ClO_4^-(aq)$

4.53 $HI(g) + H_2O \longrightarrow H_3O^+(aq) + I^-(aq)$

4.55 $N_2H_4(aq) + H_2O \rightleftharpoons N_2H_5^+(aq) + OH^-(aq)$

4.57 $HNO_2(aq) + H_2O \rightleftharpoons H_3O^+(aq) + NO_2^-(aq)$

4.59 $H_2CO_3(aq) + H_2O \rightleftharpoons H_3O^+(aq) + HCO_3^-(aq)$
$HCO_3^-(aq) + H_2O \rightleftharpoons H_3O^+(aq) + CO_3^{2-}(aq)$

4.61 (a) $H_2S(g)$: hydrogen sulfide (b) $H_2S(aq)$: hydrosulfuric acid

4.63 (a) perbromic acid (b) bromic acid
(c) bromous acid (d) hypobromous acid
(e) hydrobromic acid

4.65 (a) perbromate (b) bromate
(c) bromite (d) hypobromite
(e) bromide

4.67 (a) H_2CrO_4
(b) H_2CO_3
(c) $H_2C_2O_4$

4.69 (a) sodium bicarbonate or sodium hydrogen carbonate
(b) potassium dihydrogen phosphate
(c) ammonium hydrogen phosphate

4.71 (a) hypochlorous acid sodium hypochlorite $NaOCl$
(b) iodous acid sodium iodite $NaIO_2$
(d) perchloric acid sodium perchlorate $NaClO_4$

4.73 H_2SO_4

4.75 sodium butyrate

4.77 formic acid

4.79 The soluble ones are (a), (b), and (d).

4.81 (a) Ionic: $3Fe^{2+}(aq) + 3SO_4^{2-}(aq) + 6K^+(aq) + 2PO_4^{3-}(aq) \longrightarrow$
$Fe_3(PO_4)_2(s) + 6K^+(aq) + 3SO_4^{2-}(aq)$

Net: $3Fe^{2+}(aq) + 2PO_4^{3-}(aq) \longrightarrow Fe_3(PO_4)_2(s)$

(b) Ionic: $3Ag^+(aq) + 3C_2H_3O_2^-(aq) + Al^{3+}(aq) + 3Cl^-(aq) \longrightarrow$
$3AgCl(s) + Al^{3+}(aq) + 3C_2H_3O_2^-(aq)$

Net: $Ag^+(aq) + Cl^-(aq) \longrightarrow AgCl(s)$

(c) Ionic: $2Cr^{3+}(aq) + 6Cl^-(aq) + 3Ba^{2+}(aq) + 6OH^-(aq) \longrightarrow 2Cr(OH)_3(s) + 3Ba^{2+}(aq) + 6Cl^-(aq)$

Net: $Cr^{3+}(aq) + 3OH^-(aq) \longrightarrow Cr(OH)_3(s)$

4.83 Molecular: $NaCl(aq) + AgNO_3(aq) \longrightarrow AgCl(s) + NaNO_3(aq)$

Ionic: $Na^+(aq) + Cl^-(aq) + Ag^+(aq) + NO_3^-(aq) \longrightarrow AgCl(s) + Na^+(aq) + NO_3^-(aq)$

Net: $Cl^-(aq) + Ag^+(aq) \longrightarrow AgCl(s)$

4.85 Molecular: $Na_2S(aq) + Cu(NO_3)_2(aq) \longrightarrow CuS(s) + 2NaNO_3(aq)$

Ionic: $2Na^+(aq) + S^{2-}(aq) + Cu^{2+}(aq) + 2NO_3^-(aq) \longrightarrow CuS(s) + 2Na^+(aq) + 2NO_3^-(aq)$

Net: $Cu^{2+}(aq) + S^{2-}(aq) \longrightarrow CuS(s)$

4.87 (a) Molecular: $Ca(OH)_2(aq) + 2HNO_3(aq) \longrightarrow Ca(NO_3)_2(aq) + 2H_2O$

Ionic: $Ca^{2+}(aq) + 2OH^-(aq) + 2H^+(aq) + 2NO_3^-(aq) \longrightarrow Ca^{2+}(aq) + 2NO_3^-(aq) + 2H_2O$

Net: $H^+(aq) + OH^-(aq) \longrightarrow H_2O$

(b) Molecular: $Al_2O_3(s) + 6HCl(aq) \longrightarrow 2AlCl_3(aq) + 3H_2O$

Ionic: $Al_2O_3(s) + 6H^+(aq) + 6Cl^-(aq) \longrightarrow 2Al^{3+}(aq) + 6Cl^-(aq) + 3H_2O$

Net: $Al_2O_3(s) + 6H^+(aq) \longrightarrow 2Al^{3+}(aq) + 3H_2O$

(c) Molecular: $Zn(OH)_2(s) + H_2SO_4(aq) \longrightarrow ZnSO_4(aq) + 2H_2O$

Ionic: $Zn(OH)_2(s) + 2H^+(aq) + SO_4^{2-}(aq) \longrightarrow Zn^{2+}(aq) + SO_4^{2-}(aq) + 2H_2O$

Net: $Zn(OH)_2(s) + 2H^+(aq) \longrightarrow Zn^{2+}(aq) + 2H_2O$

4.89 The electrical conductivity would decrease regularly, until one solution had neutralized the other, forming a nonelectrolyte:

$Ba^{2+}(aq) + 2OH^-(aq) + 2H^+(aq) + SO_4^{2-}(aq) \longrightarrow BaSO_4(s) + 2H_2O$

Once the point of neutralization had been reached, the addition of excess sulfuric acid would cause the conductivity to increase, because sulfuric acid is a strong electrolyte itself.

4.91 (a) $2H^+(aq) + CO_3^{2-}(aq) \longrightarrow H_2O + CO_2(g)$

(b) $NH_4^+(aq) + OH^-(aq) \longrightarrow NH_3(g) + H_2O$

4.93 $Na_2S(aq) + 2HCl(aq) \longrightarrow 2NaCl(aq) + H_2S(g)$

Net: $S^{2-}(aq) + 2H^+(aq) \longrightarrow H_2S(g)$

4.95 These reactions have the following "driving forces":

(a) formation of insoluble $Cr(OH)_3$

(b) formation of water, a weak electrolyte

4.97 (a) Molecular: $3HNO_3(aq) + Cr(OH)_3(s) \longrightarrow Cr(NO_3)_3(aq) + 3H_2O$

Ionic: $3H^+(aq) + 3NO_3^-(aq) + Cr(OH)_3(s) \longrightarrow Cr^{3+}(aq) + 3NO_3^-(aq) + 3H_2O$

Net: $3H^+(aq) + Cr(OH)_3(s) \longrightarrow Cr^{3+}(aq) + 3H_2O$

(b) Molecular: $HClO_4(aq) + NaOH(aq) \longrightarrow NaClO_4(aq) + H_2O$

Ionic: $H^+(aq) + ClO_4^-(aq) + Na^+(aq) + OH^-(aq) \longrightarrow Na^+(aq) + ClO_4^-(aq) + H_2O$

Net: $H^+(aq) + OH^-(aq) \longrightarrow H_2O$

(c) Molecular: $Cu(OH)_2(s) + 2HC_2H_3O_2(aq) \longrightarrow Cu(C_2H_3O_2)_2(aq) + 2H_2O$

Ionic: $Cu(OH)_2(s) + 2HC_2H_3O_2^-(aq) \longrightarrow Cu^{2+}(aq) + 2C_2H_3O_2^-(aq) + 2H_2O$

Net: $Cu(OH)_2(s) + 2HC_2H_3O_2^-(aq) \longrightarrow Cu^{2+}(aq) + 2H_2O$

(d) Molecular: $ZnO(s) + H_2SO_4(aq) \longrightarrow ZnSO_4(aq) + H_2O$

Ionic: $ZnO(s) + 2H^+(aq) + SO_4^{2-}(aq) \longrightarrow Zn^{2+}(aq) + SO_4^{2-}(aq) + H_2O$

Net: $ZnO(s) + 2H^+(aq) \longrightarrow Zn^{2+}(aq) + H_2O$

4.99 (a) Molecular: $Na_2SO_3(aq) + Ba(NO_3)_2(aq) \longrightarrow BaSO_3(s) + 2NaNO_3(aq)$

Ionic: $2Na^+(aq) + SO_3^{2-}(aq) + Ba^{2+}(aq) + 2NO_3^-(aq) \longrightarrow BaSO_3(s) + 2Na^+(aq) + 2NO_3^-(aq)$

Net: $Ba^{2+}(aq) + SO_3^{2-}(aq) \longrightarrow BaSO_3(s)$

(b) Molecular: $2HCHO_2(aq) + K_2CO_3(aq) \longrightarrow CO_2(g) + H_2O + 2KCHO_2(aq)$

Ionic: $2HCHO_2(aq) + 2K^+(aq) + CO_3^{2-}(aq) \longrightarrow CO_2(g) + H_2O + 2K^+(aq) + 2CHO_2^-(aq)$

Net: $2HCHO_2(aq) + CO_3^{2-}(aq) \longrightarrow CO_2(g) + H_2O + 2CHO_2^-(aq)$

(c) Molecular: $2NH_4Br(aq) + Pb(C_2H_3O_2)_2(aq) \longrightarrow 2NH_4C_2H_3O_2(aq) + PbBr_2(s)$

Ionic: $2NH_4^+(aq) + 2Br^-(aq) + Pb^{2+}(aq) + 2C_2H_3O_2^-(aq) \longrightarrow$
$2NH_4^+(aq) + 2C_2H_3O_2^-(aq) + PbBr_2(s)$

Net: $Pb^{2+}(aq) + 2Br^-(aq) \longrightarrow PbBr_2(s)$

(d) Molecular: $2NH_4ClO_4(aq) + Cu(NO_3)_2(aq) \longrightarrow Cu(ClO_4)_2(aq) + 2NH_4NO_3(aq)$

Ionic: $2NH_4^+(aq) + 2ClO_4^-(aq) + Cu^{2+}(aq) + 2NO_3^-(aq) \longrightarrow$
$Cu^{2+}(aq) + 2ClO_4^-(aq) + 2NO_3^-(aq) + 2NH_4^+(aq)$

Net: N.R.

4.101 There are numerous possible answers. One of many possible sets of answers would be:

(c) $NaHCO_3(aq) + HCl(aq) \longrightarrow NaCl(aq) + CO_2(g) + H_2O$

(d) $FeCl_2(aq) + 2NaOH(aq) \longrightarrow Fe(OH)_2(s) + 2NaCl(aq)$

(e) $Ba(NO_3)_2(aq) + K_2SO_3(aq) \longrightarrow BaSO_3(s) + 2KNO_3(aq)$

(f) $2AgNO_3(aq) + Na_2S(aq) \longrightarrow Ag_2S(s) + 2NaNO_3(aq)$

(g) $ZnO(s) + 2HCl(aq) \longrightarrow ZnCl_2(aq) + H_2O$

4.103 (a) $$M = \left(\frac{0.17 \text{ mol } H_2SO_4}{85 \text{ mL solution}}\right)\left(\frac{1000 \text{ mL solution}}{1 \text{ L solution}}\right) = 2.0\ M\ H_2SO_4$$

(b) $$M = \left(\frac{7.5\times10^{-3} \text{ mol } NaNO_3}{13.5 \text{ mL solution}}\right)\left(\frac{1000 \text{ mL solution}}{1 \text{ L solution}}\right) = 0.56\ M\ NaNO_3$$

4.105 (a) $NaOH \longrightarrow Na^+ + OH^-$

$$\text{mol NaOH} = (4.00\text{ g NaOH})\left(\frac{1\text{ mol NaOH}}{40.00\text{ g NaOH}}\right) = 0.100\text{ mol NaOH}$$

$$M\text{ NaOH solution} = \left(\frac{0.100\text{ mol NaOH}}{100.0\text{ mL NaOH}}\right)\left(\frac{1000\text{ mL NaOH}}{1\text{ L NaOH}}\right) = 1.00\ M\text{ NaOH}$$

(b) $CaCl_2 \longrightarrow Ca^{2+} + 2Cl^-$

$$\text{mol CaCl}_2 = (16.0\text{ g CaCl}_2)\left(\frac{1\text{ mol CaCl}_2}{110.98\text{ g CaCl}_2}\right) = 0.144\text{ mol CaCl}_2$$

$$M\text{ CaCl}_2\text{ solution} = \left(\frac{0.144\text{ mol CaCl}_2}{250.0\text{ mL CaCl}_2}\right)\left(\frac{1000\text{ mL CaCl}_2}{1\text{ L CaCl}_2}\right) = 0.577\ M\text{ CaCl}_2$$

4.107 $$\text{mL NaC}_2\text{H}_3\text{O}_2 = (14.3\text{ g NaC}_2\text{H}_3\text{O}_2)\left(\frac{1\text{ mol NaC}_2\text{H}_3\text{O}_2}{82.03\text{ g NaC}_2\text{H}_3\text{O}_2}\right)\left(\frac{1\text{ L NaC}_2\text{H}_3\text{O}_2}{0.265\text{ mol NaC}_2\text{H}_3\text{O}_2}\right) \times \left(\frac{1000\text{ mL NaC}_2\text{H}_3\text{O}_2}{1\text{ L NaC}_2\text{H}_3\text{O}_2}\right) = 658\text{ mL NaC}_2\text{H}_3\text{O}_2$$

4.109 (a) $$\text{g NaCl} = 125\text{ mL soln}\left(\frac{1\text{ L}}{1000\text{ mL}}\right)\left(\frac{0.200\text{ mol NaCl}}{1\text{ L soln}}\right)\left(\frac{58.44\text{ g NaCl}}{1\text{ mol NaCl}}\right) = 1.46\text{ g NaCl}$$

(b) $$\text{g C}_6\text{H}_{12}\text{O}_6 = 250\text{ mL soln}\left(\frac{1\text{ L}}{1000\text{ mL}}\right)\left(\frac{0.360\text{ mol C}_6\text{H}_{12}\text{O}_6}{1\text{ L soln}}\right)\left(\frac{180.2\text{ g C}_6\text{H}_{12}\text{O}_6}{1\text{ mol C}_6\text{H}_{12}\text{O}_6}\right)$$

$$= 16.2\text{ g C}_6\text{H}_{12}\text{O}_6$$

(c) $$\text{g H}_2\text{SO}_4 = 250\text{ mL soln}\left(\frac{1\text{ L}}{1000\text{ mL}}\right)\left(\frac{0.250\text{ mol H}_2\text{SO}_4}{1\text{ L soln}}\right)\left(\frac{98.08\text{ g H}_2\text{SO}_4}{1\text{ mol H}_2\text{SO}_4}\right) = 6.13\text{ g H}_2\text{SO}_4$$

4.111 $$\text{mol of H}_2\text{SO}_4 = 25.0\text{ mL H}_2\text{SO}_4\left(\frac{1\text{ L soln}}{1000\text{ mL soln}}\right)\left(\frac{0.56\text{ mol H}_2\text{SO}_4}{1\text{ L soln}}\right) = 0.014\text{ mol H}_2\text{SO}_4$$

$$M\text{ of final solution} = \left(\frac{0.014\text{ mol H}_2\text{SO}_4}{125\text{ mL H}_2\text{SO}_4}\right)\left(\frac{1000\text{ mL H}_2\text{SO}_4}{1\text{ L H}_2\text{SO}_4}\right) = 0.11\ M\text{ H}_2\text{SO}_4$$

4.113 $M_1V_1 = M_2V_2$ $\quad V_2 = \left(\frac{M_1V_1}{M_2}\right)$

$$V_2 = \frac{(18.0\ M\text{ H}_2\text{SO}_4)(25.0\text{ mL})}{1.50\ M\text{ H}_2\text{SO}_4} = 300\text{ mL H}_2\text{SO}_4$$

The 25.0 mL of H_2SO_4 must be diluted to 300 mL.

4.115 $M_1V_1 = M_2V_2$ $\quad V_2 = \left(\frac{M_1V_1}{M_2}\right)$

$$V_2 = \frac{(2.50\ M\text{ KOH})(150.0\text{ mL})}{1.0\ M\text{ KOH}} = 375\text{ mL KOH}$$

The 150.0 mL of KOH must be diluted to 375 mL. The volume of water to be added is:
375 mL of V_2 – 150 mL of V_1 = 225 mL water

4.117 (a) $CaCl_2 \longrightarrow Ca^{2+} + 2Cl^-$

$$\text{mol } CaCl_2 = 0.0323 \text{ L} \times \left(\frac{0.455 \text{ mol } CaCl_2}{1 \text{ L solution}}\right) = 0.0147 \text{ mol } CaCl_2$$

$$0.0147 \text{ mol } CaCl_2\left(\frac{1 \text{ mol } Ca^{2+}}{1 \text{ mol } CaCl_2}\right) = 0.0147 \text{ mol } Ca^{2+}$$

$$0.0147 \text{ mol } CaCl_2\left(\frac{2 \text{ mol } Cl^-}{1 \text{ mol } CaCl_2}\right) = 0.0294 \text{ mol } Cl^-$$

(b) $AlCl_3 \longrightarrow Al^{3+} + 3Cl^-$

$$\text{mol } AlCl_3 = 0.0500 \text{ L} \times \left(\frac{0.408 \text{ mol } AlCl_3}{1 \text{ L solution}}\right) = 0.0204 \text{ mol } AlCl_3$$

$$0.0204 \text{ mol } AlCl_3\left(\frac{1 \text{ mol } Al^{3+}}{1 \text{ mol } AlCl_3}\right) = 0.0204 \text{ mol } Al^{3+}$$

$$0.0204 \text{ mol } AlCl_3\left(\frac{3 \text{ mol } Cl^-}{1 \text{ mol } AlCl_3}\right) = 0.0612 \text{ mol } Cl^-$$

4.119 (a) $Cr(NO_3)_2 \longrightarrow Cr^{2+} + 2NO_3^-$

$$M\ Cr^{2+} = \left(\frac{0.25 \text{ mol } Cr(NO_3)_2}{1 \text{ L } Cr(NO_3)_2 \text{ soln}}\right)\left(\frac{1 \text{ mol } Cr^{2+}}{1 \text{ mol } Cr(NO_3)_2}\right) = 0.25\ M\ Cr^{2+}$$

$$M\ NO_3^- = \left(\frac{0.25 \text{ mol } Cr(NO_3)_2}{1 \text{ L } Cr(NO_3)_2 \text{ soln}}\right)\left(\frac{2 \text{ mol } NO_3^-}{1 \text{ mol } Cr(NO_3)_2}\right) = 0.50\ M\ NO_3^-$$

(b) $CuSO_4 \longrightarrow Cu^{2+} + SO_4^{2-}$

$$M\ Cu^{2+} = \left(\frac{0.10 \text{ mol } CuSO_4}{1 \text{ L } CuSO_4 \text{ soln}}\right)\left(\frac{1 \text{ mol } Cu^{2+}}{1 \text{ mol } CuSO_4}\right) = 0.10\ M\ Cu^{2+}$$

$$M\ SO_4^{2-} = \left(\frac{0.10 \text{ mol } CuSO_4}{1 \text{ L } CuSO_4 \text{ soln}}\right)\left(\frac{1 \text{ mol } SO_4^{2-}}{1 \text{ mol } CuSO_4}\right) = 0.10\ M\ SO_4^{2-}$$

(c) $Na_3PO_4 \longrightarrow 3Na^+ + PO_4^{3-}$

$$M\ Na^+ = \left(\frac{0.16 \text{ mol } Na_3PO_4}{1 \text{ L } Na_3PO_4 \text{ soln}}\right)\left(\frac{3 \text{ mol } Na^+}{1 \text{ mol } Na_3PO_4}\right) = 0.48\ M\ Na^+$$

$$M\ PO_4^{3-} = \left(\frac{0.16 \text{ mol } Na_3PO_4}{1 \text{ L } Na_3PO_4 \text{ soln}}\right)\left(\frac{1 \text{ mol } PO_4^{3-}}{1 \text{ mol } Na_3PO_4}\right) = 0.16\ M\ PO_4^{3-}$$

(d) $Al_2(SO_4)_3 \longrightarrow 2Al^{3+} + 3SO_4^{2-}$

$$M\ Al^{3+} = \left(\frac{0.075 \text{ mol } Al_2(SO_4)_3}{1 \text{ L } Al_2(SO_4)_3 \text{ soln}}\right)\left(\frac{2 \text{ mol } Al^{3+}}{1 \text{ mol } Al_2(SO_4)_3}\right) = 0.15\ M\ Al^{3+}$$

4.121 $$\text{g } Al_2(SO_4)_3 = (50.0 \text{ mL soln})\left(\frac{0.125 \text{ mol } Al^{3+}}{1000 \text{ mL soln}}\right)\left(\frac{1 \text{ mol } Al_2(SO_4)_3}{2 \text{ mol } Al^{3+}}\right)\left(\frac{342.14 \text{ g } Al_2(SO_4)_3}{1 \text{ mol } Al_2(SO_4)_3}\right)$$
$$= 1.07 \text{ g } Al_2(SO_4)_3$$

4.123 $$\text{mL } NiCl_2 \text{ soln} = 20.0 \text{ mL soln}\left(\frac{0.153 \text{ mol } Na_2CO_3}{1000 \text{ mL soln}}\right)\left(\frac{1 \text{ mol } NiCl_2}{1 \text{ mol } Na_2CO_3}\right)\left(\frac{1000 \text{ mL soln}}{0.258 \text{ mol } NiCl_2}\right)$$
$$= 11.9 \text{ mL } NiCl_2 \text{ soln}$$

$$\text{g } NiCO_3 = 11.9 \text{ mL } NiCl_2 \text{ soln}\left(\frac{0.258 \text{ mol } NiCl_2}{1000 \text{ mL soln}}\right)\left(\frac{1 \text{ mol } NiCO_3}{1 \text{ mol } NiCl_2}\right)\left(\frac{118.7 \text{ g } NiCO_3}{1 \text{ mol } NiCO_3}\right) = 0.36 \text{ g } NiCO_3$$

4.125 $$M \text{ KOH} = \frac{(20.78 \text{ mL HCl soln})\left(\frac{1 \text{ L HCl soln}}{1000 \text{ mL HCl soln}}\right)\left(\frac{0.116 \text{ mol HCl}}{1 \text{ L HCl}}\right)\left(\frac{1 \text{ mol KOH}}{1 \text{ mol HCl}}\right)}{21.34 \text{ mL KOH}\left(\frac{1 \text{ L KOH}}{1000 \text{ mL KOH}}\right)} = 0.113\ M \text{ KOH}$$

$$KOH(aq) + HCl(aq) \longrightarrow KCl(aq) + H_2O$$

4.127 $$Al_2(SO_4)_3(aq) + 3Ba(OH)_2(aq) \longrightarrow 2Al(OH)_3(s) + 3BaSO_4(s)$$

$$\text{g } Al_2(SO_4)_3 = 85.0 \text{ mL soln}\left(\frac{0.0500 \text{ mol } Ba(OH)_2}{1000 \text{ mL soln}}\right)\left(\frac{1 \text{ mol } Al_2(SO_4)_3}{3 \text{ mol } Ba(OH)_2}\right)\left(\frac{342.2 \text{ g } Al_2(SO_4)_3}{1 \text{ mol } Al_2(SO_4)_3}\right)$$
$$= 0.485 \text{ g } Al_2(SO_4)_3$$

4.129 $$\text{mL } FeCl_3 \text{ soln} = 20.0 \text{ mL } AgNO_3\left(\frac{0.0450 \text{ mol } AgNO_3}{1000 \text{ mL soln}}\right)\left(\frac{1 \text{ mol } Ag^+}{1 \text{ mol } AgNO_3}\right)\left(\frac{1 \text{ mol } Cl^-}{1 \text{ mol } Ag^+}\right)$$
$$\times\left(\frac{1 \text{ mol } FeCl_3}{3 \text{ mol } Cl^-}\right)\left(\frac{1000 \text{ mL soln}}{0.150 \text{ mol } FeCl_3}\right) = 2.00 \text{ mL } FeCl_3 \text{ soln}$$

$$\text{g AgCl} = (20.0 \text{ mL } AgNO_3) \times$$
$$\left(\frac{1 \text{ L } AgNO_3}{1000 \text{ mL } AgNO_3}\right)\left(\frac{0.0450 \text{ mol } AgNO_3}{1 \text{ L } AgNO_3}\right)\left(\frac{3 \text{ mol AgCl}}{3 \text{ mol } AgNO_3}\right)\left(\frac{143.32 \text{ AgCl}}{1 \text{ mol AgCl}}\right) = 0.129 \text{ g AgCl}$$

4.131 $$Ag^+ + Cl^- \longrightarrow AgCl(s)$$

$$\text{mL } AlCl_3 = (20.0 \text{ mL } AgC_2H_3O_2)\left(\frac{0.500 \text{ mol } AgC_2H_3O_2}{1000 \text{ mL } AgC_2H_3O_2}\right)\left(\frac{1 \text{ mol } Ag^+}{1 \text{ mol } AgC_2H_3O_2}\right)\left(\frac{1 \text{ mol } Cl^-}{1 \text{ mol } Ag^+}\right)$$
$$\times\left(\frac{1 \text{ mol } AlCl_3}{3 \text{ mol } Cl^-}\right)\left(\frac{1000 \text{ mL } AlCl_3}{0.250 \text{ moles } AlCl_3}\right) = 13.3 \text{ mL } AlCl_3$$

4.133 $$Fe_2O_3 + 6HCl \longrightarrow 2FeCl_3 + 3H_2O$$

$$0.0250 \text{ L HCl} \times \left(\frac{0.500 \text{ mol HCl}}{1 \text{ L HCl}}\right) = 1.25 \times 10^{-2} \text{ mol HCl}$$

$$\text{mol } Fe^{3+} = (1.25 \times 10^{-2} \text{ mol HCl})\left(\frac{1 \text{ mol } Fe_2O_3}{6 \text{ mol HCl}}\right)\left(\frac{2 \text{ mol } Fe^{3+}}{1 \text{ mol } Fe_2O_3}\right) = 4.17 \times 10^{-3} \text{ mol } Fe^{3+}$$

$$M\ Fe^{3+} = \frac{4.17 \times 10^{-3} \text{ mol } Fe^{3+}}{0.0250 \text{ L soln}} = 0.167\ M\ Fe^{3+}$$

$$g\ Fe_2O_3 = 4.17 \times 10^{-3}\ mol\ Fe^{3+}\left(\frac{1\ mol\ Fe_2O_3}{2\ mol\ Fe^{3+}}\right)\left(\frac{159.69\ g\ Fe_2O_3}{1\ mol\ Fe_2O_3}\right) = 0.333\ g\ Fe_2O_3$$

Therefore, the mass of Fe_2O_3 that remains unreacted is:
(4.00 g – 0.333 g) = 3.67 g

4.135 First, calculate the number of moles HCl based on the titration using the following equation:

$$NaOH(aq) + HCl(aq) \longrightarrow NaCl(aq) + H_2O$$

$$mol\ HCl = (23.25\ mL\ NaOH)\left(\frac{0.105\ mol\ NaOH}{1000\ mL\ NaOH}\right)\left(\frac{1\ mol\ HCl}{1\ mol\ NaOH}\right) = 2.44 \times 10^{-3}\ mol\ HCl$$

Next, determine the concentration of the HCl solution:

$$\frac{2.44 \times 10^{-3}\ mol\ HCl}{0.02145\ L\ solution} = 0.114\ M\ HCl$$

4.137 Since lactic acid is monoprotic, it reacts with sodium hydroxide on a one to one mole basis:

(h) $$mol\ HC_3H_5O_3 = (17.25\ mL\ NaOH)\left(\frac{0.155\ mol\ NaOH}{1000\ mL\ NaOH}\right)\left(\frac{1\ mol\ HC_3H_5O_3}{1\ mol\ NaOH}\right)$$

$$= 2.67 \times 10^{-3}\ mol\ HC_3H_5O_3$$

(b) $$g\ HC_3H_5O_3 = 2.67 \times 10^{-3}\ mol \times \left(\frac{90.08\ g\ HC3H5O3}{1\ mol\ HC3H5O3}\right) = 0.241\ g$$

4.139 (a) $$mol\ Pb = (29.22\ mL)\left(\frac{1\ L\ Na_2SO_4}{1000\ mL\ Na_2SO_4}\right)\left(\frac{0.122\ mol\ Na_2SO_4}{1\ L\ Na_2SO_4}\right) \times$$

$$\left(\frac{1\ mol\ PbSO_4}{1\ mol\ Na_2SO_4}\right)\left(\frac{1\ mol\ Pb^{2+}}{1\ mol\ PbSO_4}\right) = 3.56 \times 10^{-3}\ mol\ Pb$$

(b) $$g\ Pb = (29.22\ mL)\left(\frac{1\ L\ Na_2SO_4}{1000\ mL\ Na_2SO_4}\right)\left(\frac{0.122\ mol\ Na_2SO_4}{1\ L\ Na_2SO_4}\right)\left(\frac{1\ mol\ PbSO_4}{1\ mol\ Na_2SO_4}\right) \times$$

$$\left(\frac{1\ mol\ Pb^{2+}}{1\ mol\ PbSO_4}\right)\left(\frac{207.2\ g\ Pb^{2+}}{1\ mol\ Pb^{2+}}\right) = 0.7386\ g\ Pb$$

(c) The percentage of Pb in the sample can be calculated as

$$\left(\frac{0.7386\ g\ Pb}{1.526\ g\ sample}\right) \times 100\% = 48.40\%\ \text{Pb in the sample.}$$

(d) The mass of dried $PbSO_4$ should give the same result.

Chapter Five
Oxidation-Reduction Reactions

Practice Exercises

5.1 $2Na(s) + O_2(g) \longrightarrow Na_2O_2(s)$
Oxygen is reduced since it gains electrons.
Sodium is oxidized since it loses electrons.

5.2 $2Al(s) + 3Cl_2(g) \longrightarrow 2AlCl_3(aq)$
Aluminum is oxidized and is, therefore, the reducing agent.
Chlorine is reduced and is, therefore, the oxidizing agent.

5.3 ClO_2^-: O –2 Cl +3

5.4 (a) Ni +2; Cl –1
(b) Mg +2; Ti +4; O –2
(c) K +1; Cr +6; O –2
(d) H +1; P +5, O –2
(e) V +3; C 0; H +1; O –2
(f) N –3; H +1

5.5 First the oxidation numbers of all atoms must be found.

$N_2O_5 + 3NaHCO_3 \longrightarrow 2NaNO_3 + 2CO_2 + H_2O$

Reactants:	Products:
N = +5	N = +5
O = –2	O = –2
Na = +1	Na = +1
H = +1	H = +1
C = +4	C = +4
O = –2	O= –2
	O = –2

None of the oxidation numbers change, therefore it is not a redox reaction.

$KClO_3 + 3HNO_2 \longrightarrow KCl + 3HNO_3$

Reactants:	Products:
K = +1	K = +1
Cl = +5	Cl = –1
O = –2	O = –2
H = +1	H = +1
N = +3	N = +5
O = –2	O = –2

The oxidation numbers for K and Na do not change. However, the oxidation numbers for the chlorine atom decreases. The oxidation numbers for nitrogen increase.
Therefore, $KClO_3$ is reduced and HNO_2 is oxidized.
This means $KClO_3$ is the oxidizing agent and HNO_2 is the reducing agent.
This reaction is the redox reaction.

5.6 First the oxidation numbers of all atoms must be found.

$Cl_2 + 2NaClO_2 \longrightarrow 2ClO_2 + 2NaCl$

Reactants:	Products:
Cl = 0	Cl = +4
Na = +1	Na = +1
Cl = +3	Cl = −1
O = −2	O = −2

The oxidation numbers for O and Na do not change. However, the oxidation numbers for all chlorine atoms change, and we cannot tell which chlorines are reduced and which are oxidized. One analysis would have the Cl in Cl_2 end up as the Cl in NaCl, while the Cl in $NaClO_2$ ends up as the Cl in ClO_2. In this case Cl_2 is reduced and is the oxidizing agent, while $NaClO_2$ is oxidized and is the reducing agent.

5.7 $Al(s) + Cu^{2+}(aq) \longrightarrow Al^{3+}(aq) + Cu(s)$

It is not balanced because the charges are not balanced.

First, we break the reaction above into half-reactions:

$Al(s) \longrightarrow Al^{3+}(aq)$

$Cu^{2+}(aq) \longrightarrow Cu(s)$

Each half-reaction is already balanced with respect to atoms, so next we add electrons to balance the charges on both sides of the equations:

$Al(s) \longrightarrow Al^{3+}(aq) + 3e^-$

$2e^- + Cu^{2+}(aq) \longrightarrow Cu(s)$

Next, we multiply both equations so that the electrons gained equals the electrons lost,

$2(Al(s) \longrightarrow Al^{3+}(aq) + 3e^-)$

$3(2e^- + Cu^{2+}(aq) \longrightarrow Cu(s))$

which gives us:

$2Al(s) \longrightarrow 2Al^{3+}(aq) + 6e^-$

$6e^- + 3Cu^{2+}(aq) \longrightarrow 3Cu(s)$

Now, by adding the half-reactions back together, we have our balanced equation:

$2Al(s) + 3Cu^{2+}(aq) \longrightarrow 2Al^{3+}(aq) + 3Cu(s)$

5.8 $TcO_4^- + Sn^{2+} \longrightarrow Tc^{4+} + Sn^{4+}$

First, we break the reaction above into half-reactions:

$TcO_4^- \longrightarrow Tc^{4+}$ $\quad$ $Sn^{2+} \longrightarrow Sn^{4+}$

Each half-reaction is already balanced with respect to atoms other than O and H, so next we balance the O atoms by using water:

$TcO_4^- \longrightarrow Tc^{4+} + 4H_2O$

$Sn^{2+} \longrightarrow Sn^{4+}$

Now we balance H by using H^+:

$8H^+ + TcO_4^- \longrightarrow Tc^{4+} + 4H_2O$

$Sn^{2+} \longrightarrow Sn^{4+}$

Next, we add electrons to balance the charges on both sides of the equations:

$3e^- + 8H^+ + TcO_4^- \longrightarrow Tc^{4+} + 4H_2O$

$Sn^{2+} \longrightarrow Sn^{4+} + 2e^-$

We multiply the equations so that the electrons gained equals the electrons lost,

$2(3e^- + 8H^+ + TcO_4^- \longrightarrow Tc^{4+} + 4H_2O)$

$3(Sn^{2+} \longrightarrow Sn^{4+} + 2e^-)$

which gives us:

$6e^- + 16H^+ + 2TcO_4^- \longrightarrow 2Tc^{4+} + 8H_2O$

$3Sn^{2+} \longrightarrow 3Sn^{4+} + 6e^-$

Now, by adding the half-reactions back together, we have our balanced equation:

$3Sn^{2+} + 16H^+ + 2TcO_4^- \longrightarrow 2Tc^{4+} + 8H_2O + 3Sn^{4+}$

5.9 $2H_2O + SO_2 \longrightarrow SO_4^{2-} + 4H^+ + 2e^-$

$4OH^- + 2H_2O + SO_2 \longrightarrow SO_4^{2-} + 4H^+ + 2e^- + 4OH^-$

$4OH^- + 2H_2O + SO_2 \longrightarrow SO_4^{2-} + 2e^- + 4H_2O$

$4OH^- + SO_2 \longrightarrow SO_4^{2-} + 2e^- + 2H_2O$

5.10 Follow the ten steps to balance a redox reaction.

$MnO_4^- \longrightarrow MnO_2$

$C_2O_4^{2-} \longrightarrow CO_3^{2-}$

$MnO_4^- \longrightarrow MnO_2$

$C_2O_4^{2-} \longrightarrow 2CO_3^{2-}$

$MnO_4^- \longrightarrow MnO_2 + 2H_2O$

$C_2O_4^{2-} + 2H_2O \longrightarrow 2CO_3^{2-}$

$MnO_4^- + 4H^+ \longrightarrow MnO_2 + 2H_2O$

$C_2O_4^{2-} + 2H_2O \longrightarrow 2CO_3^{2-} + 4H^+$

$MnO_4^- + 4H^+ + 3e^- \longrightarrow MnO_2 + 2H_2O$

$C_2O_4^{2-} + 2H_2O \longrightarrow 2CO_3^{2-} + 4H^+ + 2e^-$

$(MnO_4^- + 4H^+ + 3e^- \longrightarrow MnO_2 + 2H_2O) \times 2$

$(C_2O_4^{2-} + 2H_2O \longrightarrow 2CO_3^{2-} + 4H^+ + 2e^-) \times 3$

$2MnO_4^- + 3C_2O_4^{2-} + 2H_2O \longrightarrow 2MnO_2 + 6CO_3^{2-} + 4H^+$

Adding $4OH^-$ to both sides of the above equation we get:

$2MnO_4^- + 3C_2O_4^{2-} + 2H_2O + 4OH^- \longrightarrow 2MnO_2 + 6CO_3^{2-} + 4H^+ + 4OH^-$

$2MnO_4^- + 3C_2O_4^{2-} + 2H_2O + 4OH^- \longrightarrow 2MnO_2 + 6CO_3^{2-} + 4H_2O$

$2MnO_4^- + 3C_2O_4^{2-} + 4OH^- \longrightarrow 2MnO_2 + 6CO_3^{2-} + 2H_2O$

5.11 $Zn + H^+ \longrightarrow Zn^{2+} + H_2$

Divide the reaction into two half reactions and balance the number of atoms

$Zn \longrightarrow Zn^{2+}$ $\qquad$ $2H^+ \longrightarrow H_2$

Balance the charges with electrons

$Zn \longrightarrow Zn^{2+} + 2e^-$

$2H^+ + 2e^- \longrightarrow H_2$

5.12 (a) molecular: $Mg(s) + 2HCl(aq) \longrightarrow MgCl_2(aq) + H_2(g)$

ionic: $Mg(s) + 2H^+(aq) + 2Cl^-(aq) \longrightarrow Mg^{2+}(aq) + 2Cl^-(aq) + H_2(g)$

net ionic: $Mg(s) + 2H^+(aq) \longrightarrow Mg^{2+}(aq) + H_2(g)$

(b) molecular: $2Al(s) + 6HCl(aq) \longrightarrow 2AlCl_3(aq) + 3H_2(g)$

ionic: $2Al(s) + 6H^+(aq) + 6Cl^-(aq) \longrightarrow 2Al^{3+}(aq) + 6Cl^-(aq) + 3H_2(g)$

net ionic: $2Al(s) + 6H^+(aq) \longrightarrow 2Al^{3+}(aq) + 3H_2(g)$

5.13 $Cu^{2+}(aq) + Mg(s) \longrightarrow Cu(s) + Mg^{2+}(aq)$

5.14 (a) $2Al(s) + 3Cu^{2+}(aq) \longrightarrow 2Al^{3+}(aq) + 3Cu(s)$

(b) $Ag(s) + Mg^{2+}(aq) \longrightarrow$ No reaction

5.15 $C_4H_4S(s) + 6O_2(g) \longrightarrow 4CO_2(g) + 2H_2O + SO_2(g)$

5.16 $C_5H_8(g) + 7O_2(g) \longrightarrow 5CO_2(g) + 4H_2O(g)$

5.17 $2Sr(s) + O_2(g) \longrightarrow 2SrO(s)$

5.18 $4Fe(s) + 3O_2(g) \longrightarrow 2Fe_2O_3(s)$

Review Problems

5.29 The sum of the oxidation numbers should be equal to the total charge:

(a) ClO_4^-: Cl +7, O −2

(b) Cl^-: Cl −1

(c) SF_6: S +6, F −1

(d) $Au(NO_3)_3$: Au +3, N +5,O −2

5.31 The sum of the oxidation numbers should be equal to the total charge:

(a)	O:	−2	(c)	O:	−2
	Na:	+1		Na:	+1
	Br:	+1		Br:	+5
(b)	O:	−2	(d)	O:	−2
	Na:	+1		Na:	+1
	Br:	+3		Br:	+7

5.33 The sum of the oxidation numbers should be zero:

(a)	S:	−2	(c)	Cs	+1	
	Bi:	+3		O	−1/2	(The Cs can only have an oxidation number of +1 or 0.)
(b)	Cl:	−1	(d)	F	−1	
	Ce:	+4		O	+1	

5.35 Ti +3; N −3

5.37 O_3; oxidation number of O is 0

5.39 (a) substance reduced (and oxidizing agent): HNO_3

substance oxidized (and reducing agent): H_3AsO_3

(b) substance reduced (and oxidizing agent): HOCl

substance oxidized (and reducing agent): NaI

(c) substance reduced (and oxidizing agent): $KMnO_4$

substance oxidized (and reducing agent): $H_2C_2O_4$

(d) substance reduced (and oxidizing agent): H_2SO_4
substance oxidized (and reducing agent): Al

5.41 $Cl_2(aq) + H_2O \longrightarrow H^+(aq) + Cl^-(aq) + HOCl(aq)$
In the forward direction: The oxidation number of the chlorine atoms decreases from 0 to –1. Therefore Cl_2 is reduced. However, in HOCl, chlorine has an oxidation number of +1, so Cl_2 also oxidized! (One atom is reduced, the other is oxidized.)
In the reverse direction: The Cl^- ion begins with an oxidation number of –1 and ends with an oxidation number of 0. Therefore the Cl^- ion is oxidized: This means Cl^- is the reducing agent. Since the oxidation number of H^+ does not change, HOCl must be the oxidizing agent.

5.43 (a)
$2S_2O_3^{2-} \longrightarrow S_4O_6^{2-}$
$OCl^- \longrightarrow Cl^-$
$2S_2O_3^{2-} \longrightarrow S_4O_6^{2-}$
$OCl^- + 2H^+ \longrightarrow Cl^- + H_2O$
$2S_2O_3^{2-} \longrightarrow S_4O_6^{2-} + 2e^-$
$OCl^- + 2H^+ + 2e^- \longrightarrow Cl^- + H_2O$
$OCl^- + 2S_2O_3^{2-} + 2H^+ \longrightarrow S_4O_6^{2-} + Cl^- + H_2O$

(b)
$NO_3^- \longrightarrow NO_2$
$Cu \longrightarrow Cu^{2+}$
$NO_3^- + 2H^+ \longrightarrow NO_2 + H_2O$
$Cu \longrightarrow Cu^{2+}$
$NO_3^- + 2H^+ + e^- \longrightarrow NO_2 + H_2O$
$Cu \longrightarrow Cu^{2+} + 2e^-$
$(NO_3^- + 2H^+ + e^- \longrightarrow NO_2 + H_2O) \times 2$
$Cu \longrightarrow Cu^{2+} + 2e^-$
$2NO_3^- + Cu + 4H^+ \longrightarrow 2NO_2 + Cu^{2+} + 2H_2O$

(c)
$IO_3^- \longrightarrow I^-$
$H_3AsO_3 \longrightarrow H_3AsO_4$
$IO_3^- + 6H^+ \longrightarrow I^- + 3H_2O$
$H_2O + H_3AsO_3 \longrightarrow H_3AsO_4 + 2H^+$
$IO_3^- + 6H^+ + 6e^- \longrightarrow I^- + 3H_2O$
$H_2O + H_3AsO_3 \longrightarrow H_3AsO_4 + 2H^+ + 2e^-$
$IO_3^- + 6H^+ + 6e^- \longrightarrow I^- + 3H_2O$
$(H_2O + H_3AsO_3 \longrightarrow H_3AsO_4 + 2H^+ + 2e^-) \times 3$
$IO_3^- + 3H_3AsO_3 + 6H^+ + 3H_2O \longrightarrow I^- + 3H_3AsO_4 + 3H_2O + 6H^+$
which simplifies to give:
$IO_3^- + 3H_3AsO_3 \longrightarrow I^- + 3H_3AsO_4$

(d)
$SO_4^{2-} \longrightarrow SO_2$
$Zn \longrightarrow Zn^{2+}$
$SO_4^{2-} + 4H^+ \longrightarrow SO_2 + 2H_2O$
$Zn \longrightarrow Zn^{2+}$

$SO_4^{2-} + 4H^+ + 2e^- \longrightarrow SO_2 + 2H_2O$

$Zn \longrightarrow Zn^{2+} + 2e^-$

$Zn + SO_4^{2-} + 4H^+ \longrightarrow Zn^{2+} + SO_2 + 2H_2O$

5.45 (a) $(Sn + 2H_2O \longrightarrow SnO_2 + 4H^+ + 4e^-) \times 3$

$(NO_3^- + 4H^+ + 3e^- \longrightarrow NO + 2H_2O) \times 4$

$3Sn + 4NO_3^- + 16H^+ + 6H_2O \longrightarrow 3SnO_2 + 12H^+ + 4NO + 8H_2O$

which simplifies to: $3Sn + 4NO_3^- + 4H^+ \longrightarrow 3SnO_2 + 4NO + 2H_2O$

(b) $PbO_2 + 2Cl^- + 4H^+ + 2e^- \longrightarrow PbCl_2 + 2H_2O$

$2Cl^- \longrightarrow Cl_2 + 2e^-$

$PbO_2 + 4Cl^- + 4H^+ \longrightarrow PbCl_2 + Cl_2 + 2H_2O$

(c) $Ag \longrightarrow Ag^+ + e^-$

$NO_3^- + 2H^+ + e^- \longrightarrow NO_2 + H_2O$

$Ag + 2H^+ + NO_3^- \longrightarrow Ag^+ + NO_2 + H_2O$

(d) $(Fe^{3+} + e^- \longrightarrow Fe^{2+}) \times 4$

$2NH_3OH^+ \longrightarrow N_2O + H_2O + 6H^+ + 4e^-$

$4Fe^{3+} + 2NH_3OH^+ \longrightarrow 4Fe^{2+} + N_2O + 6H^+ + H_2O$

5.47 For redox reactions in basic solution, we proceed to balance the half reactions as if they were in acid solution, and then add enough OH^- to each side of the resulting equation in order to neutralize (titrate) all of the H^+.

(a) $(CrO_4^{2-} + 4H^+ + 3e^- \longrightarrow CrO_2^- + 2H_2O) \times 2$

$(S^{2-} \longrightarrow S + 2e^-) \times 3$

$2CrO_4^{2-} + 3S^{2-} + 8H^+ \longrightarrow 2CrO_2^- + 3S + 4H_2O$

Adding $8OH^-$ to both sides of the above equation we obtain:

$2CrO_4^{2-} + 3S^{2-} + 8H^+ + 8OH^- \longrightarrow 2CrO_2^- + 3S + 4H_2O + 8OH^-$

$2CrO_4^{2-} + 3S^{2-} + 8H_2O \longrightarrow 2CrO_2^- + 8OH^- + 3S + 4H_2O$

which simplifies to: $2CrO_4^{2-} + 3S^{2-} + 4H_2O \longrightarrow 2CrO_2^- + 3S + 8OH^-$

(b) $(C_2O_4^{2-} \longrightarrow 2CO_2 + 2e^-) \times 3$

$(MnO_4^- + 4H^+ + 3e^- \longrightarrow MnO_2 + 2H_2O) \times 2$

$3C_2O_4^{2-} + 2MnO_4^- + 8H^+ \longrightarrow 6CO_2 + 2MnO_2 + 4H_2O$

Adding $8OH^-$ to both sides of the above equation we get:

$3C_2O_4^{2-} + 2MnO_4^- + 8H^+ + 8OH^- \longrightarrow 6CO_2 + 2MnO_2 + 4H_2O + 8OH^-$

$3C_2O_4^{2-} + 2MnO_4^- + 8H_2O \longrightarrow 6CO_2 + 2MnO_2 + 4H_2O + 8OH^-$

which simplifies to give:

$3C_2O_4^{2-} + 2MnO_4^- + 4H_2O \longrightarrow 6CO_2 + 2MnO_2 + 8OH^-$

(c) $(ClO_3^- + 6H^+ + 6e^- \longrightarrow Cl^- + 3H_2O) \times 4$

$(N_2H_4 + 2H_2O \longrightarrow 2NO + 8H^+ + 8e^-) \times 3$

$4ClO_3^- + 3N_2H_4 + 24H^+ + 6H_2O \longrightarrow 4Cl^- + 6NO + 12H_2O + 24H^+$

which needs no OH^-, because it simplifies directly to:

$4ClO_3^- + 3N_2H_4 \longrightarrow 4Cl^- + 6NO + 6H_2O$

(d) $(SO_3^{2-} + H_2O \longrightarrow SO_4^{2-} + 2H^+ + 2e^-) \times 3$

$(MnO_4^- + 4H^+ + 3e^- \longrightarrow MnO_2 + 2H_2O) \times 2$

$3SO_3^{2-} + 3H_2O + 8H^+ + 2MnO_4^- \longrightarrow 3SO_4^{2-} + 6H^+ + 2MnO_2 + 4H_2O$

Adding $8OH^-$ to both sides of the equation we obtain:

$3SO_3^{2-} + 3H_2O + 8H^+ + 2MnO_4^- + 8OH^- \longrightarrow 3SO_4^{2-} + 6H^+ + 2MnO_2 + 4H_2O + 8OH^-$

$3SO_3^{2-} + 11H_2O + 2MnO_4^- \longrightarrow 3SO_4^{2-} + 10H_2O + 2MnO_2 + 2OH^-$

which simplifies to:

$3SO_3^{2-} + 2MnO_4^- + H_2O \longrightarrow 3SO_4^{2-} + 2MnO_2 + 2OH^-$

5.49 $NO_3^- + 10H^+ + 8e^- \longrightarrow NH_4^+ + 3H_2O$

$(Mg \longrightarrow Mg^{2+} + 2e^-) \times 4$

$10H^+(aq) + NO_3^-(aq) + 4Mg(s) \longrightarrow NH_4^+(aq) + 4Mg^{2+}(aq) + 3H_2O(aq)$

5.51 $H_2O + C \longrightarrow CO + H_2$

5.53 $(H_2C_2O_4 \longrightarrow 2CO_2 + 2H^+ + 2e^-) \times 3$

$K_2Cr_2O_7 + 14H^+ + 6e^- \longrightarrow 2K^+ + 2Cr^{3+} + 7H_2O$

$3H_2C_2O_4 + K_2Cr_2O_7 + 14H^+ \longrightarrow 6CO_2 + 2K^+ + 2Cr^{3+} + 6H^+ + 7H_2O$

which simplifies to:

$3H_2C_2O_4 + K_2Cr_2O_7 + 8H^+ \longrightarrow 6CO_2 + 2K^+ + 2Cr^{3+} + 7H_2O$

5.55 $O_3 + 6H^+ + 6e^- \longrightarrow 3H_2O$

$Br^- + 3H_2O \longrightarrow BrO_3^- + 6H^+ + 6e^-$

$O_3 + Br^- + 3H_2O + 6H^+ \longrightarrow BrO_3^- + 3H_2O + 6H^+$

which simplifies to:

$O_3 + Br^- \longrightarrow BrO_3^-$

5.57 (a) Molecular: $Mn(s) + 2HCl(aq) \longrightarrow MnCl_2(aq) + H_2(g)$

Ionic: $Mn(s) + 2H^+(aq) + 2Cl^-(aq) \longrightarrow Mn^{2+}(aq) + 2Cl^-(aq) + H_2(g)$

Net Ionic: $Mn(s) + 2H^+(aq) \longrightarrow Mn^{2+}(aq) + H_2(g)$

(b) M: $Cd(s) + 2HCl(aq) \longrightarrow CdCl_2(aq) + H_2(g)$

I: $Cd(s) + 2H^+(aq) + 2Cl^-(aq) \longrightarrow Cd^{2+}(aq) + 2Cl^-(aq) + H_2(g)$

Net Ionic: $Cd(s) + 2H^+(aq) \longrightarrow Cd^{2+}(aq) + H_2(g)$

(c) M: $Sn(s) + 2HCl(aq) \longrightarrow SnCl_2(aq) + H_2(g)$

I: $Sn(s) + 2H^+(aq) + 2Cl^-(aq) \longrightarrow Sn^{2+}(aq) + 2Cl^-(aq) + H_2(g)$

Net Ionic: $Sn(s) + 2H^+(aq) \longrightarrow Sn^{2+}(aq) + H_2(g)$

5.59 (a) Molecular: $Ni(s) + H_2SO_4(aq) \longrightarrow NiSO_4(aq) + H_2(g)$

Ionic: $Ni(s) + 2H^+(aq) + SO_4^{2-}(aq) \longrightarrow Ni^{2+}(aq) + SO_4^{2-}(aq) + H_2(g)$

Net Ionic: $Ni(s) + 2H^+(aq) \longrightarrow Ni^{2+}(aq) + H_2(g)$

(b) Molecular: $2Cr(s) + 3H_2SO_4(aq) \longrightarrow Cr_2(SO_4)_3(aq) + 3H_2(g)$

Ionic: $2Cr(s) + 6H^+(aq) + 3SO_4^{2-}(aq) \longrightarrow 2Cr^{3+}(aq) + 3SO_4^{2-}(aq) + 3H_2(g)$

Net Ionic: $2Cr(s) + 6H^+(aq) \longrightarrow 2Cr^{3+}(aq) + 3H_2(g)$

5.61 (a) Dilute HNO_3: $3Ag(s) + 4HNO_3(aq) \longrightarrow 3AgNO_3(aq) + 2H_2O + NO(g)$

(b) Concentrated HNO_3: $Ag(s) + 2HNO_3(aq) \longrightarrow AgNO_3(aq) + H_2O + NO_2(aq)$

5.63 $Cu(s) \longrightarrow Cu^{2+}(aq) + 2e^-$

$H_2SO_4(aq) + 2H^+(aq) + 2e^- \longrightarrow SO_2(g) + 2H_2O$

$Cu(s) + H_2SO_4(aq) + 2H^+(aq) \longrightarrow Cu^{2+}(aq) + SO_2(g) + 2H_2O$

The net ionic equation is the molecular equation.

5.65 In each case, the reaction should proceed to give the less reactive of the two metals, together with the ion of the more reactive of the two metals. The reactivity is taken from the reactivity series Table 5.3.

(a) $Mn(s) + Fe^{2+}(aq) \longrightarrow Mn^{2+}(aq) + Fe(s)$

(b) N.R.

(c) $Mg(s) + Co^{2+}(aq) \longrightarrow Mg^{2+}(aq) + Co(s)$

(d) $2Cr(s) + 3Sn^{2+}(aq) \longrightarrow 2Cr^{3+}(aq) + 3Sn(s)$

5.67 Increasing ease of oxidation: Pt, Ru, Tl, Pu

5.69 The equation given shows that Cd is more active than Ru. Coupled with the information in Review Problem 5.67, we also see that Cd is more active than Tl. This means that in a mixture of Cd and Tl^+, Cd will be oxidized and Tl^+ will be reduced:

$Cd(s) + 2TlCl(aq) \longrightarrow CdCl_2(aq) + 2Tl(s)$

(The $Tl(s)$ and the $Cd(NO_3)_2(aq)$ will not react.)

5.71 $Mg(s) + Zn^{2+}(aq) + 2Cl^-(aq) \longrightarrow Mg^{2+}(aq) + 2Cl^-(aq) + Zn(s)$

5.73 $2Li(s) + 2H_2O \longrightarrow H_2(g) + 2OH^-(aq) + 2Li^+(aq)$

5.75 (a) $2C_6H_6(l) + 15O_2(g) \longrightarrow 12CO_2(g) + 6H_2O(g)$

(b) $2C_4H_{10}(g) + 13O_2(g) \longrightarrow 8CO_2(g) + 10H_2O(g)$

(c) $C_{21}H_{44}(s) + 32O_2(g) \longrightarrow 21CO_2(g) + 22H_2O(g)$

5.77 (a) $2C_6H_6(l) + 9O_2(g) \longrightarrow 12CO(g) + 6H_2O(g)$

$2C_4H_{10}(g) + 9O_2(g) \longrightarrow 8CO(g) + 10H_2O(g)$

$2C_{21}H_{44}(s) + 43O_2(g) \longrightarrow 42CO(g) + 44H_2O(g)$

(b) $2C_6H_6(l) + 3O_2(g) \longrightarrow 12C(s) + 6H_2O(g)$

$2C_4H_{10}(g) + 5O_2(g) \longrightarrow 8C(s) + 10H_2O(g)$

$C_{21}H_{44}(s) + 11O_2(g) \longrightarrow 21C(s) + 22H_2O(g)$

5.79 $C_6H_{12}O_6(s) + 6O_2(g) \longrightarrow 6CO_2(g) + 6H_2O(g)$

5.81 $2(CH_3)_2S(g) + 9O_2(g) \longrightarrow 4CO_2(g) + 6H_2O(g) + 2SO_2(g)$

5.83 (a) $2Zn(s) + O_2(g) \longrightarrow 2ZnO(s)$

(b) $4Al(s) + 3O_2(g) \longrightarrow 2Al_2O_3(s)$

(c) $2Mg(s) + O_2(g) \longrightarrow 2MgO(s)$

(d) $4Fe(s) + 3O_2(g) \longrightarrow 2Fe_2O_3(s)$

5.85 $Cu + 2Ag^+ \longrightarrow Cu^{2+} + 2Ag$

$$\text{g Cu} = (12.0\text{ g Ag})\left(\frac{1\text{ mol Ag}}{107.868\text{ g Ag}}\right)\left(\frac{1\text{ mol Cu}}{2\text{ mol Ag}}\right)\left(\frac{63.546\text{ g Cu}}{1\text{ mol Cu}}\right) = 3.53\text{ g Cu}$$

5.87 (a) $[MnO_4^- + 8H^+ + 5e^- \longrightarrow Mn^{2+} + 4H_2O] \times 2$

$[Sn^{2+} \longrightarrow Sn^{4+} + 2e^-] \times 5$

$2MnO_4^- + 5Sn^{2+} + 16H^+ \longrightarrow 2Mn^{2+} + 5Sn^{4+} + 8H_2O$

(b)

$$\text{mL KMnO}_4 = (40.0\text{ mL SnCl}_2)\left(\frac{0.250\text{ mol SnCl}_2}{1000\text{ mL SnCl}_2}\right)\left(\frac{1\text{ mol Sn}^{2+}}{1\text{ mol SnCl}_2}\right)\left(\frac{2\text{ mol MnO}_4^-}{5\text{ mol Sn}^{2+}}\right)$$

$$\times\left(\frac{1\text{ mol KMnO}_4}{1\text{ mol MnO}_4^-}\right)\left(\frac{1000\text{ mL KMnO}_4}{0.230\text{ mol KMnO}_4}\right) = 17.4\text{ mL KMnO}_4$$

5.89 (a) $IO_3^- + 6H^+ + 6e^- \longrightarrow I^- + 3H_2O$

$[SO_3^{2-} + H_2O \longrightarrow SO_4^{2-} + 2H^+ + 2e^-] \times 3$

$IO_3^- + 3SO_3^{2-} + 6H^+ + 3H_2O \longrightarrow I^- + 3SO_4^{2-} + 3H_2O + 6H^+$

Which simplifies to:

$IO_3^- + 3SO_3^{2-} \longrightarrow I^- + 3SO_4^{2-}$

(b)

$$\text{g Na}_2\text{SO}_3 = (5.00\text{ g NaIO}_3)\left(\frac{1\text{ mol NaIO}_3}{197.9\text{ g NaIO}_3}\right)\left(\frac{3\text{ mol Na}_2\text{SO}_3}{1\text{ mol NaIO}_3}\right)\left(\frac{126.0\text{ g Na}_2\text{SO}_3}{1\text{ mol Na}_2\text{SO}_3}\right)$$

$= 9.55\text{ g Na}_2\text{SO}_3$

5.91 (a)

$$\text{mol of I}_3^- = 0.0421\text{ g NaIO}_3\left(\frac{1\text{ mol NaIO}_3}{197.89\text{ g NaIO}_3}\right)\left(\frac{1\text{ mol IO}_3^-}{1\text{ mol NaIO}_3}\right)\left(\frac{3\text{ mol I}_3^-}{1\text{ mol IO}_3^-}\right)$$

$= 6.38 \times 10^{-4}\text{ mol I}_3^-$

$$\text{Molarity of I}_3^- = \left(\frac{6.38\times10^{-4}\text{ mol I}_3^-}{100\text{ mL}}\right)\left(\frac{1000\text{ mL}}{1\text{ L}}\right) = 6.38 \times 10^{-3}\ M\text{ I}_3^-$$

(b)

$$\text{g SO}_2 = 2.47\text{ mL I}_3^-\left(\frac{1\text{ L I}_3^-}{1000\text{ mL I}_3^-}\right)\left(\frac{6.38\times10^{-3}\text{ mol I}_3^-}{1\text{ L I}_3^-}\right)\left(\frac{1\text{ mol SO}_2}{1\text{ mol I}_3^-}\right)\left(\frac{64.07\text{ g SO}_2}{1\text{ mol SO}_2}\right)$$

$= 1.01 \times 10^{-3}\text{ g SO}_2$

(c) The density of the wine was 0.96 g/mL and the SO_2 concentration was 1.01×10^{-3} g SO_2/in 50 mL

$$\text{concentration SO}_2 = \frac{1.01 \times 10^{-3}\text{ g SO}_2}{50\text{ mL}} = 2.02 \times 10^{-5}\text{ g SO}_2\text{/mL}$$

In 1 mL of solution there are 0.96 g of wine and 2.02×10^{-5} g SO_2.

Therefore the percentage of SO_2 in the wine is

$$\frac{2.02 \times 10^{-5}\text{ g SO}_2}{0.96\text{ g wine}} \times 100\% = 2.10 \times 10^{-3}\,\%$$

(d)

$$\text{ppm SO}_2 = \frac{2.02 \times 10^{-5}\text{ g SO}_2}{0.96\text{ g wine}} \times 10^6\text{ ppm} = 21\text{ ppm}$$

5.93 (a) $$\text{mol Cu}^{2+} = (29.96 \text{ mL S}_2\text{O}_3^{2-})\left(\frac{0.02100 \text{ mol S}_2\text{O}_3^{2-}}{1000 \text{ mL S}_2\text{O}_3^{2-}}\right)\left(\frac{1 \text{ mol I}_3^-}{2 \text{ mol S}_2\text{O}_3^{2-}}\right)\left(\frac{2 \text{ mol Cu}^{2+}}{1 \text{ mol I}_3^-}\right)$$
$$= 6.292 \times 10^{-4} \text{ mol Cu}^{2+}$$

$$\text{g Cu} = (6.292 \times 10^{-4} \text{ mol Cu}) \times \left(\frac{63.546 \text{ g Cu}}{\text{mol Cu}}\right) = 3.998 \times 10^{-2} \text{ g Cu}$$

$$\% \text{ Cu} = \left(\frac{3.998 \times 10^{-2} \text{ g Cu}}{0.4225 \text{ g sample}}\right) \times 100\% = 9.463\%$$

(b) $$\text{g CuCO}_3 = (6.292 \times 10^{-4} \text{ mol Cu})\left(\frac{1 \text{ mol CuCO}_3}{1 \text{ mol Cu}}\right)\left(\frac{123.56 \text{ g CuCO}_3}{1 \text{ mol CuCO}_3}\right) = 0.07774 \text{ g CuCO}_3$$

$$\% \text{ CuCO}_3 = \left(\frac{0.07774 \text{ g CuCO}_3}{0.4225 \text{ g sample}}\right) \times 100\% = 18.40\%$$

5.95 (a) $$\text{g H}_2\text{O}_2 = (17.60 \text{ mL KMnO}_4)\left(\frac{0.02000 \text{ mol KMnO}_4}{1000 \text{ mL KMnO}_4}\right)\left(\frac{1 \text{ mol MnO}_4^-}{1 \text{ mol KMnO}_4}\right)$$
$$\times \left(\frac{5 \text{ mol H}_2\text{O}_2}{2 \text{ mol MnO}_4^-}\right)\left(\frac{34.02 \text{ g H}_2\text{O}_2}{1 \text{ mol H}_2\text{O}_2}\right) = 0.02994 \text{ g H}_2\text{O}_2$$

(b) $$\left(\frac{0.02994 \text{ g H}_2\text{O}_2}{1.000 \text{ g sample}}\right) \times 100\% = 2.994\% \text{ H}_2\text{O}_2$$

5.97 (a) $$2\text{CrO}_4^{2-} + 3\text{SO}_3^{2-} + \text{H}_2\text{O} \longrightarrow 2\text{CrO}_2^- + 3\text{SO}_4^{2-} + 2\text{OH}^-$$

(b) $$\text{mol CrO}_4^{2-} = (3.18 \text{ g Na}_2\text{SO}_3)\left(\frac{1 \text{ mol Na}_2\text{SO}_3}{126.04 \text{ g Na}_2\text{SO}_3}\right)\left(\frac{1 \text{ mol SO}_3^{2-}}{1 \text{ mol Na}_2\text{SO}_3}\right)\left(\frac{2 \text{ mol CrO}_4^{2-}}{3 \text{ mol SO}_3^{2-}}\right)$$
$$= 1.68 \times 10^{-2} \text{ mol CrO}_4^{2-}$$

Since there is one mole of Cr in each mole of CrO_4^{2-}, then the above number of moles of CrO_4^{2-} is also equal to the number of moles of Cr that were present:

$$0.0168 \text{ mol Cr} \times \left(\frac{52.00 \text{ g Cr}}{\text{mol Cr}}\right) = 0.875 \text{ g Cr in the original alloy.}$$

(c) $$\frac{0.875 \text{ g Cr}}{3.450 \text{ g sample}} \times 100\% = 25.4\% \text{ Cr}$$

5.99 (a) $$\text{mol C}_2\text{O}_4^{2-} = (21.62 \text{ mL KMnO}_4)\left(\frac{0.1000 \text{ mol KMnO}_4}{1000 \text{ mL KMnO}_4}\right)\left(\frac{5 \text{ mol C}_2\text{O}_4^{2-}}{2 \text{ mol KMnO}_4}\right)$$
$$= 5.405 \times 10^{-3} \text{ mol C}_2\text{O}_4^{2-}$$

(b) The stoichiometry for calcium is as follows:

$1 \text{ mol C}_2\text{O}_4^{2-} = 1 \text{ mol Ca}^{2+} = 1 \text{ mol CaCl}_2$

Thus the number of grams of $CaCl_2$ is given simply by:

$5.405 \times 10^{-3} \text{ mol CaCl}_2 \times 110.98 \text{ g/mol} = 0.5999 \text{ g CaCl}_2$

(c) $$\frac{0.5999 \text{ g CaCl}_2}{2.2463 \text{ g sample}} \times 100\% = 26.71\% \text{ CaCl}_2$$

Chapter Six
Energy and Chemical Change

Practice Exercises

6.1 $J = 140 \text{ Cal}\left(\frac{1 \text{ kcal}}{1 \text{ Cal}}\right)\left(\frac{1000 \text{ cal}}{1 \text{ kcal}}\right)\left(\frac{4.184 \text{ J}}{1 \text{ cal}}\right) = 5.86 \times 10^5 \text{ J} = 5.9 \times 10^5 \text{ J}$

$5.9 \times 10^5 \text{ J}\left(\frac{1 \text{ kJ}}{1000 \text{ J}}\right) = 590 \text{ kJ}$

6.2 $\text{kcal} = 160 \text{ Cal}\left(\frac{1 \text{ kcal}}{1 \text{ Cal}}\right) = 160 \text{ kcal}$

$\text{cal} = 160 \text{ Cal}\left(\frac{1000 \text{ cal}}{1 \text{ Cal}}\right) = 1.6 \times 10^5 \text{ cal}$

$\text{kJ} = 160 \text{ Cal}\left(\frac{4.184 \text{ kJ}}{1 \text{ Cal}}\right) = 670 \text{ kJ}$

$J = 160 \text{ Cal}\left(\frac{4.184 \text{ kJ}}{1 \text{ Cal}}\right)\left(\frac{1000 \text{ J}}{1 \text{ kJ}}\right) = 6.7 \times 10^5 \text{ J}$

6.3 $\Delta T = T_{\text{final}} - T_{\text{initial}} = 32\ °\text{C} - 25\ °\text{C} = 7\ °\text{C}$

6.4 $\Delta T_{\text{water}} = T_{\text{final}} - T_{\text{initial}} = 30.0\ °\text{C} - 20.0\ °\text{C} = 10.0\ °\text{C}$

$q_{\text{water}} = (10.0\ °\text{C})(250 \text{ g } H_2O)\left(4.184 \frac{\text{J}}{\text{g } °\text{C}}\right) = 10{,}460 \text{ J}$

$q_{\text{ball bearing}} = -q_{\text{water}} = -10{,}460 \text{ J}$

$C = \frac{q_{\text{ball bearing}}}{\Delta T_{\text{ball bearing}}} = \frac{-10{,}460 \text{ J}}{30.0\ °\text{C} - 220\ °\text{C}} = 55.1 \text{ J } °\text{C}^{-1}$

6.5 The amount of heat transferred into the water is:
$J = (255 \text{ g } H_2O)(4.184 \text{ J g}^{-1}\ °\text{C}^{-1})(30.0\ °\text{C} - 25.0\ °\text{C}) = 5330 \text{ J}$

$\text{kJ} = (5330 \text{ J})\left(\frac{1 \text{ kJ}}{1000 \text{ J}}\right) = 5.33 \text{ kJ}$

$\text{cal} = (5335 \text{ J})\left(\frac{1 \text{ cal}}{4.184 \text{ J}}\right) = 1280 \text{ cal}$

$\text{kcal} = (1280 \text{ cal})\left(\frac{1 \text{ kcal}}{1000 \text{ cal}}\right) = 1.28 \text{ kcal}$

6.6 $q = ms\Delta t$
$549 \text{ J} = (7.54 \text{ g Si})(0.705 \text{ J g}^{-1}\ °\text{C}^{-1})(t_{°\text{C}_f} - 25.0\ °\text{C})$
$t_{°\text{C}_f} = 128\ °\text{C}$

6.7 The reaction of hydrogen and oxygen is an explosion, as stated, explosions are exothermic because they release heat.

6.8 Since the container becomes cold, the reaction is absorbing heat, and is endothermic.

6.9 Both reactions are exothermic, as stated. The first reaction releases some of the energy as work, the second reaction releases all of the energy as heat since it is run at constant volume. The second run releases more heat.

6.10 The first reaction releases some of the energy as work. The second reaction releases all of the energy as heat. The second reaction will have a larger increase in temperature.

6.11 $\text{mol} = (2.85\text{ g})\left(\dfrac{1\text{ mol}}{122.12\text{ g}}\right) = 0.0233\text{ mol benzoic acid}$

$q_{\text{benzoic acid}} = \Delta H_{\text{combustion}} \times n_{\text{benzoic acid}} = (-3227\text{ kJ mol}^{-1})(0.0233\text{ mol}) = -75.3\text{ kJ}$

$q_{\text{benzoic acid}} = -q_{\text{calorimeter}}$ $\qquad -q_{\text{calorimeter}} = 75.3\text{ kJ}$

$-q_{\text{calorimeter}} = C\Delta t$

$75.3\text{ kJ} = C(29.19\text{ °C} - 24.05\text{ °C})$

$C = \dfrac{75.3\text{ kJ}}{29.19\text{ °C} - 24.05\text{ °C}} = \dfrac{75.3\text{ kJ}}{5.14\text{ °C}} = 14.6\text{ kJ °C}^{-1}$

6.12 $C_{12}H_{22}O_{11}$ MM = (12 mol C × 12.01 g/mol C) + (22 mol H × 1.01 g/mol H) + (11 mol O × 16.0 g/mol O) = 342.34 g/mol $C_{12}H_{22}O_{11}$

$q_{\text{calorimeter}} = C\Delta t = 8.930\text{ kJ/°C}(26.77\text{ °C} - 24.00\text{ °C}) = 8.930\text{ kJ/°C}(2.77\text{ °C}) = 24.74\text{ kJ}$

$q_{\text{sucrose}} = -q_{\text{calorimeter}} = -(24.74\text{ kJ}) = -24.74\text{ kJ}$

$\text{heat released by 1 mole of sucrose} = \left(\dfrac{-24.7\text{ kJ}}{1.50\text{ g } C_{12}H_{22}O_{11}}\right)\left(\dfrac{342.3\text{ g } C_{12}H_{22}O_{11}}{1\text{ mol } C_{12}H_{22}O_{11}}\right) = -5636\text{ kJ mol}^{-1}$

$= -5640\text{kJ mol}^{-1}$

$\text{heat released by 1 g of sucrose in Cal} = -5640\text{ kJ mol}^{-1}\left(\dfrac{1\text{ mol } C_{12}H_{22}O_{11}}{342.3\text{ g } C_{12}H_{22}O_{11}}\right)\left(\dfrac{1\text{ Cal}}{4.184\text{ kJ}}\right)$

$= -3.94\text{ Cal/g}$

6.13 HCl and NaOH react 1:1, therefore the heat released per mole of NaOH will be the same as per mole of HCl, or –58 kJ mol^{-1} NaOH

6.14 Net ionic equation: $HC_2H_3O_2 + OH^- \longrightarrow H_2O + C_2H_3O_2^-$

Density of acid solution: 1.00 g/mL

Mass of acid solution: $50.00\text{ mL}\left(\dfrac{1.00\text{ g}}{1.00\text{ mL}}\right) = 50.0\text{ g}$

Mass of base solution: 53.0 g

Total mass = 50 g + 53 g = 103 g solution

$q_{\text{water}} = ms\Delta t = (103\text{ g})(4.184\text{ J g}^{-1}\text{ °C}^{-1})(6.6\text{ °C}) = 2840\text{ J} = 2.84\text{ kJ} = 2.8\text{ kJ}$

$q_{\text{water}} = -q_{\text{reaction}}$

$-q_{\text{reaction}} = -2.8\text{ kJ}$

$\text{mol } HC_2H_3O_2 = 50\text{ mL solution}\left(\dfrac{1.00\ M}{1000\text{ mL}}\right) = 5.0 \times 10^{-2}\text{ mol } HC_2H_3O_2$

NaOH and $HC_2H_3O_2$ react 1:1, the $\Delta H°$ is $\dfrac{-2.84\text{ kJ}}{5.0\times10^{-2}\text{ mol } HC_2H_3O_2} = -57\text{ kJ/mol } HC_2H_3O_2$

One mole of the acid reacts in the reaction, so the $\Delta H° = -57$ kJ

6.15 Multiply the reaction and the $\Delta H°$ by 3:

$3CH_4(g) + 6O_2(g) \longrightarrow 3CO_2(g) + 6H_2O(l) \quad \Delta H = -890.5 \times 3 = -2672$ kJ

6.16 From the text we are give the thermochemical equation for the formation of 2 moles of NH_3:

$N_2(g) + 3H_2(g) \longrightarrow 2NH_3(g) \quad \Delta H° = -92.38$ kJ

We can proceed by dividing both the equation and the value for $\Delta H°$ by 2 and then multiplying that by 2.5:

$\frac{1}{2}N_2(g) + \frac{3}{2}H_2(g) \longrightarrow NH_3(g) \quad \Delta H = -46.29$ kJ

$\frac{2.5}{2}N_2(g) + \frac{7.5}{2}H_2(g) \longrightarrow 2.5NH_3(g) \quad \Delta H = -115.5$ kJ

6.17

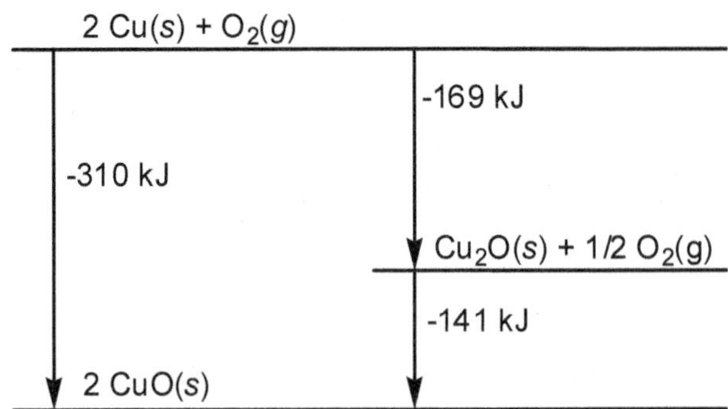

The reaction is exothermic.

6.18

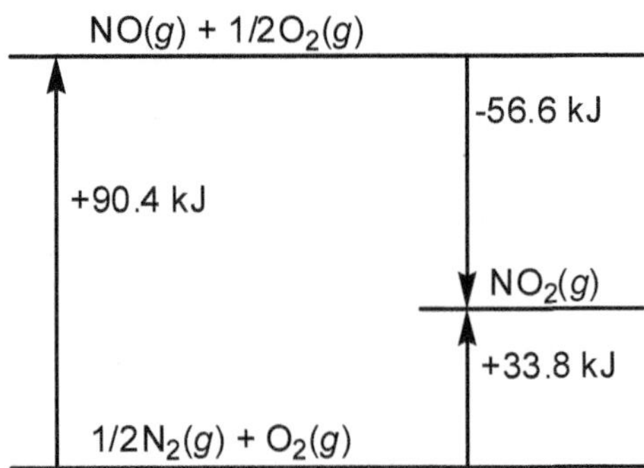

The reaction is endothermic.

6.19 For this problem: divide the second reaction by two:

$\frac{1}{2}\{2N_2O(g) + 3O_2(g) \longrightarrow 4NO_2(g)\} \quad \Delta H = \frac{1}{2}(-28.0 \text{ kJ})$

$N_2O(g) + \frac{3}{2}O_2(g) \longrightarrow 2NO_2(g)\} \quad \Delta H = -14.0$ kJ

Reverse the first reaction

$2NO_2(g) \longrightarrow 2NO(g) + O_2(g) \quad \Delta H = +113.2$ kJ

Add the reactions together:

$N_2O(g) + \frac{3}{2}O_2(g) \longrightarrow 2NO_2(g) \quad \Delta H = -14.0$ kJ

$2NO_2(g) \longrightarrow 2NO(g) + O_2(g) \quad \Delta H = +113.2$ kJ

$N_2O(g) + \frac{1}{2}O_2(g) \longrightarrow 2NO(g) \quad \Delta H = +99.2$ kJ

6.20 This problem requires that we add the reverse of the second equation (remembering to change the sign of the associated ΔH value) to the first equation:

$$C_2H_4(g) + 3O_2(g) \longrightarrow 2CO_2(g) + 2H_2O(l) \quad \Delta H° = -1411.1 \text{ kJ}$$
$$2CO_2(g) + 3H_2O(l) \longrightarrow C_2H_5OH(l) + 3O_2(g) \quad \Delta H° = +1367.1 \text{ kJ}$$
$$C_2H_4(g) + H_2O(l) \longrightarrow C_2H_5OH(l) \quad \Delta H° = -44.0 \text{ kJ}$$

6.21 $$\text{kJ} = (12.5 \text{ g } C_3H_6O)\left(\frac{1 \text{ moles } C_3H_6O}{58.077 \text{ g } C_3H_6O}\right)\left(\frac{-1790.4 \text{ kJ/mol}}{1 \text{ moles } C_8H_{18}}\right) = 385 \text{ kJ}$$

6.22 $$\text{kJ} = (480 \text{ mol } C_8H_{18})\left(\frac{-5450.5 \text{ kJ/mol}}{1 \text{ moles } C_8H_{18}}\right) = 2.62 \times 10^6 \text{ kJ}$$

6.23 $K(s) + \frac{1}{2}Br_2(l) \longrightarrow KBr(s) \quad \Delta H_f^o = -393.8 \text{ kJ}$

6.24 $Na(s) + \frac{1}{2}H_2(g) + C(s) + \frac{3}{2}O_2(g) \longrightarrow NaHCO_3(s) \quad \Delta H_f^o = -947.7 \text{ kJ/mol}$

6.25 $\Delta H° = [\text{sum } \Delta H_f^o \text{ products}] - [\text{sum } \Delta H_f^o \text{ reactants}]$

$\Delta H° = \{\Delta H_f^o\ [CaSO_4(s)] + 2\,\Delta H_f^o\ [HCl(g)]\} - \{\Delta H_f^o\ [CaCl_2(s)] + \Delta H_f^o\ [H_2SO_4(l)]\}$

$\Delta H° = \{[1 \text{ mol} \times (-1432.7 \text{ kJ/mol})] + [2 \text{ mol} \times (-92.30 \text{ kJ/mol})]\} -$
$\{[1 \text{ mol} \times (-795.0 \text{ kJ/mol})] + [1 \text{ mol} \times (-811.32 \text{ kJ/mol})]\}$

$\Delta H° = -10.98 \text{ kJ} = 11.0 \text{ kJ}$

6.26 $S(s) + \frac{3}{2}O_2(g) \longrightarrow SO_3(g) \ \Delta H_f^o = -395.2 \text{ kJ/mol}$

$S(s) + O_2(g) \longrightarrow SO_2(g) \ \Delta H_f^o = -296.9 \text{ kJ/mol}$

Reverse the first reaction and add the two reactions together to get

$SO_3(g) \longrightarrow SO_2(g) + \frac{1}{2}O_2(g) \ \Delta H_f^o = +98.3 \text{ kJ}$

$\Delta H° = [\text{sum } \Delta H_f^o \text{ products}] - [\text{sum } \Delta H_f^o \text{ reactants}]$

$\Delta H° = \{[1 \text{ mol } SO_2 \times \Delta H_f^o\ SO_2(g)] + [\frac{1}{2} \text{mol } O_2 \times \Delta H_f^o\ O_2(g)]\} - \{1 \text{ mol } SO_3 \times \Delta H_f^o\ SO_3(s)\}$

$\Delta H° = \{[1 \text{ mol} \times (-296.9 \text{ kJ/mol})] + [\frac{1}{2} \text{ mol} \times 0 \text{ kJ/mol}]\} - [1 \text{ mol} \times (-395.2 \text{ kJ/mol})]$

$\Delta H° = +98.3 \text{ kJ}$

The answers for the enthalpy of reaction are the same using either method.

6.27 $\Delta H° = [\text{sum } \Delta H_f^o \text{ products}] - [\text{sum } \Delta H_f^o \text{ reactants}]$

(a) $\Delta H° = \{2 \text{ mol } NO_2 \times \Delta H_f^o\ NO_2(g)\}$

$- \{[2 \text{ mole NO} \times \Delta H_f^o\ NO(g)] + [1 \text{ mol } O_2 \times \Delta H_f^o\ O_2(g)]\}$

$= \{2 \text{ mol} \times 33.8 \text{ kJ/mol}\} - \{[2 \text{ mol} \times 90.37 \text{ kJ/mol}] + [1 \text{ mol} \times 0 \text{ kJ/mol}]\}$

$= -113.1 \text{ kJ}$

(b) $\Delta H° = \{[1 \text{ mol } H_2O \times \Delta H_f^o\ H_2O(l)] + [1 \text{ mole NaCl} \times \Delta H_f^o\ NaCl(s)]\}$

$- \{[1 \text{ mol NaOH} \times \Delta H_f^o\ NaOH(s)] + [1 \text{ mole HCl} \times \Delta H_f^o\ HCl(g)]\}$

$= [(-285.9 \text{ kJ/mol}) + (-411.0 \text{ kJ/mol})] - [(-426.8 \text{ kJ/mol}) + (-92.30 \text{ kJ/mol})]$

$= -177.8 \text{ kJ}$

Review Problems

6.49 $KE = \frac{1}{2}mv^2$ $KE_1 = \frac{1}{2}m(30\text{ mi/h})^2$ $KE_2 = \frac{1}{2}m(60\text{ mi/h})^2$

$KE_1 = \frac{1}{2}m(900\text{ mi}^2/\text{h}^2)$ $KE_2 = \frac{1}{2}m(3600\text{ mi}^2/\text{h}^2)$

$$\frac{KE_2}{KE_1} = \frac{3600}{900} = 4$$

There is a four-fold increase in the kinetic energy due to the doubling of the speed of the car.
If the speed decreases to 10 mph:

$KE_1 = \frac{1}{2}m(10\text{ mi/h})^2$ $KE_2 = \frac{1}{2}m(30\text{ mi/h})^2$

$KE_1 = \frac{1}{2}m(100\text{ mi}^2/\text{h}^2)$ $KE_2 = \frac{1}{2}m(900\text{ mi}^2/\text{h}^2)$

$$\frac{KE_2}{KE_1} = \frac{900}{100} = 9$$

The kinetic energy decreases by 9 times due to the decreasing of the speed of the car by 10 mph.

6.51 $$\text{kg } O_2 \text{ molecule} = 1\text{ mol } O_2\left(\frac{32.00\text{ g } O_2}{1\text{ mol } O_2}\right)\left(\frac{1\text{ kg } O_2}{1000\text{ g } O_2}\right)$$

$$= 3.200 \times 10^{-3}\text{ kg}$$

$$KE = \frac{1}{2}mv^2 = \frac{1}{2}(3.200 \times 10^{-3}\text{ kg})(255\text{ m s}^{-1})^2 = 1.04 \times 10^3\text{ J}$$

$$1.04 \times 10^3\text{ J}\left(\frac{1\text{ kJ}}{10^3\text{ J}}\right) = 1.04\text{ kJ}$$

6.53 $\Delta t = 15.0\text{ °C} - 25.0\text{ °C} = -10.0\text{ °C}$

$q = ms\Delta t$

$$q = \left[1.75\text{ mol } H_2O \times \left(\frac{18.02\text{ g } H_2O}{1\text{ mol } H_2O}\right)\right] \times 4.184\text{ J g}^{-1}\text{ °C}^{-1} \times -10.0\text{ °C} = -1320\text{ J}$$

$$\text{cal} = (-1320\text{ J}) \times \left(\frac{1\text{ cal}}{4.184\text{ J}}\right) = -315\text{ cal}$$

6.55 $q_{Fe} = (85.0\text{ g Fe})(0.4498\text{ J g}^{-1}\text{ °C}^{-1})(35.0\text{ °C} - 85.0\text{ °C}) = -1912\text{ J}$

$q_{Fe} = -q_{H_2O} = 1912\text{ J}$

$$m = \frac{q}{s\Delta t}$$

$$\text{g } H_2O = \left(\frac{1912\text{ J}}{(4.184\text{ J g}^{-1}\text{ °C}^{-1})(35.0\text{ °C} - 25.0\text{ °C})}\right) = 45.7\text{ g } H_2O$$

6.57 (a) $\Delta t = 28.00\text{ °C} - 24.00\text{ °C} = 4.00\text{ °C}$

$q = ms\Delta t$

$q_{H_2O} = 4.184\text{ J g}^{-1}\text{ °C}^{-1} \times 100\text{ g} \times 4.0\text{ °C} = 1.67 \times 10^3\text{ J}$

(b) $1.67 \times 10^3\text{ J}$

(c) $\dfrac{1.67 \times 10^3\ \text{J}}{100\ °\text{C} - 28.0\ °\text{C}} = 23.2\ \text{J}\ °\text{C}^{-1}$

(d) $\dfrac{23.2\ \text{J}\ °\text{C}^{-1}}{50.0\ \text{g}} = 0.464\ \text{J g}^{-1}\ °\text{C}^{-1}$

6.59 $\dfrac{\text{J}}{\text{mol}\ °\text{C}} = \left(\dfrac{0.4498\ \text{J}}{\text{g}\ °\text{C}}\right)\left(\dfrac{55.847\ \text{g Fe}}{1\ \text{mol Fe}}\right) = 25.12\ \dfrac{\text{J}}{\text{mol}\ °\text{C}}$

6.61 $4.184\ \text{J g}^{-1}\ °\text{C}^{-1} \times (4.54 \times 10^3\ \text{g}) \times (58.65\ °\text{C} - 60.25\ °\text{C}) = -3.04 \times 10^4\ \text{J} = -30.4\ \text{kJ}$

6.63 $HNO_3(aq) + KOH(aq) \longrightarrow KNO_3(aq) + H_2O(l)$
Keep in mind that the total mass (assume the densities to be 1.00 g/mL for the solutions) must be considered in this calculation, and that both liquids, once mixed, undergo the same temperature increase:
$q_{H_2O} = (4.18\ \text{J/g}\ °\text{C}) \times (55.0\ \text{g} + 55.0\ \text{g}) \times (31.8\ °\text{C} - 23.5\ °\text{C})$
$= 3.8 \times 10^3$ J of heat energy released
Next determine the number of moles of reactant involved in the reaction:
$0.0550\ \text{L} \times 1.3\ \text{mol L}^{-1} = 0.072$ mol of acid and of base.
$q_{H_2O} = -q_{\text{reaction}} = -3.8 \times 10^3\ \text{J}$

Thus the enthalpy change is: $\dfrac{\left(-3.8 \times 10^3\ \text{J}\right)\left(\dfrac{1\ \text{kJ}}{1000\ \text{J}}\right)}{(0.072\ \text{mol})} = -53\ \text{kJ/mol}$

6.65 (a) $C_3H_8(g) + 5O_2(g) \longrightarrow 3CO_2(g) + 4H_2O(l)$
(b) $q_V = (97.1\ \text{kJ/°C})(27.282\ °\text{C} - 25.000\ °\text{C}) = 222\ \text{kJ} = 2.22 \times 10^5\ \text{J}$
(c) $\Delta H° = -222$ kJ/mol

6.67 $\Delta E = q + w = 28\ \text{J} - 45\ \text{J} = -17\ \text{J}$

6.69 Here, ΔE must = 0 in order for there to be no change in energy for the cycle.
$\Delta E = q + w$
$0 = q + (-100\ \text{J})$
$q = +100\ \text{J}$

6.71 If the engine absorbs 250 J of heat, the maximum amount of work it can do is 250 J.

6.73 (a) Divide the given equation by 4.
$NH_3(g) + 7/4\ O_2(g) \longrightarrow NO_2(g) + 3/2H_2O(g)\ \ \Delta H° = -1132\ \text{kJ}/4 = -283\ \text{kJ}$
(b) Divide the given equation by 6
$2/3NH_3(g) + 7/6\ O_2(g) \longrightarrow 2/3\ NO_2(g) + H_2O(g)\qquad \Delta H° = -1132\ \text{kJ}/6 = -189\ \text{kJ}$

6.75 $q = 6.54\ \text{g Mg}\left(\dfrac{-1203\ \text{kJ}}{2\ \text{mol Mg}}\right)\left(\dfrac{1\ \text{mol Mg}}{24.305\ \text{g Mg}}\right) = -162\ \text{kJ}$

162 kJ of heat are evolved

6.77 $\text{g}\ CH_4 = -432\ \text{kJ}\left(\dfrac{1\ \text{mol}\ CH_4}{-802\ \text{kJ}}\right)\left(\dfrac{16.04\ \text{g}}{1\ \text{mol}\ CH_4}\right) = 8.64\ \text{g}$

6.79

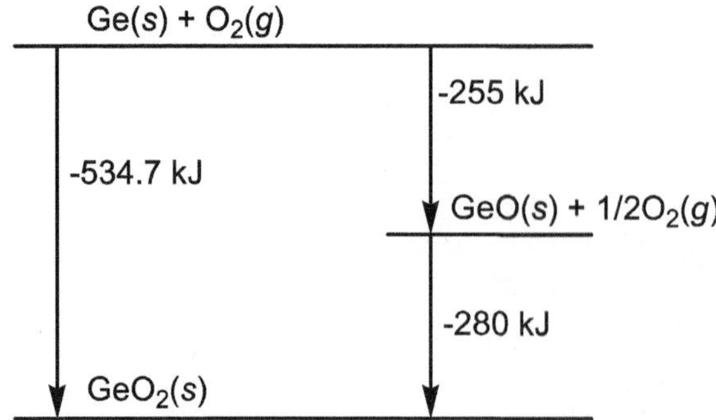

6.81 Since NO_2 does not appear in the desired overall reaction, the two steps are to be manipulated in such a manner so as to remove it by cancellation. Add the second equation to the reverse of the first, remembering to change the sign of $\Delta H°$ for the first equation, since it is to be reversed:

$2NO_2(g) \longrightarrow N_2O_4(g)$ $\quad \Delta H° = -57.93$ kJ

$2NO(g) + O_2(g) \longrightarrow 2NO_2(g)$ $\quad \Delta H° = -113.14$ kJ

Adding, we have:

$2NO(g) + O_2(g) \longrightarrow N_2O_4(g)$ $\quad \Delta H° = -171.07$ kJ

6.83 If we label the four known thermochemical equations consecutively, 1, 2, 3, and 4, then the sum is made in the following way: Divide equation #3 by two, and reverse all of the other equations (#1, #2, and #4), while also dividing each by two:

$\frac{1}{2}Na_2O(s) + HCl(g) \longrightarrow \frac{1}{2}H_2O(l) + NaCl(s)$ $\quad \Delta H° = -253.66$ kJ

$NaNO_2(s) \longrightarrow \frac{1}{2}Na_2O(s) + \frac{1}{2}NO_2(g) + \frac{1}{2}NO(g)$ $\quad \Delta H° = +213.57$ kJ

$\frac{1}{2}NO(g) + \frac{1}{2}NO_2(g) \longrightarrow \frac{1}{2}N_2O(g) + \frac{1}{2}O_2(g)$ $\quad \Delta H° = -21.34$ kJ

$\frac{1}{2}H_2O(l) + \frac{1}{2}O_2(g) + \frac{1}{2}N_2O(g) \longrightarrow HNO_2(l)$ $\quad \Delta H° = -17.18$ kJ

Adding gives:

$HCl(g) + NaNO_2(s) \longrightarrow HNO_2(l) + NaCl(s), \ \Delta H° = -78.61$ kJ

6.85 Multiply all of the equations by $\frac{1}{2}$ and add them together.

$\frac{1}{2}CaO(s) + \frac{1}{2}Cl_2(g) \longrightarrow \frac{1}{2}CaOCl_2(s)$ $\quad \Delta H° = -55.5$ kJ

$\frac{1}{2}H_2O(l) + \frac{1}{2}CaOCl_2(s) + NaBr(s) \longrightarrow NaCl(s) + \frac{1}{2}Ca(OH)_2(s) + \frac{1}{2}Br_2(l)$ $\quad \Delta H° = -30.1$ kJ

$\frac{1}{2}Ca(OH)_2(s) \longrightarrow \frac{1}{2}CaO(s) + \frac{1}{2}H_2O(l)$ $\quad \Delta H° = +32.6$ kJ

Adding gives:

$\frac{1}{2}Cl_2(g) + NaBr(s) \longrightarrow NaCl(s) + \frac{1}{2}Br_2(l)$ $\quad \Delta H° = -53$ kJ

6.87 Multiply the second equation by two and add them together:

$3Mg(s) + 2NH_3(g) \longrightarrow Mg_3N_2(s) + 3H_2(g)$ $\quad \Delta H° = -371$ kJ

$N_2(g) + 3H_2(g) \longrightarrow 2NH_3(g)$ $\quad \Delta H° = 2(-46$ kJ)

Adding gives:

$3Mg(s) + N_2(g) \longrightarrow Mg_3N_2(s)$ $\quad \Delta H° = -463$ kJ

6.89 The heat of formation is defined as the enthalpy change when one mole of a compound is produced from its elements in their standard states.
Only (c) satisfies this requirement.
For choice (a,) reactant $CO(NH_2)_2$ is not an element
For choice (b), O and H are not the standard states for oxygen and nitrogen gas. The state for C is not given as graphite or as a solid.
For choice (d), the reaction is not balanced

6.91 (a) $2C(s, \textit{graphite}) + 2H_2(g) + O_2(g) \longrightarrow HC_2H_3O_2(l) \quad \Delta H_f^o = -487.0 \text{ kJ}$

(b) $2C(s, \textit{graphite}) + \frac{1}{2}O_2(g) + 3H_2(g) \longrightarrow C_2H_5OH(l) \quad \Delta H_f^o = -277.63 \text{ kJ}$

(c) $Ca(s) + S(s) + 3O_2(g) + 2H_2(g) \longrightarrow CaSO_4 \cdot 2H_2O(s) \quad \Delta H_f^o = -2021.1 \text{ kJ}$

(d) $2Na(s) + S(s) + 2\,O_2(g) \longrightarrow Na_2SO_4(s) \quad \Delta H_f^o = -\,1384.5 \text{ kJ}$

6.93 (a) $\Delta H^o = \{[1 \text{ mol } O_2(g) \times \Delta H_f^o\, O_2(g)] + [2 \text{ mol } H_2O(l) \times \Delta H_f^o\, H_2O(l)]\}$
$- \{2 \text{ mol } H_2O_2(l) \times \Delta H_f^o\, H_2O_2(l)\}$
$= 0 \text{ kJ/mol} + [2 \text{ mol} \times (-285.9 \text{ kJ/mol})] - [2 \times (-187.6 \text{ kJ/mol})]$
$= -196.6 \text{ kJ}$

(b) $\Delta H^o = \{[1 \text{ mol } H_2O(l) \times \Delta H_f^o\, H_2O(l)] + [1 \text{ mol } NaCl(s) \times \Delta H_f^o\, NaCl(s)]\}$
$- \{[1 \text{ mol } HCl(g) \times \Delta H_f^o\, HCl(g)] + [1 \text{ mole } NaOH \times \Delta H_f^o\, NaOH(s)]\}$
$= [1 \text{ mol} \times (-285.9 \text{ kJ/mol})] + [1 \text{ mol} \times (-411.0 \text{ kJ/mol})]$
$- [1 \text{ mol} \times (-92.30 \text{ kJ/mol})] - [1 \text{ mol} \times (-426.8 \text{ kJ/mol})]$
$= -177.8 \text{ kJ}$

6.95 $C_{12}H_{22}O_{11}(s) + 12O_2(g) \longrightarrow 12CO_2(g) + 11H_2O(l) \quad \Delta H^o_{combustion} = -5.65 \times 10^3 \text{ kJ/mol}$

$\Delta H^o_{combustion} = [\text{sum } \Delta H_f^o \text{ products}] - [\text{sum } \Delta H_f^o \text{ reactants}]$
$= \{[12 \text{ mol } CO_2(g) \times \Delta H_f^o\, CO_2(g)] + [11 \text{ mol } H_2O(l) \times \Delta H_f^o\, H_2O(l)]\}$
$- \{[1 \text{ mol } C_{12}H_{22}O_{11}(s) \times \Delta H_f^o\, C_{12}H_{22}O_{11}(s)] + [12 \text{ mol } O_2(g) \times \Delta H_f^o\, O_2(g)]\}$

Rearranging and realizing the $\Delta H_f^o O_2(g) = 0$ we get
$\Delta H_f^o\, C_{12}H_{22}O_{11}(s) = [12 \text{ mol} \times \Delta H_f^o\, CO_2(g)] + [11 \text{ mol } H_2O(l) \times \Delta H_f^o\, H_2O(l)] - \Delta H^o_{combustion}$
$= 12(-393.5 \text{ kJ}) + 11(-285.9 \text{ kJ}) - (-5.65 \times 10^3 \text{ kJ}) = -2.22 \times 10^3 \text{ kJ}$

Chapter Seven
The Quantum Mechanical Atom

Practice Exercises

7.1 $\lambda = 588 \text{ nm}\left(\dfrac{1 \times 10^{-9}\text{m}}{1 \text{ nm}}\right) = 5.88 \times 10^{-7} \text{ m}$

$\nu = \dfrac{c}{\lambda} = \dfrac{2.998 \times 10^{8} \text{m/s}}{5.88 \times 10^{-7} \text{m}} = 5.10 \times 10^{14} \text{ s}^{-1} = 5.10 \times 10^{14} \text{ Hz}$

7.2 $\lambda = \dfrac{c}{\nu} = \dfrac{2.998 \times 10^{8} \text{ m s}^{-1}}{92.3 \times 10^{6} \text{ s}^{-1}} = 3.25 \text{ m}$

7.3 $\dfrac{1}{\lambda} = 109{,}678 \text{ cm}^{-1} \times \left(\dfrac{1}{4^2} - \dfrac{1}{6^2}\right) = 109{,}678 \text{ cm}^{-1} \times (0.0625 - 0.02778)$

$\dfrac{1}{\lambda} = 3.808 \times 10^{3} \text{ cm}^{-1}$

$\lambda = 2.63 \times 10^{-4} \text{ cm} = 2.63 \text{ μm}$

This is in the infrared portion of the electromagnetic spectrum.

7.4 $\dfrac{1}{\lambda} = 109{,}678 \text{ cm}^{-1} \times \left(\dfrac{1}{2^2} - \dfrac{1}{3^2}\right) = 109{,}678 \text{ cm}^{-1} \times (0.2500 - 0.1111)$

$\dfrac{1}{\lambda} = 1.5233 \times 10^{4} \text{ cm}^{-1}$

$\lambda = 6.565 \times 10^{-5} \text{ cm} = 656.5 \text{ nm}$, which is red.

7.5 $\Delta E = b\left(\dfrac{1}{n_{\text{low}}^{2}} - \dfrac{1}{n_{\text{high}}^{2}}\right)$

$\Delta E = 2.18 \times 10^{-18} \text{ J}\left(\dfrac{1}{4^2} - \dfrac{1}{6^2}\right) = (2.18 \times 10^{-18} \text{ J})(0.0625 - 0.0278) = (2.18 \times 10^{-18} \text{ J})(0.0347)$

$= 7.57 \times 10^{-20} \text{ J}$

7.6 $\Delta E = b\left(\dfrac{1}{n_{\text{low}}^{2}} - \dfrac{1}{n_{\text{high}}^{2}}\right)$

$2.04 \times 10^{-18} \text{ J} = 2.18 \times 10^{-18} \text{ J}\left(\dfrac{1}{n_{\text{low}}^{2}} - \dfrac{1}{4^2}\right)$

$2.04 \times 10^{-18} \text{ J} = 2.18 \times 10^{-18} \text{ J}\left(\dfrac{1}{n_{\text{low}}^{2}} - 0.0625\right)$

$\dfrac{2.04 \times 10^{-18} \text{ J}}{2.18 \times 10^{-18} \text{ J}} + 0.0625 = \dfrac{1}{n_{\text{low}}^{2}}$

$n_{\text{low}}^{2} = \dfrac{1}{0.9983} = 1.00 \qquad n_{\text{low}} = 1$

7.7 $E = \dfrac{n^2h^2}{8mL^2}$

$n = 1$, $h = 6.626 \times 10^{-34}$ J s $= 6.626 \times 10^{-34}$ kg m^2 s^{-1}, $m = 9.109 \times 10^{-31}$ kg, $L = 1$ nm $= 1 \times 10^{-9}$ m

$$E = \frac{1^2\left(6.626\times10^{-34}\text{ kg m}^2\text{ s}^{-1}\right)^2}{8\left(9.109\times10^{-31}\text{ kg}\right)\left(1\times10^{-9}\text{ m}\right)^2} = 6.02 \times 10^{-20}\text{ kg m}^2\text{ s}^{-2} = 6.02 \times 10^{-20}\text{ J}$$

For $L = 2$ nm:

$$E = \frac{1^2\left(6.626\times10^{-34}\text{ kg m}^2\text{ s}^{-1}\right)^2}{8\left(9.109\times10^{-31}\text{ kg}\right)\left(2\times10^{-9}\text{ m}\right)^2} = 1.51 \times 10^{-20}\text{ kg m}^2\text{ s}^{-2} = 1.51 \times 10^{-20}\text{ J}$$

7.8 $\Delta E = \dfrac{n_{\text{high}}{}^2h^2}{8mL^2} - \dfrac{n_{\text{low}}{}^2h^2}{8mL^2} = \dfrac{h^2}{8mL^2}\left(n_{\text{high}}{}^2 - n_{\text{low}}{}^2\right)$

$h = 6.626 \times 10^{-34}$ J s $= 6.626 \times 10^{-34}$ kg m^2 s^{-1}, $m = 9.109 \times 10^{-31}$ kg, $L = 1$ nm $= 1 \times 10^{-9}$ m

$$\Delta E = \frac{\left(6.626 \times 10^{-34}\text{ kg m}^2\text{ s}^{-1}\right)^2}{8\left(9.109 \times 10^{-31}\text{ kg}\right)\left(1 \times 10^{-9}\text{ m}\right)^2}\left(3^2 - 2^2\right) = 3.01 \times 10^{-19}\text{ J}$$

$$E = \frac{hc}{\lambda}$$

$$\lambda = \frac{hc}{E} = \frac{\left(6.626 \times 10^{-34}\text{ J s}\right)\left(3.00 \times 10^8\text{ m s}^{-1}\right)}{3.01 \times 10^{-19}\text{ J}} = 6.60 \times 10^{-7}\text{ m} = 660\text{ nm}$$

7.9 (a) $n = 4$, $\ell = 2$
(b) $n = 5$, $\ell = 3$
(c) $n = 7$, $\ell = 0$

7.10 When $n = 2$, $\ell = 0, 1$. Thus we have *s*, and *p* subshells.
When $n = 5$, $\ell = 0, 1, 2, 3, 4$. Thus we have *s*, *p*, *d*, *f*, and *g* subshells.
The number of subshells spans the values: 0,1,2,3, …, $n - 1$, and the number of subshells is equal to the value of n.

7.11 There are five *d* orbitals, and each orbital can hold two electrons, so a *d* subshell can hold ten (10) electrons.

7.12 For $n = 5$, the subshells are *s*, *p*, *d*, *f*, *g*.

s holds 2 electrons

p holds 6 electrons

d holds 10 electrons

f holds 14 electrons

g holds 18 electrons

The total number of electrons is 50.

7.13 (a) Na

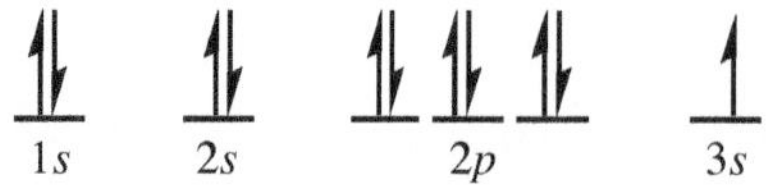

$1s$ ↑↓ $2s$ ↑↓ $2p$ ↑↓ ↑↓ ↑↓ $3s$ ↑

(b) S

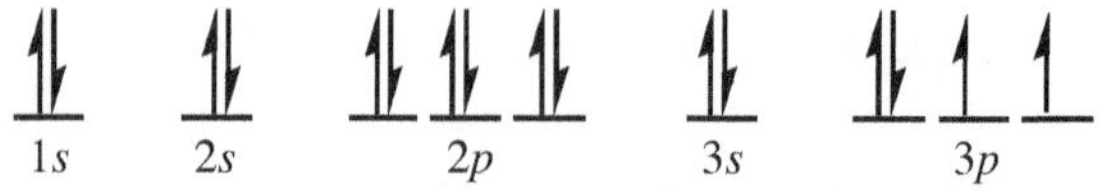

$1s$ ↑↓ $2s$ ↑↓ $2p$ ↑↓ ↑↓ ↑↓ $3s$ ↑↓ $3p$ ↑↓ ↑ ↑

(c) Ar

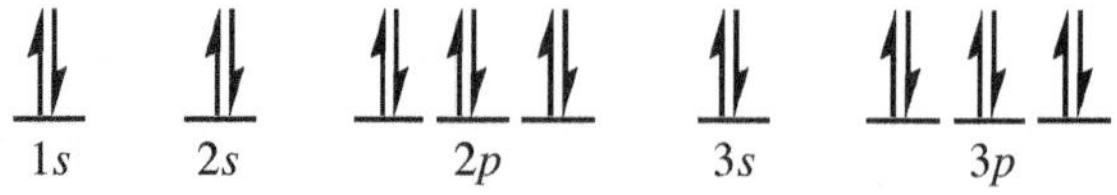

$1s$ ↑↓ $2s$ ↑↓ $2p$ ↑↓ ↑↓ ↑↓ $3s$ ↑↓ $3p$ ↑↓ ↑↓ ↑↓

7.14 (a) Si $1s^22s^22p^63s^23p^2$

(b) Ti $1s^22s^22p^63s^23p^64s^23d^2$

(c) Se $1s^22s^22p^63s^23p^64s^23d^{10}4p^4$

7.15 Yes, Ti, Cr, Fe, Ni, and the elements in their groups have even numbers of electrons and are paramagnetic. Additionally, oxygen has eight electrons, but it is paramagnetic since it has two unpaired electrons in the $2p$ orbitals.

7.16 (a) Mg

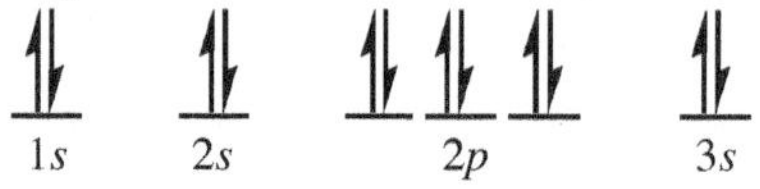

$1s$ ↑↓ $2s$ ↑↓ $2p$ ↑↓ ↑↓ ↑↓ $3s$ ↑↓

0 unpaired electrons

(b) Ge

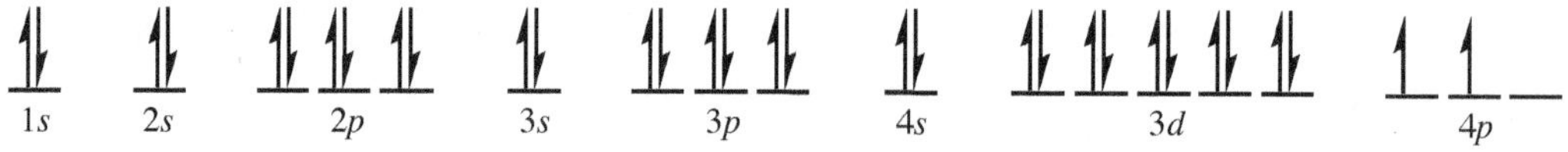

$1s$ ↑↓ $2s$ ↑↓ $2p$ ↑↓ ↑↓ ↑↓ $3s$ ↑↓ $3p$ ↑↓ ↑↓ ↑↓ $4s$ ↑↓ $3d$ ↑↓ ↑↓ ↑↓ ↑↓ ↑↓ $4p$ ↑ ↑ _

2 unpaired electrons

(c) Cd

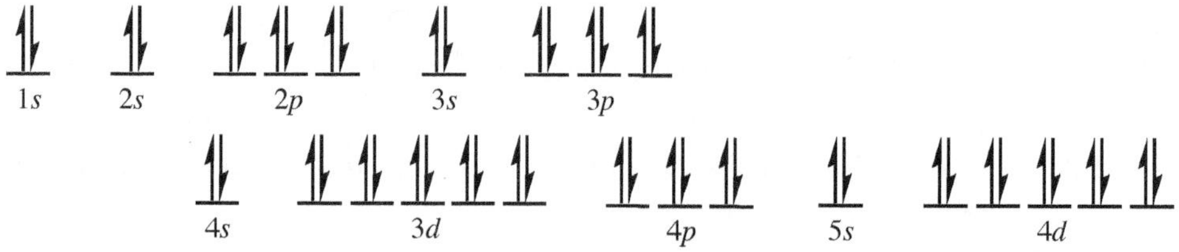

$1s$ ↑↓ $2s$ ↑↓ $2p$ ↑↓ ↑↓ ↑↓ $3s$ ↑↓ $3p$ ↑↓ ↑↓ ↑↓

$4s$ ↑↓ $3d$ ↑↓ ↑↓ ↑↓ ↑↓ ↑↓ $4p$ ↑↓ ↑↓ ↑↓ $5s$ ↑↓ $4d$ ↑↓ ↑↓ ↑↓ ↑↓ ↑↓

0 unpaired electrons

(d) Gd

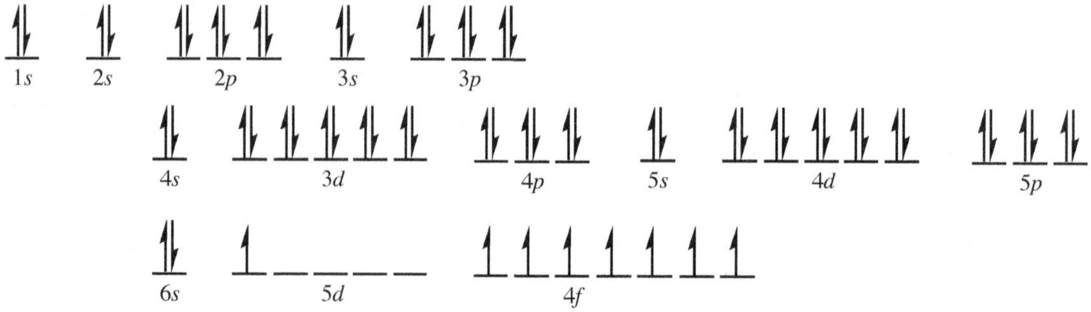

$1s$ ↑↓ $2s$ ↑↓ $2p$ ↑↓ ↑↓ ↑↓ $3s$ ↑↓ $3p$ ↑↓ ↑↓ ↑↓

$4s$ ↑↓ $3d$ ↑↓ ↑↓ ↑↓ ↑↓ ↑↓ $4p$ ↑↓ ↑↓ ↑↓ $5s$ ↑↓ $4d$ ↑↓ ↑↓ ↑↓ ↑↓ ↑↓ $5p$ ↑↓ ↑↓ ↑↓

$6s$ ↑↓ $5d$ ↑ _ _ _ _ $4f$ ↑ ↑ ↑ ↑ ↑ ↑ ↑

8 unpaired electrons

7.17 (a) Mg $1s^22s^22p^63s^2$
(b) Ge $1s^22s^22p^63s^23p^63d^{10}4s^24p^2$
(c) Cd $1s^22s^22p^63s^23p^63d^{10}4s^24p^64d^{10}5s^2$
(d) Gd $1s^22s^22p^63s^23p^63d^{10}4s^24p^64d^{10}4f^75s^25p^65d^16s^2$

7.18 (a) O $1s^22s^22p^4$
S $1s^22s^22p^63s^23p^4$
Se $1s^22s^22p^63s^23p^63d^{10}4s^24p^4$
(b) P $1s^22s^22p^63s^23p^3$
N $1s^22s^22p^3$
Sb $1s^22s^22p^63s^23p^63d^{10}4s^24p^64d^{10}5s^25p^3$
The elements have the same number of electrons in the valence shell, and the only differences between the valence shells are the energy levels.

7.19 (a) Zr [Kr] $4d^25s^2$
[Kr] 4d 5s
(b) Po [Xe] $4f^{14}5d^{10}6s^26p^4$
[Xe] 4f 5d 6s 6p

7.20 (a) P [Ne]$3s^23p^3$
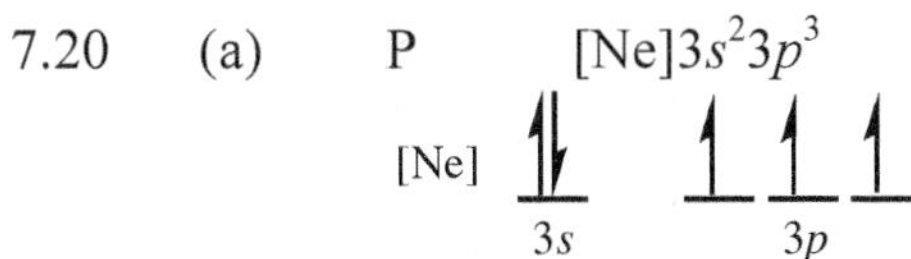

3 unpaired electrons
(b) Sn [Kr]$4d^{10}5s^25p^2$
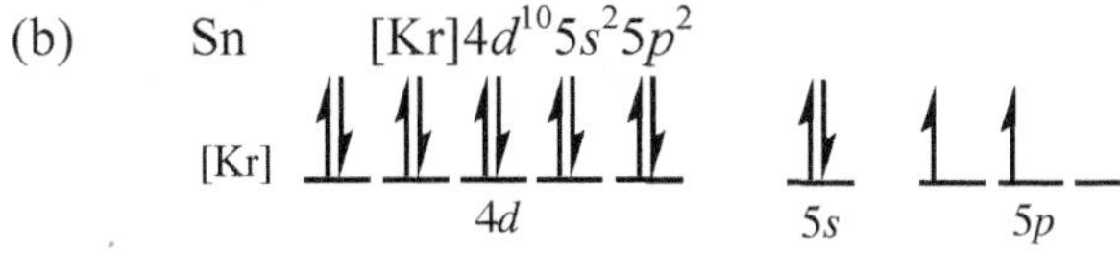

2 unpaired electrons

7.21 Based on the definition of valence, there are no examples where more than 8 electrons would occupy the valence shell. For representative elements the valence shell is defined as the occupied shell with the highest value of *n*. In the ground state atom, only *s* and *p* electrons fit this definition. The transition elements have outer electron configurations: $(n-1)d^a\ ns^b$ so the valence shell is the *ns* subshell.

7.22 (a) Se $4s^24p^4$ (b) Sn $5s^25p^2$ (c) I $5s^25p^5$

7.23 (a) Sn (b) Ga (c) Cr (d) S^{2-}

7.24 (a) P (b) Fe^{3+} (c) Fe (d) Cl^-

7.25 (a) Be (b) C

7.26 (a) C^{2+} (b) Mg^{2+}

Review Problems

The number used for the speed of light, *c*, depends on the number of significant figures. For one to three significant figures, the value for *c* is 3.00×10^8 m/s, for four significant figures, the value for *c* is 2.998×10^8 m/s.

7.79 $$\nu = \frac{c}{\lambda} = \frac{3.00 \times 10^8 \text{ m s}^{-1}}{436 \times 10^{-9} \text{ m}} = 6.88 \times 10^{14} \text{ s}^{-1} = 6.88 \times 10^{14} \text{ Hz}$$

7.81 $295 \text{ nm} = 295 \times 10^{-9}$ m

$$\nu = \frac{c}{\lambda} = \frac{3.00 \times 10^8 \text{ m/s}}{295 \times 10^{-9} \text{ m}} = 1.02 \times 10^{15} \text{ s}^{-1} = 1.02 \times 10^{15} \text{ Hz}$$

7.83 $101.1 \text{ MHz} = 101.1 \times 10^6 \text{ Hz} = 101.1 \times 10^6 \text{ s}^{-1}$

$$\lambda = \frac{c}{\nu} = \frac{2.998 \times 10^8 \text{ m/s}}{101.1 \times 10^6 \text{ s}^{-1}} = 2.965 \text{ m}$$

7.85 $E = h\nu = 6.63 \times 10^{-34} \text{ J s} \times (4.0 \times 10^{14} \text{ s}^{-1}) = 2.7 \times 10^{-19}$ J/photon

$$E \text{ per mol} = \left(\frac{2.7 \times 10^{-19} \text{ J}}{1 \text{ photon}}\right)\left(\frac{6.02 \times 10^{23} \text{ photons}}{1 \text{ mol}}\right) = 1.6 \times 10^5 \text{ J mol}^{-1}$$

7.87 (a) violet (see Figure 7.7)

(b) $$\nu = \frac{c}{\lambda} = \frac{2.998 \times 10^8 \text{ m s}^{-1}}{410.3 \times 10^{-9} \text{ m}} = 7.307 \times 10^{14} \text{ s}^{-1}$$

(c) $E = h\nu = (6.626 \times 10^{-34} \text{ J s}) \times (7.307 \times 10^{14} \text{ s}^{-1}) = 4.842 \times 10^{-19}$ J

7.89 $$\frac{1}{\lambda} = \left(109{,}678 \text{ cm}^{-1}\right)\left(\frac{1}{3^2} - \frac{1}{6^2}\right) = \left(109{,}678 \text{ cm}^{-1}\right)\left(0.1111 - 0.02778\right) = 9.140 \times 10^3 \text{ cm}^{-1}$$

$\lambda = 1.094 \times 10^{-4} \text{ cm} = 1094$ nm

This light is in the infrared region. We would not expect to see the light since it is not in the visible region.

7.91 $$\frac{1}{\lambda} = \left(109{,}678 \text{ cm}^{-1}\right)\left(\frac{1}{4^2} - \frac{1}{10^2}\right) = 5.758 \times 10^3 \text{ cm}^{-1}$$

$\lambda = 1.74 \times 10^{-6}$ m,

$$E = \frac{hc}{\lambda} = \frac{\left(6.626 \times 10^{-34} \text{ J s}\right)\left(3.00 \times 10^8 \text{ m s}^{-1}\right)}{1.737 \times 10^{-6} \text{ m}} = 1.14 \times 10^{-19} \text{ J}$$

This is in the infrared region.

7.93 (a) p (b) f

7.95 (a) 3 (b) 2

7.97 (a) $n = 3, \ell = 0$ (b) $n = 5, \ell = 2$

7.99 $\ell = 0, 1, 2, 3, 4,$ or 5

7.101 $n = 8$

7.103 (a) $m_\ell = 1, 0, \text{or } -1$ (b) $m_\ell = 3, 2, 1, 0, -1, -2, \text{or } -3$

7.105 When $m_\ell = -4$ the minimum value of ℓ is 4 and the minimum value of n is 5.

7.107

n	ℓ	m_ℓ	m_s
2	1	−1	+1/2
2	1	−1	−1/2
2	1	0	+1/2
2	1	0	−1/2
2	1	+1	+1/2
2	1	+1	−1/2

7.109 21 electrons have $\ell = 1$
4 electrons have $m_\ell = 2$

7.111 (a) S $1s^22s^22p^63s^23p^4$
(b) K $1s^22s^22p^63s^23p^64s^1$
(c) Ti $1s^22s^22p^63s^23p^63d^24s^2$
(d) Sn $1s^22s^22p^63s^23p^63d^{10}4s^24p^64d^{10}5s^25p^2$

7.113 (a) Mn $[Ar]4s^23d^5$ five unpaired electrons, paramagnetic
(b) As $[Ar]\,3d^{10}4s^24p^3$ three unpaired electrons, paramagnetic
(c) S $[Ne]3s^23p^4$ two unpaired electrons, paramagnetic
(d) Sr $[Kr]5s^2$ zero unpaired electrons, not paramagnetic
(e) Ar $1s^22s^22p^63s^23p^6$ zero unpaired electrons, not paramagnetic

7.115 (a) Mg $1s^22s^22p^63s^2$ zero unpaired electrons
(b) P $1s^22s^22p^63s^23p^3$ three unpaired electrons
(c) V $1s^22s^22p^63s^23p^63d^34s^2$ three unpaired electrons

7.117 (a) Ni $[Ar]3d^84s^2$
(b) Cs $[Xe]6s^1$
(c) Ge $[Ar]\,3d^{10}4s^24p^2$
(d) Br $[Ar]\,3d^{10}4s^24p^5$
(e) Bi $[Xe]\,4f^{14}5d^{10}6s^26p^3$

7.119 (a) Mg

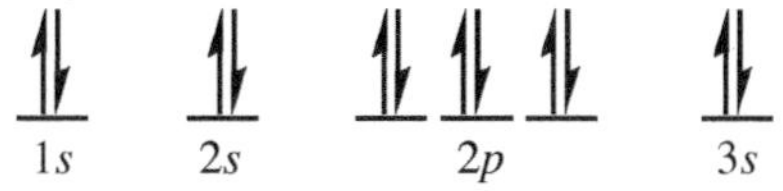

↑↓ (1s) ↑↓ (2s) ↑↓ ↑↓ ↑↓ (2p) ↑↓ (3s)

(b) Ti

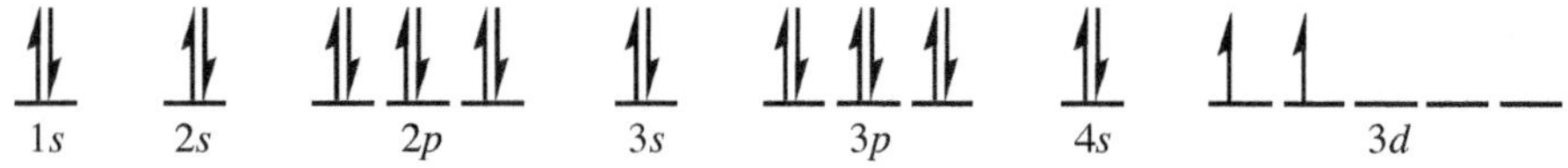

↑↓ (1s) ↑↓ (2s) ↑↓ ↑↓ ↑↓ (2p) ↑↓ (3s) ↑↓ ↑↓ ↑↓ (3p) ↑↓ (4s) ↑ ↑ _ _ _ (3d)

7.121 (a) Ni

[Ar] ↑↓ (4s) ↑↓ ↑↓ ↑↓ ↑ ↑ (3d)

(b) Cs

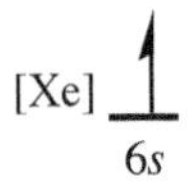

[Xe] ↑ (6s)

(c) Ge

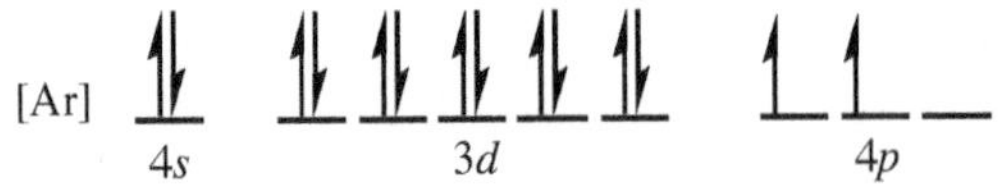

[Ar] ↑↓ (4s) ↑↓ ↑↓ ↑↓ ↑↓ ↑↓ (3d) ↑ ↑ _ (4p)

(d) Br

[Ar] ↑↓ (4s) ↑↓ ↑↓ ↑↓ ↑↓ ↑↓ (3d) ↑↓ ↑↓ ↑ (4p)

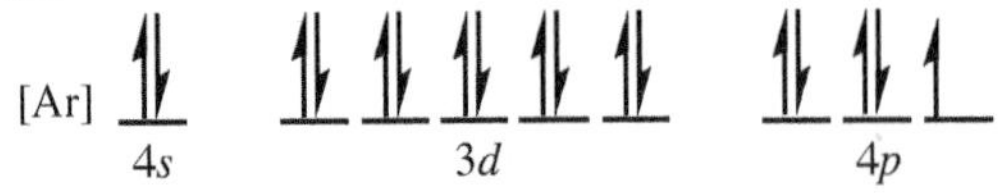

7.123 The value corresponds to the row in which the element resides:
(a) 5 (b) 4 (c) 4 (d) 6

7.125 (a) Na $3s^1$ (b) Al $3s^2 3p^1$ (c) Ge $4s^2 4p^2$ (d) P $3s^2 3p^3$

7.127 (a) Na

↑ (3s)

(b) Al

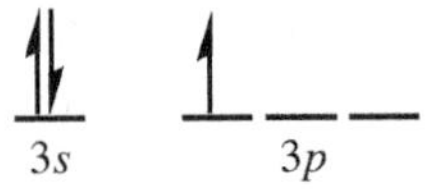

↑↓ (3s) ↑ _ _ (3p)

(c) Ge

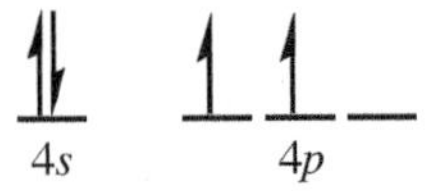

↑↓ (4s) ↑ ↑ _ (4p)

(d) P

↑↓ (3s) ↑ ↑ ↑ (3p)

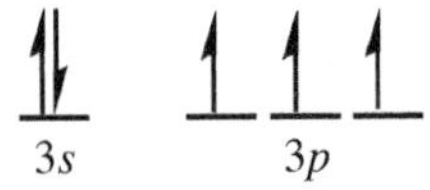

7.129 (a) There are 10 core electrons in Na so the valence electron would see 11 – 10 or +1 as the effective nuclear charge.

(b) There are 10 core electrons in S so the valence electrons would see 16 – 10 or +6 as the effective nuclear charge.

(c) There are 10 core electrons in Cl so the valence electrons would see 17 – 10 or +7 as the effective nuclear charge.

7.131 (a) Mg (b) Bi

7.133 As < Ge < Sb < Sn
Sb is a bit larger than Sn, according to Figure 7.30. (Based upon trends, Sn would be predicted to be larger, but this is one area in which an exception to the trend exists. See Figure 7.30.)

7.135 Cations are generally smaller than the corresponding atom, and anions are generally larger than the corresponding atom:
(a) Na (b) Co^{2+} (c) Cl^-

7.137 (a) N (b) S (c) Cl

7.139 (a) Br (b) As

7.141 The element with the largest difference between the second and third ionization potential would be the element with two valence electrons. The third ionization would remove an electron from the core, which is much higher in energy than removing electrons from valence levels. Mg has the valence structure:

$\underline{\uparrow\downarrow}$ $\underline{\ \ }\ \underline{\ \ }\ \underline{\ \ }$

3*s* 3*p*

Chapter Eight
The Basics of Chemical Bonding

Practice Exercises

8.1 (a) $CaBr_2$ (b) LiF (c) CaO

8.2

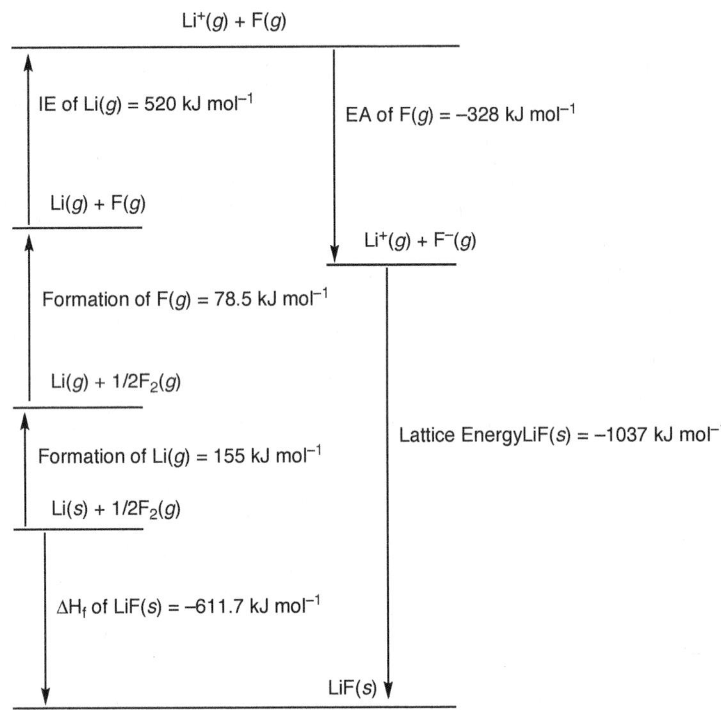

8.3 Indium should lose its $5p^1$ electron to have an electron configuration of.
$1s^12s^22p^63s^23p^63d^{10}4s^24p^64d^{10}5s^2$

8.4 Cr: $[Ar]3d^54s^1$
(a) Cr^{2+}: $[Ar]3d^4$ The 4*s* electron and one 3*d* electron are lost.
(b) Cr^{3+}: $[Ar]3d^3$ The 4*s* electron and two 3*d* electrons are lost.
(c) Cr^{6+}: [Ar] The 4*s* electron and all of the 3*d* electrons are lost.

8.5 S^{2-}: [Ne] $3s^23p^6$
Cl^-: [Ne] $3s^23p^6$
The electron configurations are identical, [Ar].

8.6

·Pb· :Te:

They are in the fourth and sixth groups, respectively.

8.7

:I· ·Ca· :I· ⟶ Ca^{2+} + 2[:I:]$^-$

8.8

$$\cdot\ddot{\underset{\cdot\cdot}{O}}\cdot \quad \cdot Mg\cdot \longrightarrow Mg^{2+} + [:\ddot{\underset{\cdot\cdot}{O}}:]^{2-}$$

8.9 (a) Sulfur will make two bonds.
(b) Phosphorous will make three bonds.
(c) Silicon will make four bonds.

8.10 Both F atoms need an additional pair of electrons.
The O with a double bond needs two more electrons.
The C with a double bond needs to lose two electrons.

8.11 $\mu = q \times r$

$$q = 0.167\ e^- \left(\frac{1.602\times10^{-19}\ C}{1\ e^-}\right) = 2.675 \times 10^{-20}\ C$$

$r = 154.6\ pm = 154.6 \times 10^{-12}\ m$

$q = (2.675 \times 10^{-20}\ C) \times (154.6 \times 10^{-12}\ m) = 4.136 \times 10^{-30}\ C\ m$

in Debye units:

$$q = 4.136 \times 10^{-30}\ C\ m\left(\frac{1\ D}{3.34\times10^{-30}\ C\ m}\right) = 1.24\ D$$

8.12 $q = \frac{\mu}{r}$

$$\mu = 9.00\ D\left(\frac{3.34\times10^{-30}\ C\ m}{1\ D}\right) = 3.006 \times 10^{-29}\ C\ m$$

$r = 236\ pm = 236 \times 10^{-12}\ m$

$$q = \left(\frac{3.006\times10^{-29}\ C\ m}{236\times10^{-12}\ m}\right) = 1.27 \times 10^{-19}\ C$$

In electron charges

$$q = 1.27 \times 10^{-19}\ C\left(\frac{1\ e^-}{1.602\ \times\ 10^{-19}\ C}\right) = 0.795\ e^-$$

On the sodium the charge is $+0.795\ e^-$ and on the chlorine, the charge is $-0.795\ e^-$.
This would be 79.5% positive charge on the Na and 79.5% negative charge on the Cl.

8.13 The bond is polar and the Cl carries the negative charge.

8.14 (a) Br (b) Cl (c) Cl
Order of increasing polarity: S—Cl < P—Br < Si—Cl

8.15

```
        O

H   O   P   O   H

        O
```

Number of valence electrons:

O 6 each and 4 O atoms total 24 electrons

P 5 electrons

H 1 each and 2 H atoms total 2 electrons

Negative charge 1 electron

Total 32 valence electrons

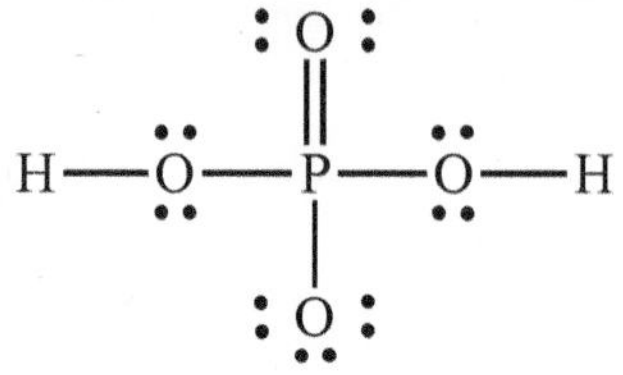

8.16

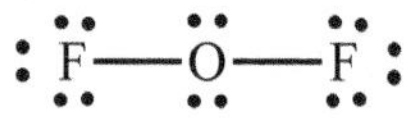

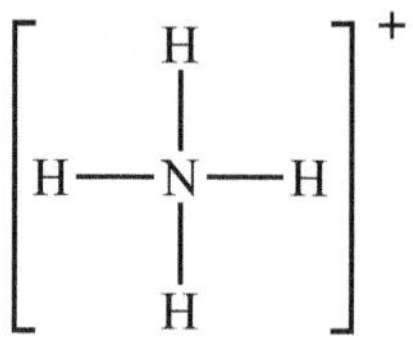

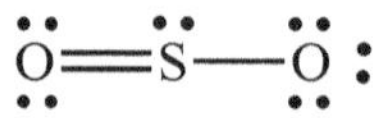

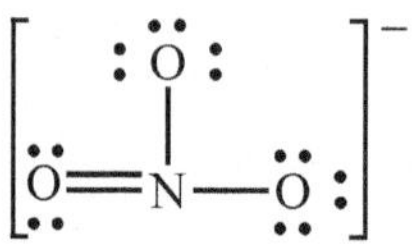

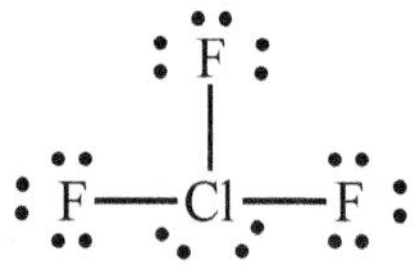

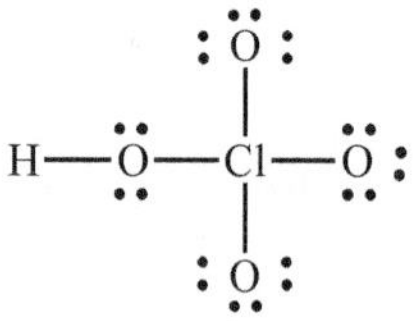

8.17

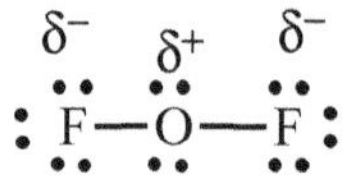

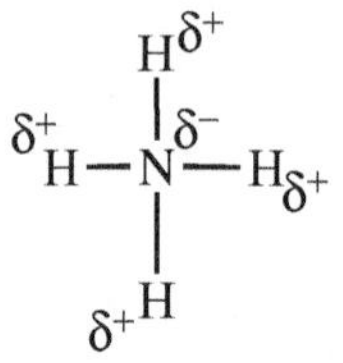

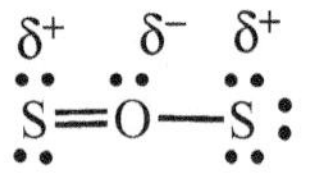

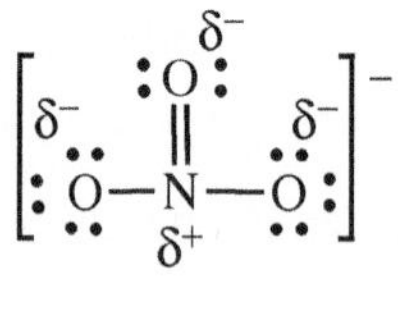

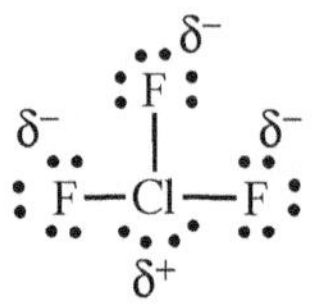

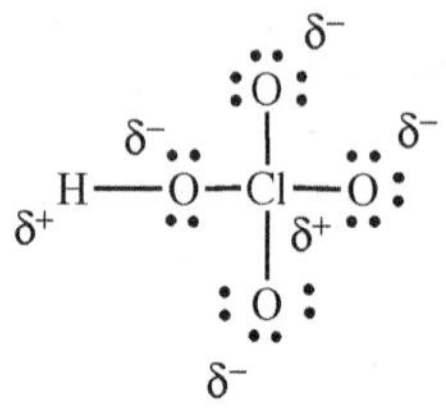

8.18 The negative sign should be on the oxygen, so two of the oxygen atoms should have a single bond and three lone pairs and the sulfur should have one double bond, two single bonds, and a lone pair. This is the best formal structure. The single bonded oxygens each have a –1 charge and the sulfur and double bonded oxygen have a zero formal charge.

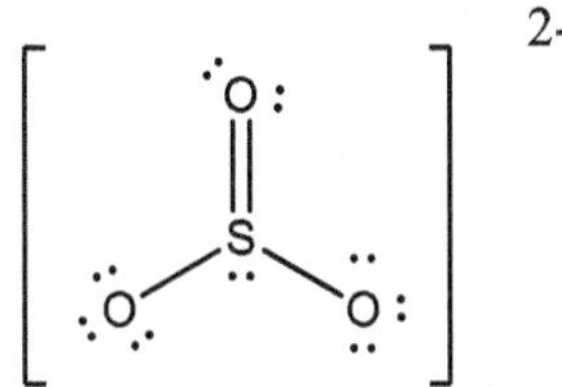

8.19 (a) 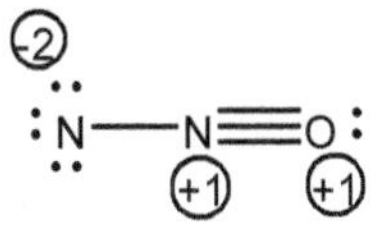 (b)

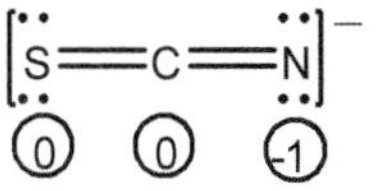

8.20 In each of these problems, we try to minimize the formal charges in order to determine the preferred Lewis structure. This frequently means violating the octet rule by expanding the octet. Of course, this can only be done for atoms beyond the second period as the atoms in the first and second periods will never expand the octet.

(a) SeO_4^{2-}

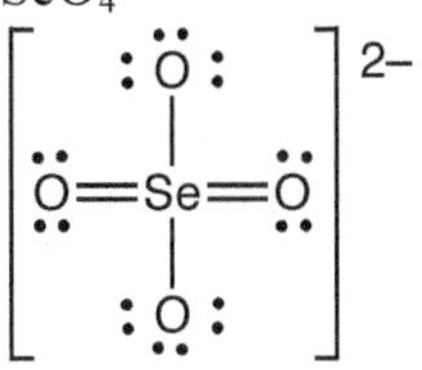

(b) $HClO_3$

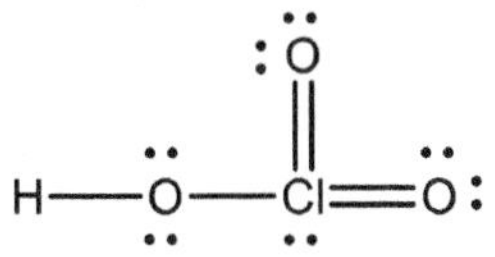

(c) H_3PO_4

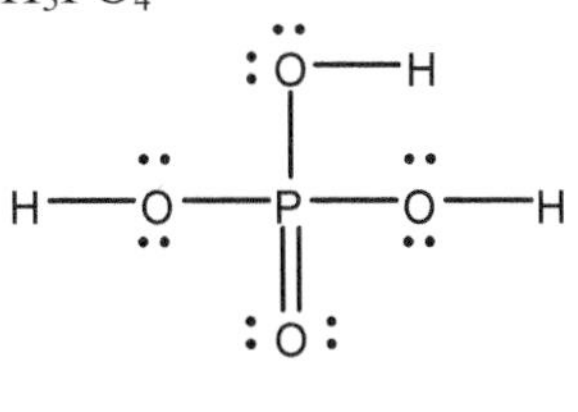

8.21

There is no difference between the coordinate covalent bond and the other covalent bonds.

8.22

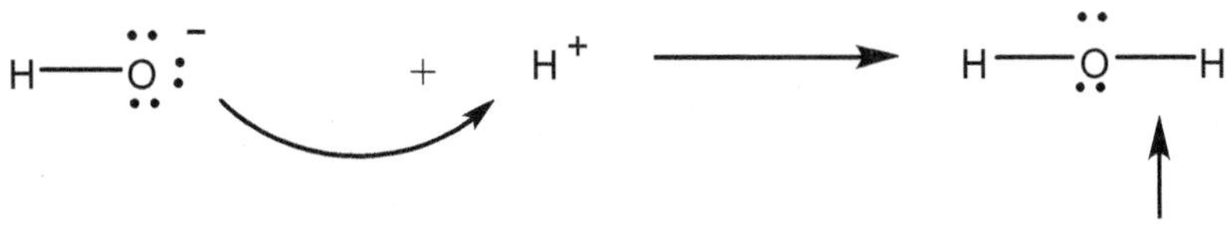

coordinate covalent bond

8.23 There are four resonance structures.

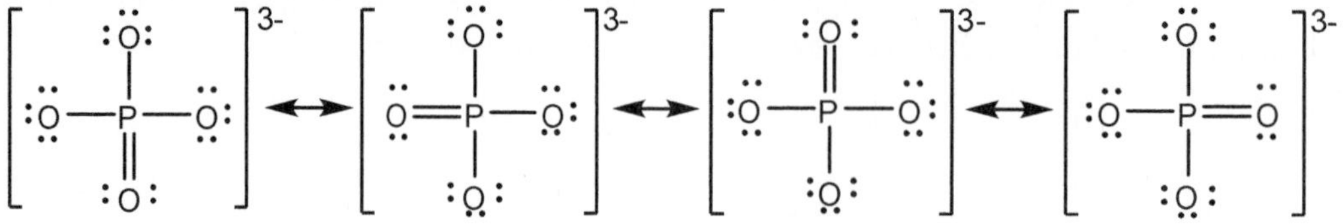

8.24

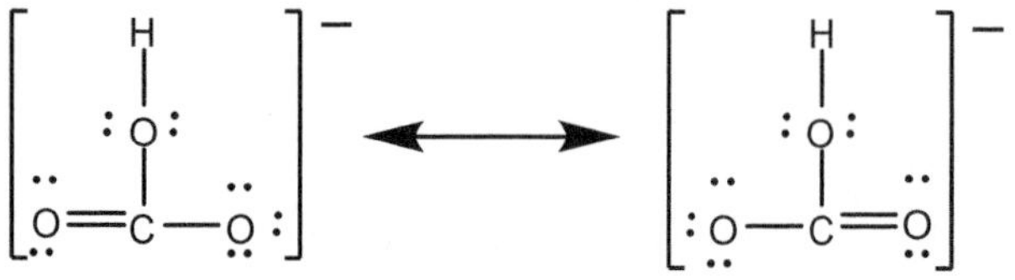

8.25

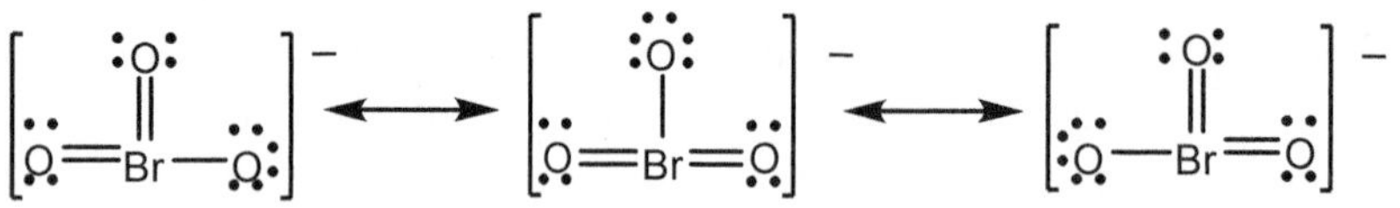

8.26

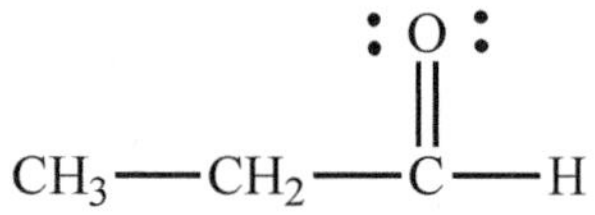

aldehyde

amine

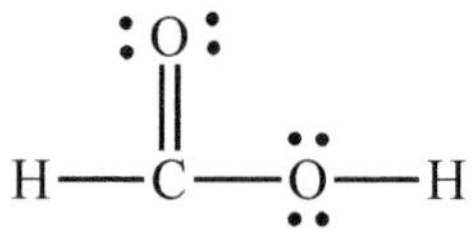

acid

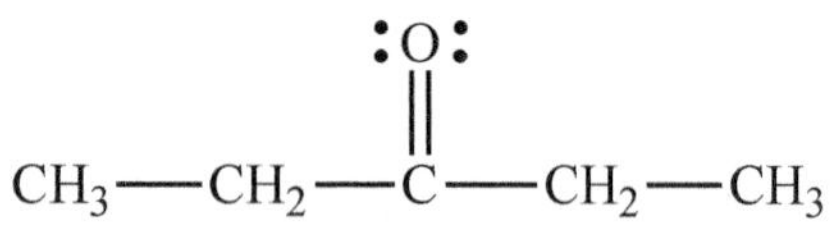

ketone

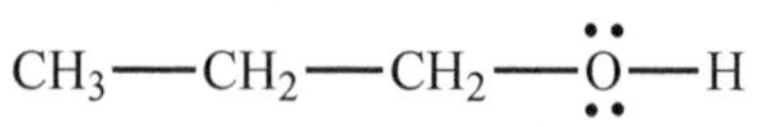

alcohol

8.27 (a) CH_3NHCH_3 will produce a basic solution

(b) HCOOH will produce an acidic solution

(c)

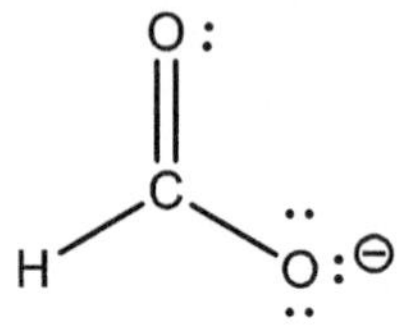

(d)

```
       H      ]+
       |
H3C—N—CH3
       |
       H
```

Review Problems

8.63 (a) Al_2O_3 (b) BeO (c) NaCl

8.65

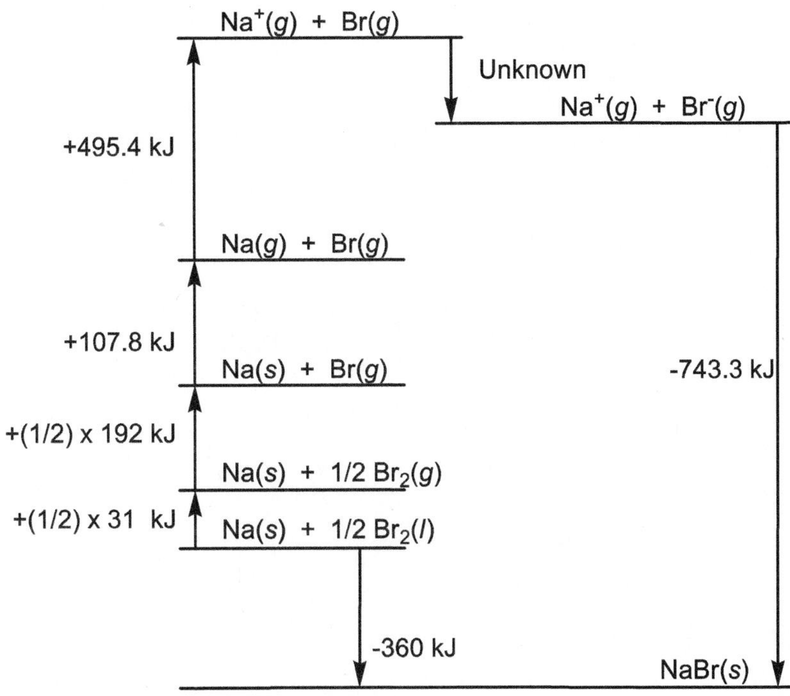

Electron affinity for Br = –331.4 kJ/mol

8.67 Magnesium loses two electrons:

$$Mg \longrightarrow Mg^{2+} + 2e^-$$

$$[Ne]3s^2 \longrightarrow [Ne]$$

Bromine gains an electron:

$$Br + e^- \longrightarrow Br^-$$

$$[Ar]3d^{10}4s^24p^5 \longrightarrow [Kr]$$

To keep the overall change of the formula unit neutral, two Br^- ions combine with one Mg^{2+} ion to form $MgBr_2$:

$$Mg^{2+} + 2Br^- \longrightarrow MgBr_2$$

8.69 Pb^{2+}: [Xe] $4f^{14}5d^{10}6s^2$
Pb^{4+}: [Xe]$4f^{14}5d^{10}$

8.71 Mn^{3+}: $[Ar]3d^4$ 4 unpaired electrons

8.73 (a)

$\cdot\dot{\underset{\cdot}{Si}}\cdot$

(b)

$:\dot{\underset{\cdot}{Sb}}\cdot$

(c)

$\cdot Ba \cdot$

(d)

$\cdot\dot{Al}\cdot$

(e)

$:\dot{\underset{\cdot}{S}}:$

8.75 (a)

$\cdot Mg \cdot \quad \cdot\ddot{\underset{\cdot\cdot}{S}}\cdot \longrightarrow Mg^{2+} + [:\ddot{\underset{\cdot\cdot}{S}}:]^{2-}$

(b)

$:\ddot{\underset{\cdot\cdot}{Cl}}\cdot \quad \cdot Mg \cdot \quad :\ddot{\underset{\cdot\cdot}{Cl}}\cdot \longrightarrow Mg^{2+} + 2[:\ddot{\underset{\cdot\cdot}{Cl}}:]^-$

(c)

$\cdot Mg \cdot \quad \cdot\ddot{\underset{\cdot}{N}}\cdot \quad \cdot Mg \cdot \quad \cdot\ddot{\underset{\cdot}{N}}\cdot \quad \cdot Mg \cdot \longrightarrow 3\ Mg^{2+} + 2[:\ddot{\underset{\cdot\cdot}{N}}:]^{3-}$

8.77 $J\text{/molecule} = (242.6 \times 10^3 \text{ J/mol})\left(\dfrac{1 \text{ mole}}{6.022 \times 10^{23} \text{ molecules}}\right)$

$= 4.029 \times 10^{-19}$ J/molecule

8.79 $E = h\nu = \dfrac{hc}{\lambda} \qquad \lambda = \dfrac{hc}{E}$

$$E = (348 \times 10^3 \text{ J/mol})\left(\frac{1 \text{ mol}}{6.022 \times 10^{23} \text{ molecules}}\right) = 5.78 \times 10^{-19} \text{ J/molecule}$$

$$\lambda = \frac{\left(6.626 \times 10^{-34} \text{ J sec}\right)\left(3.00 \times 10^8 \text{ m s}^{-1}\right)}{\left(5.78 \times 10^{-19} \text{ J}\right)} = 3.44 \times 10^{-7} \text{ m} = 344 \text{ nm}$$

Ultraviolet region

8.81 (a)

$:\ddot{\underset{\cdot\cdot}{Br}}\cdot \quad \cdot\ddot{\underset{\cdot\cdot}{Br}}: \longrightarrow :\ddot{\underset{\cdot\cdot}{Br}}-\ddot{\underset{\cdot\cdot}{Br}}:$

(b)

$H\cdot \quad \cdot\ddot{\underset{\cdot\cdot}{O}}\cdot \quad \cdot H \longrightarrow H-\ddot{\underset{\cdot\cdot}{O}}-H$

(c)

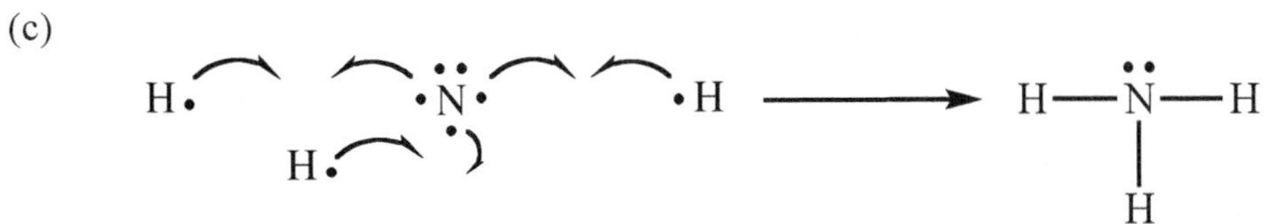

8.83 (a)

Cl
|
Cl—N—Cl

(b)

Cl
|
Cl—C—Cl
|
Cl

(c)

Cl—S—Cl

(d)

Br—Cl

8.85 (a) Each chlorine atom needs one further electron in order to achieve an octet, and the phosphorus atom requires three electrons from an appropriate number of chlorine atoms. We conclude that a phosphorus atom is bonded to three chlorine atoms and that each chlorine atom is bonded only once to the phosphorus atom: PCl_3.

(b) Since carbon needs four additional electrons, the formula must be CF_4. In this arrangement, each fluorine atom acquires the one additional electron that is needed to reach its octet.

(c) Each halogen atom needs only one additional electron from the other: ICl.

8.87 $\mu = q \times r$

$$\mu = 0.16\ \text{D}\left(\frac{3.34 \times 10^{-30}\ \text{C m}}{1\ \text{D}}\right) = q \times (115\ \text{pm})\left(\frac{1\ \text{m}}{10^{12}\ \text{pm}}\right)$$

$5.34 \times 10^{-31}\ \text{C m} = q \times (115 \times 10^{-12}\ \text{m})$

$q = 4.64 \times 10^{-21}\ \text{C}$

$$q = 4.64 \times 10^{-21}\ \text{C}\left(\frac{1\ e^-}{1.60 \times 10^{-19}\ \text{C}}\right) = 0.029\ e^-$$

The charge on the oxygen is $-0.029\ e^-$ and the charge on the nitrogen is $+0.029\ e^-$. The nitrogen atom is positive.

8.89 $\mu = q \times r$

$$\mu = 7.88\ \text{D}\left(\frac{3.34 \times 10^{-30}\ \text{C m}}{1\ \text{D}}\right) = q \times 0.255\ \text{nm}\left(\frac{1\ \text{m}}{10^{9}\ \text{nm}}\right)$$

$2.63 \times 10^{-29}\ \text{C m} = q \times (0.255 \times 10^{-9}\ \text{m})$
$q = 1.03 \times 10^{-19}\ \text{C}$

$$q = 1.03 \times 10^{-19}\ \text{C}\left(\frac{1\ e^-}{1.60 \times 10^{-19}\ \text{C}}\right) = 0.645\ e^-$$

This is 64.5% of a positive charge on the cesium and 64.5% of a negative charge on the fluorine.

8.91 Here we choose the atom with the smaller electronegativity:
(a) S (b) Si (c) Br (d) C

8.93 Here we choose the linkage that has the greatest difference in electronegativities between the atoms of the bond: N—S.

8.95 (a) (b)

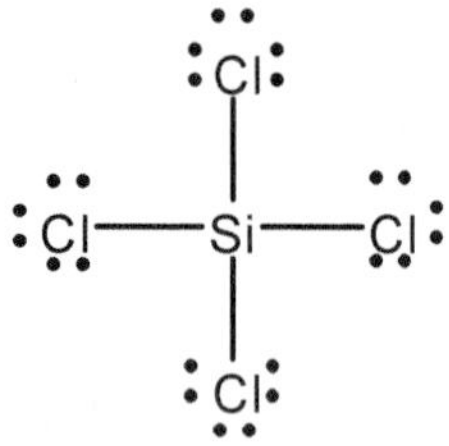

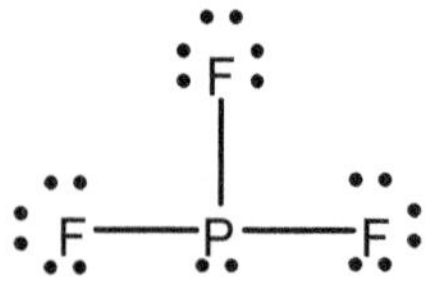

(c) 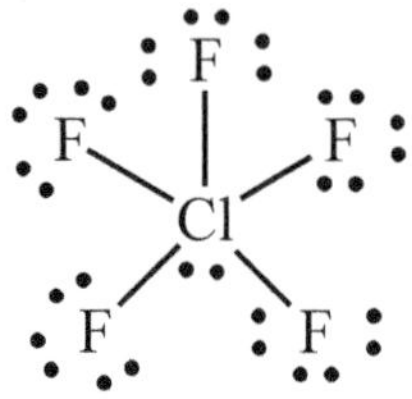 (d)

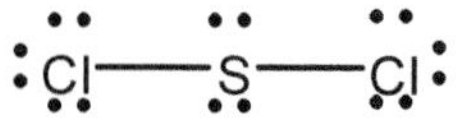

8.97 (a) 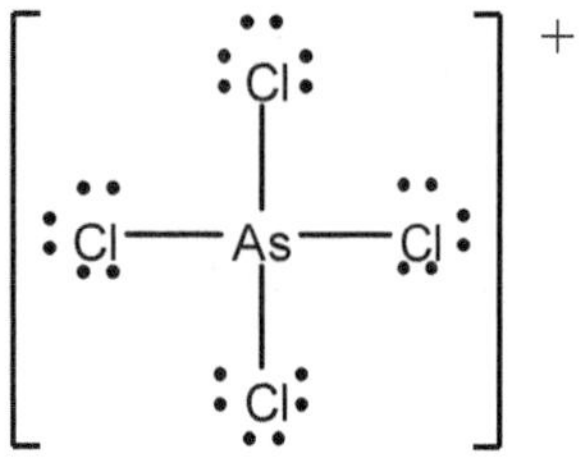 (b)

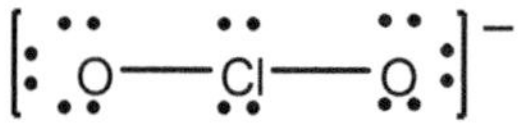

(c) 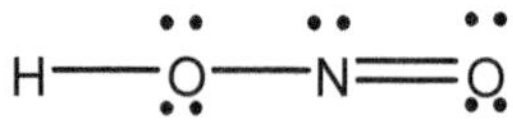 (d)

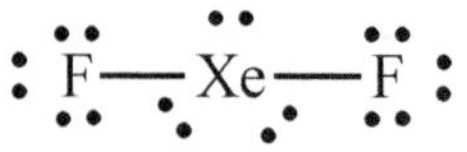

8.99 (a) 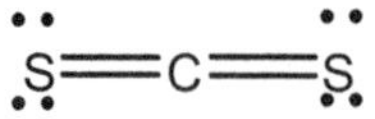 (b)

(c) (d)

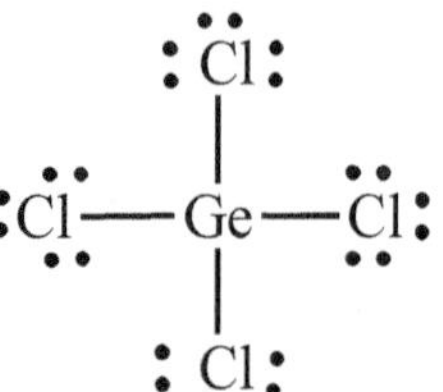

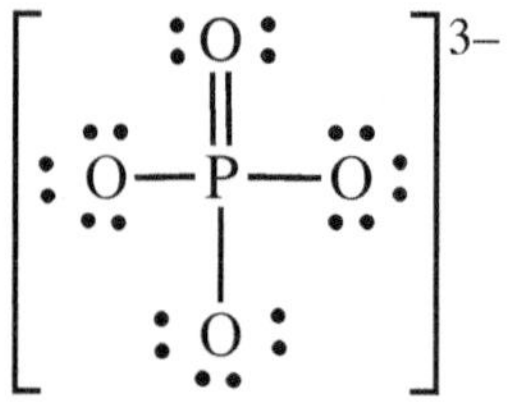

8.101 (a)

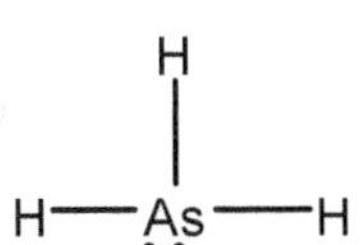

(b)

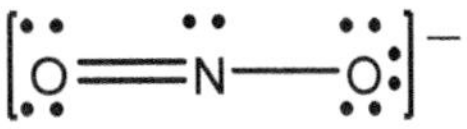

(c)

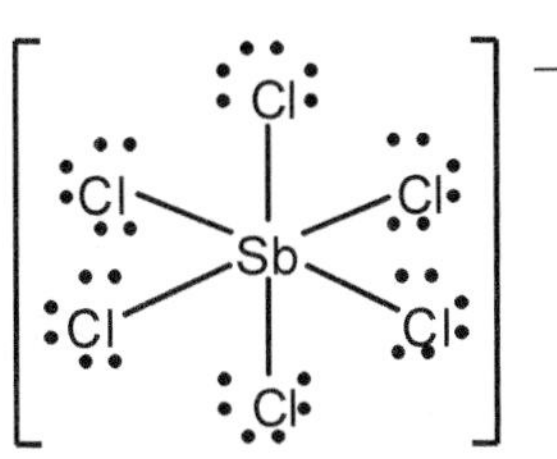

(d)

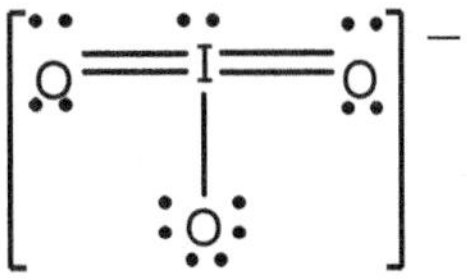

8.103 (a)

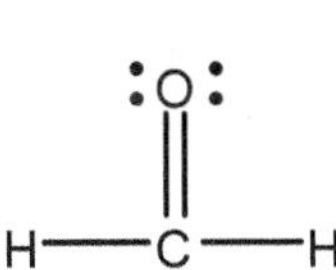

(b)

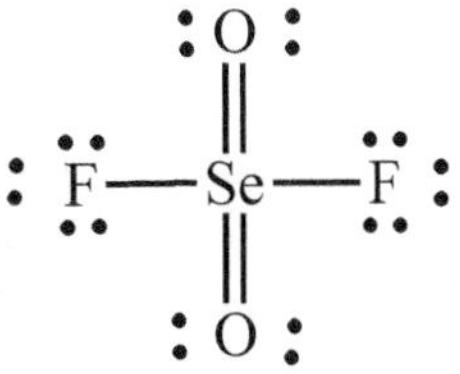

8.105 (a)

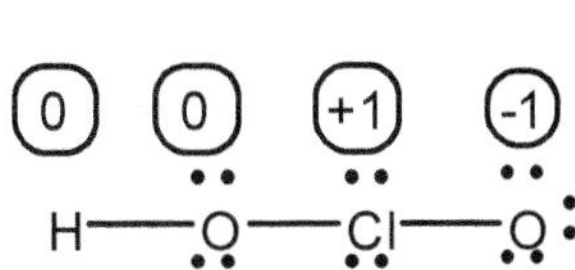

(b)

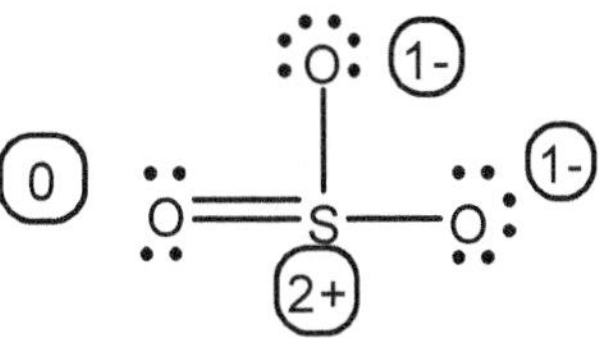

(c)

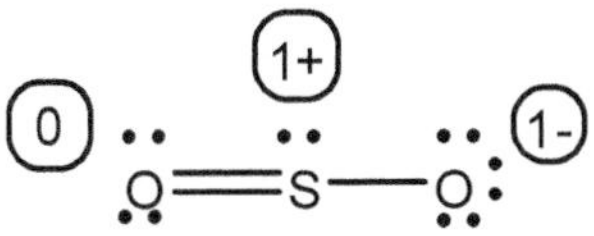

8.107

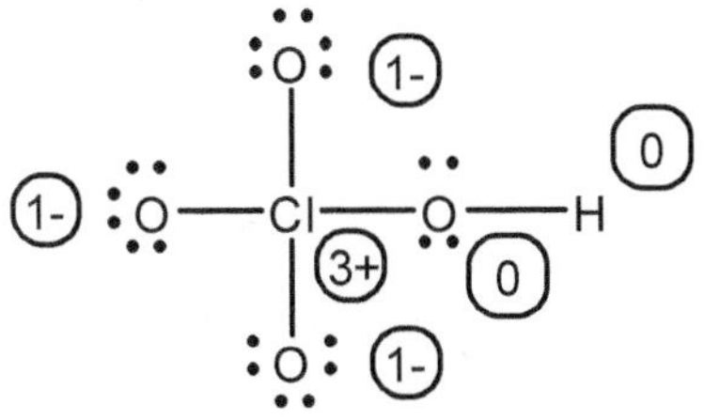

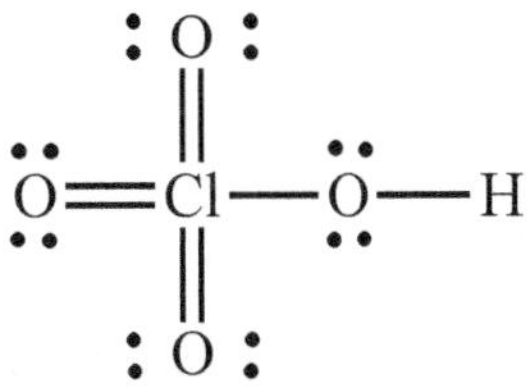

The preferred structure is the one on the right. The formal charges on the double bonded oxygen are zero and the formal charge on the chlorine is +1.

8.109 (1) The formal charges on all of the atoms of the left structure are zero, therefore, the potential energy of this molecule is lower and it is more stable. (2) Group 2 atoms normally have only two bonds.

8.111

8.113

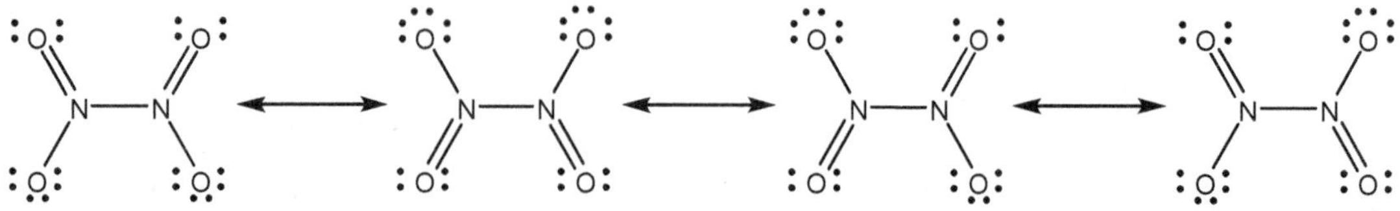

The average bond order is 3/2.

8.115 The Lewis structure for NO_3^- is given in Section 8.8, and that for NO_2^- is given below.

Resonance causes the average number of bonds in each N—O linkage of NO_3^- to be 1.33. Resonance causes the average number of electron pair bonds in each linkage of NO_2^- to be 1.5. We conclude that the N—O bond in NO_2^- should be shorter than that in NO_3^-.

8.117

These are not preferred structures, because in each Lewis diagram, one oxygen bears a formal charge of +1 whereas the other bears a formal charge of –1. The structure with the formal charges of zero has a lower potential energy and is more stable.

Chapter Nine
Theories of Bonding and Structure

Practice Exercises

9.1 (a) Carbon tetrabromide, CBr_4, should have a tetrahedral shape (Figure 9.4) because it has four bonding electron pairs around the central atom.

(b) Arsenic pentaflouride, AsF_5, should have a trigonal bipyramidal shape (Figure 9.4) because it has five bonding electron pairs around the central atom.

9.2 SeF_6 should have an octahedral shape (Figure 9.4) because it has six bonding electron pairs around the central atom.

9.3 $SbCl_5$ should have a trigonal bipyramidal shape (Figure 9.4) because, like PCl_5, it has five bonding electron pairs around the central atom.

9.4 HArF should have a linear shape (Figure 9.7) because although it has five electron pairs around the central Ar atom, only two are being used for bonding.

9.5 IBr_2^- should have a linear shape (Figure 9.7) because although it has five electron pairs around the central I atom, only two are being used for bonding.

9.6 SO_3 should have trigonal planar shape because it has three electron domains around the central S atom. Two of the domains are double bonds.

9.7 SF_4 is distorted tetrahedral and has one lone pair of electrons on the sulfur, therefore it is polar.

9.8 (a) TeF_6 is octahedral, and it is not polar.
(b) SeO_2 is bent, and it is polar.
(c) BrCl is polar because there is a difference in electronegativity between Br and Cl.
(d) AsH_3, like NH_3, is pyramidal, and it is polar.
(e) CF_2Cl_2 is polar, because there is a difference in electronegativity between F and Cl.

9.9 The H—Cl bond is formed by the overlap of the half–filled 1*s* atomic orbital of a H atom with the half–filled 3*p* valence orbital of a Cl atom:

Cl atom in HCl (x = H electron):

⇅ ⇅ ⇅ ↑x
3*s* 3*p*

The overlap that gives rise to the H—Cl bond is that of a 1*s* orbital of H with a 3*p* orbital of Cl:

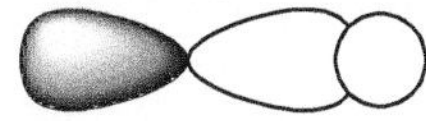

9.10 The half–filled 1*s* atomic orbital of each H atom overlaps with a half–filled 3*p* atomic orbital of the P atom, to give three P—H bonds. This should give a bond angle of 90°.

P atom in PH_3 (x = H electron):

⇅ ↑x ↑x ↑x
3*s* 3*p*

The orbital overlap that forms the P—H bond combines a 1*s* orbital of hydrogen with a 3*p* orbital of phosphorus (note: only half of each *p* orbital is shown):

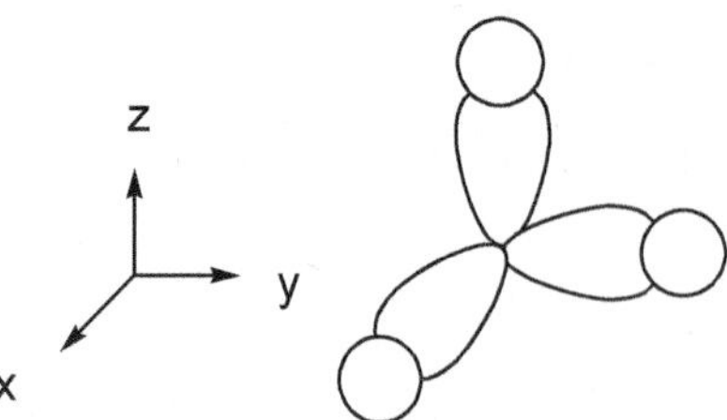

9.11 BF_3 uses sp^2 hybridized orbitals since it has only three bonding electron pairs and no lone pairs of electrons.
The sp^2 hybrid orbitals on the B, x = F electron

↿x ↿x ↿x   —
sp^2 $2p$

9.12 BeF_2 uses sp hybridized orbitals since it has only two bonding electron pairs and no lone pairs of electrons.
The sp hybrid orbitals on the Be; x = F electron

↿x ↿x   — —
sp $2p$

9.13 SiH_4 uses sp^3 hybridized orbitals since it has four bonding electron pairs and no lone pairs of electrons.

↿x ↿x ↿x ↿x
sp^3

9.14 Since there are five bonding pairs of electrons on the central phosphorous atom, we choose sp^3d hybridization for the P atom. Each of phosphorus's five sp^3d hybrid orbitals overlaps with a $3p$ atomic orbital of a chlorine atom to form a total of five PCl single bonds. Four of the $3d$ atomic orbitals of P remain unhybridized.

9.15 VSEPR theory predicts that $AsCl_5$ will be trigonal bipyramidal. Since there are five bonding pairs of electrons on the central arsenic atom, we choose sp^3d hybridization for the As atom as a trigonal bipyramid. Each of arsenic's five sp^3d hybrid orbitals overlaps with a $3p$ atomic orbital of a chlorine atom to form a total of five As—Cl single bonds. Four of the $4d$ atomic orbitals of As remain unhybridized.

9.16 Using VSEPR theory we find that XeF_4 has eight valence electrons for the Xe and seven for each F, so there are six electron domains around the Xe. To accommodate the six electron domains, the Xe needs six hybrid orbitals, so the hybridization is sp^3d^2

F   ••   F
Xe
F   ••   F

9.17 (a) PCl_3 has four pair of electrons around the central atom, P, so the hybridization is sp^3
(b) $BrCl_3$ has five pair of electrons around the central atom, Br, so the hybridization is sp^3d

9.18 NH_3 is sp^3 hybridized. Three of the electron pairs are use for bonding with the three hydrogens. The fourth pair of electrons is a lone pair of electrons. This pair of electrons is used for the formation of the bond between the nitrogen of NH_3 and the hydrogen ion, H^+.

9.19 Since there are six bonding pairs of electrons on the central phosphorous atom, we choose sp^3d^2 hybridization for the P atom. Each of phosphorus's six sp^3d^2 hybrid orbitals overlaps with a $3p$

atomic orbital of a chlorine atom to form a total of six P–Cl single bonds. Three of the 3*d* atomic orbitals of P remain unhybridized.
P atom in PCl_6^- (x = Cl electron):

↑x ↑x ↑x ↑x ↑x ↑x — — — 3*d*
sp^3d^2

The ion is octahedral because six atoms and no lone pairs surround the central atom.

$[PCl_6]^-$ (octahedral: P bonded to six Cl)

9.20 Atom 1has three electron domains: sp^2
Atom 2 has four electron domains: sp^3
Atom 3 has three electron domains: sp^2
There are 10 σ bonds and 2 π bonds in the molecule.

9.21 Atom 1has two electron domains: *sp*
Atom 2 has three electron domains: sp^2
Atom 3 has four electron domains: sp^3
There are 9 σ bonds and 3 π bonds in the molecule.

9.22 CN^- has 10 valence electrons and the MO diagram is similar to that of C and N. The bond order of the ion is 3 and this does agree with the Lewis structure.

— σ^*_{2pz}

— — π^*_{2px} π^*_{2spy}

↑↓ σ_{2pz}

↑↓ ↑↓ π_{2px} π_{2py}

↑↓ σ^*_{2s}

↑↓ σ_{2s}

9.23 NO has 11 valence electrons, and the MO diagram is similar to that shown in Table 9.1 for O_2, except that one fewer electron is employed at the highest energy level

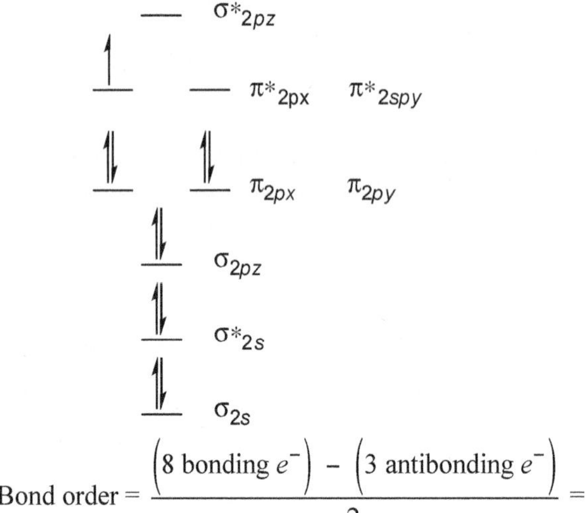

$$\text{Bond order} = \frac{\left(8 \text{ bonding } e^-\right) - \left(3 \text{ antibonding } e^-\right)}{2} = \frac{5}{2}$$

The bond order is calculated to be 5/2.

9.24 The hybridization of the N in NO_3^- is sp^2. This leaves one extra p orbital on the nitrogen which is available for π bonding. The three oxygen atoms are equivalent, and only one can form a π bond at a time with the N, so there are three resonance structures over the three N—O bonds. The atomic orbitals used in the delocalized molecular orbital are the p orbitals on the N and O that are not involved in forming the σ bond framework.

9.25 For calcium, the valence band is formed by the 4*s* orbitals and the conduction band is formed by the 3*d* orbitals.

9.26 Non-metals have the largest band gap, metals have the smallest, or non-existent band gap, and semi-conductors have a band gap between the two. Magnesium is smaller than germanium which is smaller than sulfur.

9.27 (a) The carbon atoms in diamonds have a tetrahedral structure, so the hybridization is sp^3.
(b) The carbon atoms in graphite have a trigonal planar structure, so the hybridization is sp^2.
(c) The carbon atoms in buckyballs have a trigonal planar structure, so the hybridization is sp^2.

Review Problems

9.73 (a) Bent (central N atom has two single bonds and two lone pairs)
(b) Planar triangular (central C atom has three bonding domains—a double bond and two single bonds)
(c) T-shaped (central I atom has three single bonds and two lone pairs)
(d) Linear (central Br atom has two single bonds and three lone pairs)
(e) Planar triangular (central Ga atom has three single bonds and no lone pairs)

9.75 (a) Bent (central F atom has four electron domains; two are lone pair)
(b) Trigonal bipyramidal (central As atom has five electron domains)
(c) Trigonal pyramidal (central As atom has four electron domains; one is a lone pair)
(d) Trigonal pyramidal (central Sb atom has four electron domains; one is a lone lair)
(e) Bent (central Se atom has four electron domains; two are lone pair)

9.77 (a) Tetrahedral (central I atom has four electron domains)
(b) Square planar (central I atom has six electron domains; two are lone pair)
(c) Octahedral (central Te atom has six electron domains)
(d) Tetrahedral (central Si atom has four electron domains)
(e) Linear (central I atom has five electron domains; three are lone pair)

9.79 BrF_4^+ The central Br atom has five electron domains, one is a lone pair

9.81 All angles are 120°, since there are three electron domains around each carbon atom.

9.83 (a) 109.5° (b) 109.5°
(c) 120° (d) 109.5°
(e) 120°

9.85 The ones that are polar are (a), (b), and (c). The last two have symmetrical structures, and although individual bonds in these substances are polar bonds, the geometry of the bonds serves to cause the individual dipole moments of the various bonds to cancel one another.

9.87 All are polar. (a), (b), (c) and (e) have asymmetrical structures, and (d) only has one bond, which is polar.

9.89 In SF_6, although the individual bonds in this substance are polar bonds, the geometry of the bonds is symmetrical which serves to cause the individual dipole moments of the various bonds to cancel one another. In SF_5Br, one of the six bonds has a different polarity so the individual dipole moments of the various bonds do not cancel one another.

9.91 This is shown in Figure 9.18 for F_2 . Since Cl_2 is in the same family the bonding will be similar, using the n = 3 shell rather than the n = 2 shell. We can diagram it showing the orbitals of one of the chlorine atoms:
Each Cl atom (x = an electron from the other Cl atom):

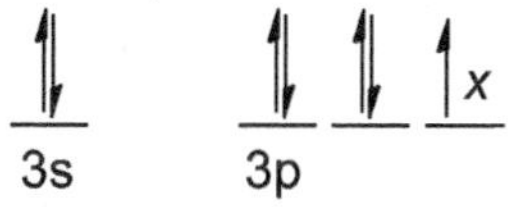

↑↓ ↑↓ ↑↓ ↑x
3s 3p

9.93 Atomic Be:

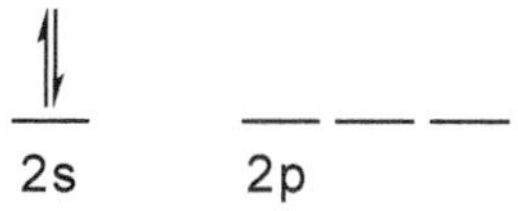

↑↓ __ __ __
2s 2p

Hybridized Be: (x = a Cl electron)

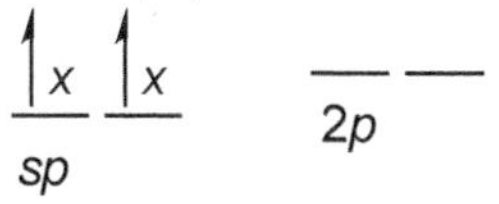

↑x ↑x __ __
sp *2p*

Beryllium uses *sp* hybrid orbitals.

9.95 Sb atom in $SbCl_5$ (x = a Cl electron)

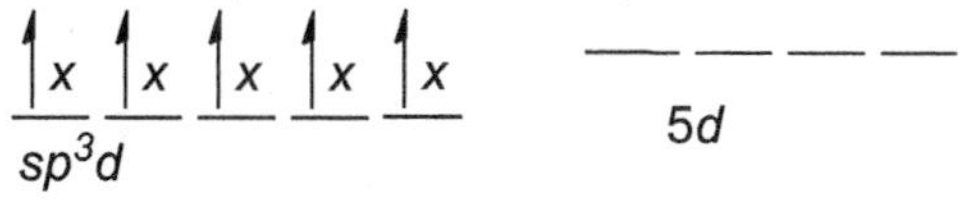

↑x ↑x ↑x ↑x ↑x __ __ __ __
sp^3d 5d

Antimony uses sp^3d hybrid orbitals.

9.97 (a) There are three bonds to the central Cl atom, plus one lone pair of electrons. The geometry of the electron pairs is tetrahedral so the Cl atom is to be sp^3 hybridized:

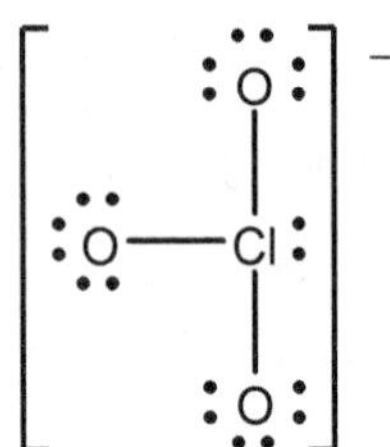

(b) There are three atoms bonded to the central sulfur atom, and no lone pairs on the central sulfur. The geometry of the electron pairs is that of a planar triangle, and the hybridization of the S atom is sp^2:

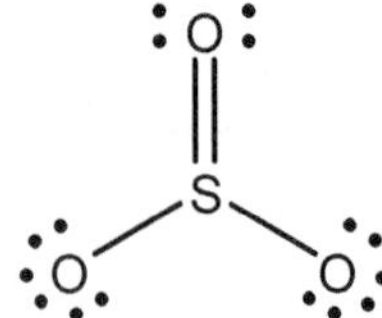

Two other resonance structures should also be drawn for SO_3.

(c) There are two bonds to the central O atom, as well as two lone pairs. The O atom is to be sp^3 hybridized, and the geometry of the electron pairs is tetrahedral.

9.99 (a) There are three bonds to As and one lone pair at As, requiring As to be sp^3 hybridized. The Lewis diagram

The hybrid orbital diagram for As: (x = a Cl electron)

(b) There are three atoms bonded to the central Cl atom, and it also has two lone pairs of electrons. The hybridization of Cl is thus sp^3d.
The Lewis diagram

The hybrid orbital diagram for Cl: (x = a F electron)

9.101 We can consider that this ion is formed by reaction of SbF_5 with F^-. The antimony atom accepts a pair of electrons from fluoride:

Sb in SbF_6^-: (*xx* = an electron pair from the donor F^-)

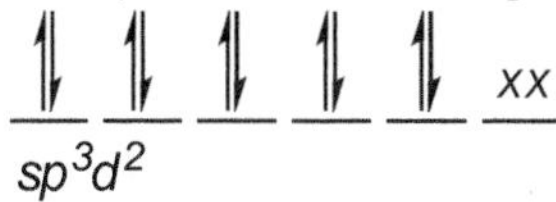

9.103 (a) N in the C=N system:

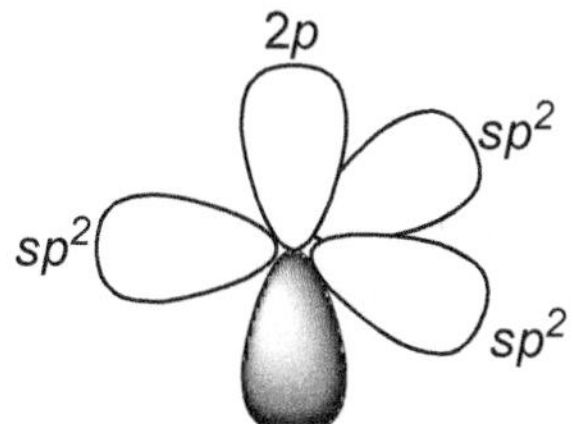

(b) sigma bond

C N

pi bond

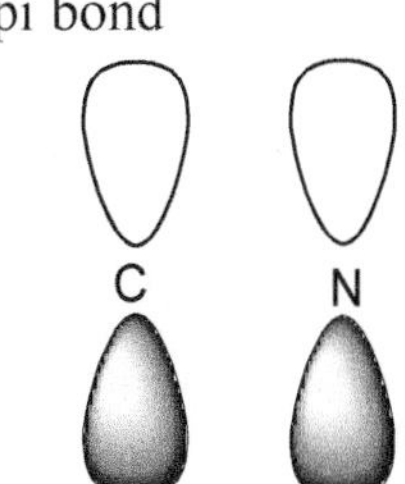

(c)

H 120°

120° C=N

H 120° 120° H

9.105 Each carbon atom is sp^2 hybridized, and each C–Cl bond is formed by the overlap of an sp^2 hybrid of carbon with a p atomic orbital of a chlorine atom. The C=C double bond consists first of a C–C σ bond formed by "head on" overlap of sp^2 hybrids from each C atom. Secondly, the C=C double bond consists of a side–to–side overlap of unhybridized p orbitals of each C atom, to give one π bond. The molecule is planar, and the expected bond angles are all 120°.

9.107 1. sp^3 2. sp 3. sp^2 4. sp^2

9.109
1. One σ bond
2. One σ bond and two π bonds
3. One σ bond
4. One σ bond and one π bond

9.111 The two unpaired electrons are located in the π antibonding molecular orbitals.

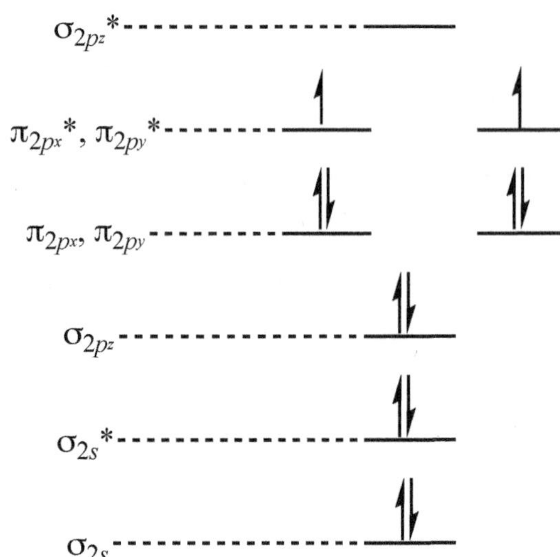

The net bond order is (8 Bonding e^- – 4 Antibonding e^-)/2 = 2.

9.113 (a) (i) O_2: BO = 2 O_2^+: BO = 2.5 (ii) O_2: BO = 2 O_2^-: BO = 1.5
(b) (i) O_2 (ii) O_2^-
(c) (i) O_2^+ (ii) O_2

9.115 NO has 11 valence electrons, and the MO diagram is similar to that shown in Table 9.1 for O_2, except that one fewer electron is employed at the highest energy level
The bond order for NO is calculated to be 5/2:

$$\text{Bond Order} = \frac{\left(8\text{ bonding } e^-\right) - \left(3\text{ antibonding } e^-\right)}{2} = \frac{5}{2}$$

The bond order for NO^+ is calculated to be 3, since one electron is removed from an antibonding orbital:

$$\text{Bond order} = \frac{\left(8\text{ bonding } e^-\right) - \left(2\text{ antibonding } e^-\right)}{2} = 3$$

If we remove one electron to form NO^+, the bond order becomes 3 (there are only two antibonding electrons). The larger bond order indicates a shorter bond length. The bond energy of NO is lower than the bond energy of NO^+.

9.117 All of the molecules or ions are paramagnetic except N_2. O_2^+, O_2, O_2^-, and NO are paramagnetic

Chapter Ten
Properties of Gases

Practice Exercises

10.1 Adding more gas increases the pressure because by adding the gas, there are more particles to bounce against the walls of the container, increasing the force per unit area.

10.2 When apples rot, they lose mass because they are emitting a gas which would be the by-product of the chemical reaction. When nails rust, the iron combines with the oxygen in the air, thus gaining mass.

10.3 $$\text{psi} = 730 \text{ mm Hg}\left(\frac{14.7 \text{ psi}}{760 \text{ mm Hg}}\right) = 14.1 \text{ psi}$$

$$\text{In. Hg} = 730 \text{ mm Hg}\left(\frac{29.921 \text{ in. Hg}}{760 \text{ mm Hg}}\right) = 28.7 \text{ in. Hg}$$

10.4 $$\text{pascals} = 888 \text{ mbar}\left(\frac{1 \text{ bar}}{1000 \text{ mbar}}\right)\left(\frac{1 \text{ atm}}{1.013 \text{ bar}}\right)\left(\frac{101{,}325 \text{ pascal}}{1 \text{ atm}}\right) = 88{,}800 \text{ pascal}$$

$$\text{torr} = 888 \text{ mbar}\left(\frac{1 \text{ bar}}{1000 \text{ mbar}}\right)\left(\frac{1 \text{ atm}}{1.013 \text{ bar}}\right)\left(\frac{760 \text{ torr}}{1 \text{ atm}}\right) = 666 \text{ torr}$$

10.5 $$\text{mm Hg} = 15 \text{ cm Hg}\left(\frac{10 \text{ mm Hg}}{1 \text{ cm Hg}}\right) = 150 \text{ mm Hg}$$

$$\text{mm Hg} = 770 \text{ torr}\left(\frac{760 \text{ mm Hg}}{760 \text{ torr}}\right) = 770 \text{ mm Hg}$$

The maximum pressure = 770 mm Hg + [(150 mm Hg) × 2] = 1070 mm Hg
The minimum pressure = 770 mm Hg – [(150 mm Hg) × 2] = 470 mmHg

10.6 The pressure of the gas in the manometer is the pressure of the atmosphere less the pressure of the mercury, 11.7 cm Hg.
Using the pressure in the atmosphere from the previous example:

$$\text{mm Hg} = 11.7 \text{ cm Hg}\left(\frac{10 \text{ mm Hg}}{1 \text{ cm Hg}}\right) = 117 \text{ mm Hg}$$

770 mm Hg – 117 mm Hg = 653 mm Hg

10.7 $$\frac{P_1V_1}{T_1} = \frac{P_2V_2}{T_2}$$

$V_2 = 3V_1$ and $T_2 = 2T_1$

$$\frac{P_1V_1}{T_1} = \frac{P_2 3V_1}{2T_1}$$

$$P_2 = \frac{2}{3}P_1$$

The pressure must change by 2/3.

10.8 In general the combined gas law equation is: $\frac{P_1V_1}{T_1} = \frac{P_2V_2}{T_2}$, and in particular, for this problem, we have:

$$P_2 = \frac{P_1V_1T_2}{T_1V_2} = \frac{(745 \text{ torr})(950 \text{ cm}^3)(333.2 \text{ K})}{(1150 \text{ cm}^3)(298.2 \text{ K})} = 688 \text{ torr}$$

10.9 When gases are held at the same temperature and pressure, as reactants, then they react in a ratio of volumes that is equal to the ratio of the coefficients (moles) in the balanced chemical equation for the given reaction. We can, therefore, directly use the stoichiometry of the balanced chemical equation to determine the combining ratio of the gas volumes:

$$\text{L } O_2 = (4.50 \text{ L } CH_4)\left(\frac{2 \text{ volume } O_2}{1 \text{ volume } CH_4}\right) = 9.00 \text{ L } O_2$$

10.10 $\text{L } O_2 = (6.75 \text{ L } CH_4)\left(\frac{2 \text{ volume } O_2}{1 \text{ volume } CH_4}\right) = 13.50 \text{ L } O_2$

$$20.9\% \ O_2 = \frac{13.50 \text{ L } O_2}{x \text{ L air}} \times 100\%$$

$$x \text{ L air} = (13.50 \text{ L } O_2)\left(\frac{100\%}{20.9\%}\right) = 64.6 \text{ L air}$$

10.11 $n = \frac{PV}{RT} = \frac{(57.8 \text{ atm})(12.0 \text{ L})}{\left(0.0821 \frac{\text{L atm}}{\text{mol K}}\right)(298 \text{ K})} = 28.3 \text{ moles gas}$

28.3 mol Ar (39.95 g Ar/mol) = 1,130 g Ar

10.12 First determine the number of moles of CO_2 in the tank:

$$n = \frac{PV}{RT}$$

$$P = 2000 \text{ psig}\left(\frac{1 \text{ atm}}{14.7 \text{ psig}}\right) = 136 \text{ atm} = 140 \text{ atm}$$

$$V = 6.0 \text{ ft}^3\left(\frac{(30.48 \text{ cm})^3}{(1 \text{ ft})^3}\right)\left(\frac{1 \text{ mL}}{1 \text{ cm}^3}\right)\left(\frac{1 \text{ L}}{1000 \text{ mL}}\right) = 170 \text{ L}$$

$$R = 0.0821 \ \frac{\text{L}\cdot\text{atm}}{\text{mol}\cdot\text{K}}$$

$$T = (22\ ^\circ\text{C} + 273\ ^\circ\text{C})\left(\frac{1 \text{ K}}{1\ ^\circ\text{C}}\right) = 295 \text{ K}$$

$$\text{mol } CO_2 \text{ in the tank} = \frac{(140 \text{ atm})(170 \text{ L})}{\left(0.0821 \frac{\text{L}\cdot\text{atm}}{\text{mol}\cdot\text{K}}\right)(295 \text{ K})} = 980 \text{ mol } CO_2$$

Then find the total number of grams of CO_2 in the tank, MW CO_2 = 44.01 g/mol

$$\text{g } CO_2 \text{ in the tank} = 980 \text{ mol } CO_2\left(\frac{44.01 \text{ g } CO_2}{1 \text{ mol } CO_2}\right) = 43{,}000 \text{ g } CO_2$$

Amount of solid CO_2 = 43,000 g CO_2 × 0.35 = 15,000 g solid CO_2

10.13 Since $PV = nRT$, then $n = PV/RT$

$$n = \frac{PV}{RT} = \frac{(685 \text{ torr})\left(\frac{1 \text{ atm}}{760 \text{ torr}}\right)(0.300 \text{ L})}{\left(0.0821 \frac{\text{L atm}}{\text{mol K}}\right)(300.2 \text{ K})} = 0.0110 \text{ moles gas}$$

$$\text{molar mass} = \frac{1.45 \text{ g}}{0.0110 \text{ mol}} = 132 \text{ g mol}^{-1}$$

The gas must be xenon.

10.14 Find the number of moles of argon

$$n = \frac{PV}{RT} = \frac{(1.0000 \text{ atm})(0.54423 \text{ L})}{\left(0.082057 \frac{\text{L atm}}{\text{mol K}}\right)(273.15 \text{ K})} = 0.024281 \text{ mol argon}$$

$$\text{Mass of the argon} = (0.024281 \text{ mol argon})\left(\frac{39.948 \text{ g argon}}{1 \text{ mol argon}}\right) = 0.96998 \text{ g argon}$$

The mass of the flask = 735.6898 g – 0.96998 g air = 734.7198 g

Mass of the organic compound = 736.13106 g – 734.7198 g = 1.4113 g

The number of moles of the organic compound equals the number of moles of air:

$$\text{MW} = \frac{1.4113 \text{ g organic compound}}{0.024281 \text{ mole organic compound}} = 58.12 \text{ g/mol}$$

The unknown could be butane, MW = 58.1 g/mol

10.15 $d = m/V$

Taking 1.00 mol SO_2:

$m = 64.1$ g

$$V = \frac{nRT}{P} = \frac{(1.00 \text{ mol})\left(0.0821 \frac{\text{L atm}}{\text{mol K}}\right)(268.2 \text{ K})}{96.5 \text{ kPa}\left(\frac{1 \text{ atm}}{101.325 \text{ kPa}}\right)} = 23.1 \text{ L} = 23{,}100 \text{ mL}$$

$$\text{density} = \frac{64.1 \text{ g}}{23.1 \text{ L}} = 2.77 \text{ g L}^{-1}$$

10.16 $$\text{Density of air} = 1 \text{ mol air}\left(\frac{28.8 \text{ g air}}{1 \text{ mol air}}\right)\left(\frac{1 \text{ mol air}}{22.4 \text{ L}}\right) = 1.29 \text{ g/L}$$

$$\text{Density of radon at STP} = 1 \text{ mol Rn}\left(\frac{222.0 \text{ g Rn}}{1 \text{ mol Rn}}\right)\left(\frac{1 \text{ mol Rn}}{22.4 \text{ L}}\right) = 9.91 \text{ g/L}$$

Since radon is almost eight times denser than air, the sensor should be in the lowest point in the house: the basement.

10.17 In general $PV = nRT$, where n = mass × formula mass. Thus

$$PV = \frac{\text{mass}}{\text{formula mass}} RT$$

We can rearrange this equation to get;

$$\text{formula mass} = \frac{\left(\frac{\text{mass}}{V}\right)RT}{P} = \frac{dRT}{P}$$

$$\text{formula mass} = \frac{\left(5.60\ \text{g L}^{-1}\right)\left(0.0821\ \text{L atm mol}^{-1}\ \text{K}^{-1}\right)\left(273.15\ \text{K} + 23.0\ °\text{C}\right)}{750\ \text{torr}\left(\dfrac{1\ \text{atm}}{760\ \text{torr}}\right)}$$

$$\text{formula mass} = \frac{\left(5.60\ \text{g L}^{-1}\right)\left(0.0821\ \frac{\text{L atm}}{\text{mol K}}\right)\left(296.2\ \text{K}\right)}{\left(750\ \text{torr}\right)\left(\dfrac{1\ \text{atm}}{760\ \text{torr}}\right)} = 138\ \text{g mol}^{-1}$$

The empirical mass is 69 g mol^{-1}. The ratio of the molecular mass to the empirical mass is

$$\frac{138\ \text{g mol}^{-1}}{69\ \text{g mol}^{-1}} = 2$$

Therefore, the molecular formula is 2 times the empirical formula, i.e., P_2F_4.

10.18 $$\text{formula mass} = \frac{\left(\dfrac{\text{mass}}{V}\right)RT}{P} = \frac{dRT}{P}$$

$$\text{formula mass} = \frac{\left(5.55\ \text{g L}^{-1}\right)\left(0.0821\ \frac{\text{L atm}}{\text{mol K}}\right)\left(313.2\ \text{K}\right)}{\left(1.25\ \text{atm}\right)} = 114\ \text{g mol}^{-1}$$

Since the compound contains C and H it could be an alkane, C_nH_{2n+2}, an alkene, C_nH_{2n}, or an alkyne, C_nH_{2n-2}.

9 C and 6 H
8 C and 18 H
7 C and 30 H
6 C and 42 H
5 C and 54 H
4 C and 66 H
3 C and 78 H
2 C and 90 H
1 C and 102 H

The most probable compound is C_8H_{18} also known as octane.

10.19 $CS_2(g) + 3O_2(g) \longrightarrow 2SO_2(g) + CO_2(g)$

$$\text{mol of } CS_2 = 11.0\ \text{g } CS_2\left(\frac{1\ \text{mol } CS_2}{76.131\ \text{g } CS_2}\right) = 0.1445\ \text{mol } CS_2$$

Since the volumes of the gases are related by stoichiometry, we will calculate the volume of the CS_2 and then calculate the volume of the products.

$$V = \frac{nRT}{P}$$

$$\text{L } CS_2 = \frac{\left(0.1445\ \text{mol } CS_2\right)\left(0.0821\ \text{L atm mol}^{-1}\ \text{K}^{-1}\right)\left(301\ \text{K}\right)}{\left(883\ \text{torr}\right)\left(\dfrac{1\ \text{atm}}{760\ \text{torr}}\right)} = 3.07\ \text{L } CS_2$$

The ratio of the total volume of CO_2 and SO_2 to the CS_2 is 3 L:1 L

$$\text{Volume of products} = 3.07\ \text{L } CS_2\left(\frac{3\ \text{L } CO_2 \text{ and } SO_2}{1\ \text{L } CS_2}\right) = 9.22\ \text{L}$$

The ratio of the volume of CO_2 to CS_2 is 1:1, so the volume of CO_2 is 3.07 L.
The ratio of the volume of SO_2 to CS_2 is 2:1, so the volume of SO_2 is 6.14 L.

10.20 $CaCO_3(s) \longrightarrow CaO(s) + CO_2(s)$

$$\text{mol } CO_2 = \frac{PV}{RT} = \frac{(738 \text{ torr})\left(\frac{1 \text{ atm}}{760 \text{ torr}}\right)(0.257 \text{ L})}{(0.0821 \text{ L atm mol}^{-1} \text{ K}^{-1})(296 \text{ K})} = 0.01027 \text{ mol } CO_2$$

$$\text{mol } CaCO_3 = 0.01027 \text{ mol } CO_2\left(\frac{1 \text{ mol } CaCO_3}{1 \text{ mol } CO_2}\right) = 0.01027 \text{ mol } CaCO_3$$

$$\text{g } CaCO_3 = 0.01027 \text{ mol } CaCO_3\left(\frac{100.09 \text{ g } CaCO_3}{1 \text{ mol } CaCO_3}\right) = 1.03 \text{ g } CaCO_3$$

10.21 $P_{\text{total}} = P_{CO} + P_{H_2O} + P_{CO_2} + P_{N_2}$

$$P = \frac{nRT}{V}$$

$$n_{CO} = 0.0250 \text{ mol } C(CH_2ONO_2)_4 \frac{2 \text{ mol CO}}{1 \text{ mol } C(CH_2ONO_2)_4} = 0.0500 \text{ mol CO}$$

$$P_{CO} = \frac{(0.0500 \text{ mol CO})(0.0821 \text{ L atm mol}^{-1} \text{ K}^{-1})(298 \text{ K})}{(30.0 \text{ L})} = 0.0408 \text{ atm CO}$$

$$n_{H_2O} = 0.0250 \text{ mol } C(CH_2ONO_2)_4 \frac{4 \text{ mol } H_2O}{1 \text{ mol } C(CH_2ONO_2)_4} = 0.100 \text{ mol } H_2O$$

$$P_{H_2O} = \frac{(0.100 \text{ mol } H_2O)(0.0821 \text{ L atm mol}^{-1} \text{ K}^{-1})(298 \text{ K})}{(30.0 \text{ L})} = 0.0816 \text{ atm } H_2O$$

$$n_{CO_2} = 0.0250 \text{ mol } C(CH_2ONO_2)_4 \frac{3 \text{ mol } CO_2}{1 \text{ mol } C(CH_2ONO_2)_4} = 0.0750 \text{ mol } CO_2$$

$$P_{CO_2} = \frac{(0.0750 \text{ mol } CO_2)(0.0821 \text{ L atm mol}^{-1} \text{ K}^{-1})(298 \text{ K})}{(30.0 \text{ L})} = 0.0612 \text{ atm } CO_2$$

$$n_{N_2} = 0.0250 \text{ mol } C(CH_2ONO_2)_4 \frac{2 \text{ mol } N_2}{1 \text{ mol } C(CH_2ONO_2)_4} = 0.0500 \text{ mol } N_2$$

$$P_{N_2} = \frac{(0.0500 \text{ mol } N_2)(0.0821 \text{ L atm mol}^{-1} \text{ K}^{-1})(298 \text{ K})}{(30.0 \text{ L})} = 0.0408 \text{ atm } N_2$$

$P_{\text{total}} = P_{CO} + P_{H_2O} + P_{CO_2} + P_{N_2}$

$P_{\text{total}} = 0.0408 \text{ atm} + 0.0816 \text{ atm} + 0.0612 \text{ atm} + 0.0408 \text{ atm} = 0.2244 \text{ atm}$

10.22 We can determine the pressure due to the oxygen since $P_{\text{total}} = P_{\text{other gas}} + P_{O_2}$. $P_{O_2} = P_{\text{total}} - P_{\text{other gas}} = 237.0 \text{ atm} - 115.0 \text{ atm} = 122.0 \text{ atm}$. We can now use the ideal gas law to determine the number of moles of O_2:

$$n = \frac{PV}{RT} = \frac{(122.0 \text{ atm})(17.00 \text{ L})}{\left(0.0821 \frac{\text{L atm}}{\text{mol K}}\right)(298 \text{ K})} = 84.8 \text{ mol } O_2$$

$$g\ O_2 = (84.8\ \text{mol}\ O_2)\left(\frac{32.0\ g\ O_2}{1\ \text{mol}\ O_2}\right) = 2713\ g\ O_2$$

10.23 First we find the partial pressure of nitrogen, using the vapor pressure of water at 15 °C:
$P_{N_2} = P_{total} - P_{water} = 745$ torr – 12.79 torr = 732 torr.
To calculate the volume of the nitrogen we can use the combined gas law

$$\frac{P_1V_1}{T_1} = \frac{P_2V_2}{T_2}$$

For this problem,

$$V_2 = \frac{P_1V_1T_2}{P_2T_1} = \frac{(732.2\ \text{torr})(0.317\ \text{L})(273\ \text{K})}{(760\ \text{torr})(288\ \text{K})} = 0.289\ \text{L} = 289\ \text{mL}$$

10.24 The total pressure is the pressure of the methane and the pressure of the water. We can determine the pressure of the methane by subtracting the pressure of the water from the total pressure.
The pressure of the water is determined by the temperature of the sample. At 28 °C, the partial pressure of water is 28.3 torr.
$P_{CH_4} = T_{total} - P_{water} = 775$ torr – 28.3 torr = 747 torr
The pressure in the flask is 747 torr.

$$\text{mol}\ CH_4 = \frac{PV}{RT} = \frac{(747\ \text{torr})\left(\frac{1\ \text{atm}}{760\ \text{torr}}\right)(2.50\ \text{L})}{\left(0.0821\ \frac{\text{L atm}}{\text{mol K}}\right)(301\ \text{K})} = 0.0994\ \text{mol}\ CH_4$$

10.25 Find the number of moles of both the H_2 and NO then find the mol fractions.

$$\text{mol}\ H_2 = 2.15\ g\ H_2\left(\frac{1\ \text{mol}\ H_2}{2.016\ g\ H_2}\right) = 1.07\ \text{mol}\ H_2$$

$$\text{mol NO} = 34.0\ \text{g NO}\left(\frac{1\ \text{mol NO}}{30.01\ \text{g NO}}\right) = 1.13\ \text{mol NO}$$

$$X_{H_2} = \frac{1.07\ \text{mol}\ H_2}{1.13\ \text{mol NO} + 1.07\ \text{mol}\ H_2} = 0.486$$

$$X_{NO} = \frac{1.13\ \text{mol NO}}{1.13\ \text{mol NO} + 1.07\ \text{mol}\ H_2} = 0.514$$

$P_{H_2} = (P_{total})(X_{H_2}) = (2.05\ \text{atm})(0.486) = 0.996$ atm
$P_{NO} = (P_{total})(X_{NO}) = (2.05\ \text{atm})(0.514) = 1.05$ atm

10.26 The stoichiometric ratio of the SO_2 to O_2 is 2 mol SO_2 to 1 mol O_2

$$P_{O_2}\ \text{added to flask} = 0.750\ \text{atm}\ SO_2\left(\frac{1\ \text{atm}\ O_2}{2\ \text{atm}\ SO_2}\right) = 0.375\ \text{atm}\ O_2$$

The total pressure, before the reaction, of SO_2 and O_2 = 0.750 atm + 0.375 atm = 1.125 atm

The total pressure, after the reaction will be

$$P_{SO_3} = 0.750\ \text{atm}\ SO_2\left(\frac{2\ \text{atm}\ SO_3}{2\ \text{atm}\ SO_2}\right) = 0.750\ \text{atm}\ SO_3$$

10.27 $\dfrac{\text{effusion rate}_{^{81}\text{Br}^{81}\text{Br}}}{\text{effusion rate}_{^{81}\text{Br}^{79}\text{Br}}} = \sqrt{\dfrac{159.8}{161.8}} = 0.994$

$\dfrac{\text{effusion rate}_{^{81}\text{Br}^{81}\text{Br}}}{\text{effusion rate}_{^{79}\text{Br}^{79}\text{Br}}} = \sqrt{\dfrac{157.8}{161.8}} = 0.988$

10.28 $\dfrac{\text{effusion rate}_{HX}}{\text{effusion rate}_{\text{HCl}}} = \sqrt{\dfrac{M_{\text{HCl}}}{M_{HX}}}$

Rearrange the equation

$$M_{HX} = M_{\text{HCl}}\left(\frac{\text{effusion rate}_{HX}}{\text{effusion rate}_{\text{HCl}}}\right)^2 = 36.46 \text{ g mol}^{-1} \times (1.88)^2 = 128.9 \text{ g mol}^{-1}$$

The unknown gas must be HI.

Review Problems

10.35 (a) $\text{torr} = 1.26 \text{ atm}\left(\dfrac{760 \text{ torr}}{1 \text{ atm}}\right) = 958 \text{ torr}$

(b) $\text{atm} = 740 \text{ torr}\left(\dfrac{1 \text{ atm}}{760 \text{ torr}}\right) = 0.974 \text{ atm}$

(c) $\text{mm Hg} = 738 \text{ torr}\left(\dfrac{760 \text{ torr}}{760 \text{ mm Hg}}\right) = 738 \text{ mm Hg}$

(d) $\text{torr} = 1.45 \times 10^3 \text{ Pa}\left(\dfrac{760 \text{ torr}}{1.01325 \times 10^5 \text{ Pa}}\right) = 10.9 \text{ torr}$

10.37 (a) $\text{torr} = 0.329 \text{ atm}\left(\dfrac{760 \text{ torr}}{1 \text{ atm}}\right) = 250 \text{ torr}$

(b) $\text{torr} = 0.460 \text{ atm}\left(\dfrac{760 \text{ torr}}{1 \text{ atm}}\right) = 350 \text{ torr}$

10.39 765 torr – 720 torr = 45 torr $\quad 45 \text{ torr}\left(\dfrac{760 \text{ mm Hg}}{760 \text{ torr}}\right) = 45 \text{ mm Hg}$

$\text{cm Hg} = 45 \text{ mm Hg}\left(\dfrac{1 \text{ cm}}{10 \text{ mm}}\right) = 4.5 \text{ cm Hg}$

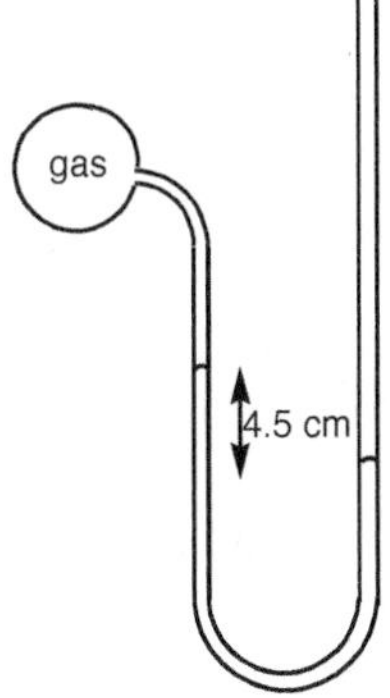

10.41 $65 \text{ mm Hg}\left(\dfrac{760 \text{ torr}}{760 \text{ mm Hg}}\right) = 65 \text{ torr}$ $\quad$ $748 \text{ torr} + 65 \text{ torr} = 813 \text{ torr}$

10.43 In a closed-end manometer the difference in height of the mercury levels in the two arms corresponds to the pressure of the gas. Therefore, the pressure of the gas is

$$12.5 \text{ cm Hg}\left(\frac{10 \text{ mm Hg}}{1 \text{ cm Hg}}\right) = 125 \text{ mm Hg}.$$

$$125 \text{ mm Hg}\left(\frac{760 \text{ torr}}{760 \text{ mm Hg}}\right) = 125 \text{ torr}$$

10.45 Since volume is to decrease, pressure must increase, and we multiply the starting pressure by a volume ratio that is larger than one. Also, since $P_1V_1 = P_2V_2$, we can solve for P_2:

$$P_2 = \frac{P_1V_1}{V_2} = \frac{(740 \text{ torr})(880 \text{ mL})}{(550 \text{ mL})} = 1180 \text{ torr}$$

10.47 Use Charles' Law to solve for the second volume:

$$T_1 = (22\ °\text{C} + 273\ °\text{C})\left(\frac{1 \text{ K}}{1\ °\text{C}}\right) = 295 \text{ K}$$

$$T_2 = (-15\ °\text{C} + 273\ °\text{C})\left(\frac{1 \text{ K}}{1\ °\text{C}}\right) = 258 \text{ K}$$

$$V_2 = \frac{V_1T_2}{T_1} = \frac{(2.50 \text{ L})(258 \text{ K})}{295 \text{ K}} = 2.19 \text{ L}$$

10.49 $P_2 = 2P_1 = 2(854 \text{ torr}) = 1708 \text{ torr}$

$$T_1 = (285\ °\text{C} + 273\ °\text{C})\left(\frac{1 \text{ K}}{1\ °\text{C}}\right) = 558 \text{ K}$$

Compare pressure change to temperature to solve for temperature change:

$$T_2 = \frac{P_2T_1}{P_1} = \frac{(1708 \text{ torr})(558 \text{ K})}{854 \text{ torr}} = 1116 \text{ K} \qquad 1116 \text{ K} - 273 \text{ K} = 843\ °\text{C}$$

10.51 In general the combined gas law equation is: $\dfrac{P_1V_1}{T_1} = \dfrac{P_2V_2}{T_2}$, and in particular, for this problem, we have:

$$T_1 = (24.0\ °\text{C} + 273.2\ °\text{C})\left(\frac{1 \text{ K}}{1\ °\text{C}}\right) = 297.2 \text{ K}$$

$$T_2 = (75.0\ °\text{C} + 273.2\ °\text{C})\left(\frac{1 \text{ K}}{1\ °\text{C}}\right) = 348.2 \text{ K}$$

$$P_2 = \frac{P_1V_1T_2}{T_1V_2} = \frac{(745 \text{ torr})(2.58 \text{ L})(348.2 \text{ K})}{(297.2 \text{ K})(2.81 \text{ L})} = 801 \text{ torr}$$

10.53 In general the combined gas law equation is $\dfrac{P_1V_1}{T_1} = \dfrac{P_2V_2}{T_2}$, and in particular, for this problem, we have:

$$T_1 = (24.0\ °\text{C} + 273.2\ °\text{C})\left(\frac{1 \text{ K}}{1\ °\text{C}}\right) = 297.2 \text{ K}$$

$$T_2 = (375.0\ °C + 273.2\ °C)\left(\frac{1\ K}{1\ °C}\right) = 648.2\ K$$

$$V_2 = \frac{P_1V_1T_2}{T_1P_2} = \frac{(745\ torr)(2.68\ L)(648.2\ K)}{(297.2\ K)(765\ torr)} = 5.69\ L$$

10.55 In general the combined gas law equation is: $\frac{P_1V_1}{T_1} = \frac{P_2V_2}{T_2}$, and in particular, for this problem, we have:

$$T_1 = (20.0\ °C + 273.2\ °C)\left(\frac{1\ K}{1\ °C}\right) = 293.2\ K$$

$$T_2 = \frac{P_2V_2T_1}{P_1V_1} = \frac{(373\ torr)(9.45\ L)(293.2\ K)}{(761\ torr)(6.18\ L)} = 219.8\ K = -53\ °C$$

10.57 The balanced equation is

$$2C_6H_{14}(g) + 19O_2(g) \longrightarrow 12CO_2(g) + 14H_2O(g)$$

$$mL\ O_2 = 855\ mL\ CO_2\left(\frac{19\ mL\ O_2}{12\ mL\ CO_2}\right) = 1.35 \times 10^3\ mL\ O_2$$

10.59 $CH_4 + 2O_2 \longrightarrow CO_2 + 2H_2O$

$$n_{CH_4} = \frac{PV}{RT} = \frac{(725\ torr)\left(\frac{1\ atm}{760\ torr}\right)(16.8 \times 10^{-3}\ L)}{\left(0.0821\ \frac{L\ atm}{mol\ K}\right)(308.2\ K)} = 6.34 \times 10^{-4}\ mol\ CH_4$$

$$mol\ O_2 = (6.34 \times 10^{-4}\ mol\ CH_4)\left(\frac{2\ mol\ O_2}{1\ mol\ CH_4}\right) = 1.27 \times 10^{-3}\ mol\ O_2$$

$$V_{O_2} = \frac{nRT}{P} = \frac{(1.27 \times 10^{-3}\ moles)\left(0.0821\ \frac{L\ atm}{mol\ K}\right)(300.2\ K)}{(654\ torr)\left(\frac{1\ atm}{760\ torr}\right)} = 3.63 \times 10^{-2}\ L = 36.3\ mL\ O_2$$

10.61 $2CO(g) + O_2(g) \longrightarrow 2CO_2(g)$

$$moles\ CO = \frac{(683\ torr)\left(\frac{1\ atm}{760\ torr}\right)(0.300\ L)}{\left(0.0821\ \frac{L\ atm}{mol\ K}\right)(298.2\ K)} = 1.10 \times 10^{-2}\ moles$$

$$moles\ O_2 = \frac{(715\ torr)\left(\frac{1\ atm}{760\ torr}\right)(0.155\ L)}{\left(0.0821\ \frac{L\ atm}{mol\ K}\right)(398.2\ K)} = 4.46 \times 10^{-3}\ moles$$

$$moles\ of\ CO\ from\ O_2 = 4.46 \times 10^{-3}\ mol\ O_2\left(\frac{2\ mol\ CO}{1\ mol\ O_2}\right) = 8.92 \times 10^{-3}\ mol\ CO$$

Since we have 1.10×10^{-2} mol CO, O_2 is the limiting reactant.

$$V = \frac{\left(8.92 \times 10^{-3} \text{ mol}\right)\left(0.0821 \frac{\text{L atm}}{\text{mol K}}\right)(300.2 \text{ K})}{(745 \text{ torr})\left(\frac{1 \text{ atm}}{760 \text{ torr}}\right)} = 2.24 \times 10^{-1} \text{ L} \Rightarrow 224 \text{ mL}$$

10.63 $$R = \left(0.0821 \frac{\text{L atm}}{\text{mol K}}\right)\left(\frac{1000 \text{ mL}}{1 \text{ L}}\right)\left(\frac{760 \text{ torr}}{1 \text{ atm}}\right) = 6.24 \times 10^4 \frac{\text{mL torr}}{\text{mol K}}$$

10.65 $$V = \frac{nRT}{P} = \frac{\left(0.136 \text{ g}\left(\frac{1 \text{ mol}}{32.0 \text{ g}}\right)\right)\left(0.0821 \frac{\text{L atm}}{\text{mol K}}\right)(293.2 \text{ K})}{748 \text{ torr}\left(\frac{1 \text{ atm}}{760 \text{ torr}}\right)} = 0.104 \text{ L}$$

10.67 $$P = \frac{nRT}{V} = \frac{10.0 \text{ g}\left(\frac{1 \text{ mol}}{32.0 \text{ g}}\right)\left(0.0821 \frac{\text{L atm}}{\text{mol K}}\right)(300.2 \text{ K})}{(2.50 \text{ L})} = 3.08 \text{ atm}\left(\frac{760 \text{ torr}}{1 \text{ atm}}\right) = 2340 \text{ torr}$$

10.69 $$n = \frac{PV}{RT} = \frac{624 \text{ torr}\left(\frac{\text{atm}}{760 \text{ torr}}\right)(0.0265 \text{ L})}{\left(0.0821 \frac{\text{L atm}}{\text{mol K}}\right)(293.2 \text{ K})} = 9.04 \times 10^{-4} \text{ mol}\left(\frac{44.0 \text{ g}}{1 \text{ mol}}\right) = 0.0398 \text{ g } CO_2$$

10.71 First determine the number of moles from the ideal gas law:

$$n = \frac{PV}{RT} = \frac{(10.0 \text{ torr})\left(\frac{1 \text{ atm}}{760 \text{ torr}}\right)(255 \text{ mL})\left(\frac{1 \text{ L}}{1000 \text{ mL}}\right)}{\left(0.0821 \frac{\text{L atm}}{\text{mol K}}\right)(298.2 \text{ K})} = 1.37 \times 10^{-4} \text{ mol}$$

Now calculate the molecular mass:

$$\text{molecular mass} = \frac{\text{mass}}{\text{\# of moles}} = \frac{(12.1 \text{ mg})\left(\frac{1 \text{ g}}{1000 \text{ mg}}\right)}{1.37 \times 10^{-4} \text{ mol}} = 88.2 \text{ g/mol}$$

10.73 (a) $$\text{density } C_2H_6 = \left(\frac{30.1 \text{ g } C_2H_6}{1 \text{ mol } C_2H_6}\right)\left(\frac{1 \text{ mol}}{22.4 \text{ L}}\right) = 1.34 \text{ g L}^{-1}$$

(b) $$\text{density } N_2 = \left(\frac{28.0 \text{ g } N_2}{1 \text{ mol } N_2}\right)\left(\frac{1 \text{ mol}}{22.4 \text{ L}}\right) = 1.25 \text{ g L}^{-1}$$

(c) $$\text{density } Cl_2 = \left(\frac{70.9 \text{ g } Cl_2}{1 \text{ mol } Cl_2}\right)\left(\frac{1 \text{ mol}}{22.4 \text{ L}}\right) = 3.17 \text{ g L}^{-1}$$

(d) $$\text{density Ar} = \left(\frac{39.9 \text{ g Ar}}{1 \text{ mol Ar}}\right)\left(\frac{1 \text{ mol}}{22.4 \text{ L}}\right) = 1.78 \text{ g L}^{-1}$$

10.75 In general $PV = nRT$, where $n = \frac{\text{mass}}{\text{formula mass}}$

$$PV = \frac{\text{mass}}{\text{formula mass}} RT$$

and we arrive at the formula for the density (mass divided by volume) of a gas:

$$d = \frac{P \times (\text{formula mass})}{RT}$$

$$d = \frac{(742 \text{ torr})\left(\frac{1 \text{ atm}}{760 \text{ torr}}\right)(32.0 \text{ g/mol})}{\left(0.0821 \frac{\text{L atm}}{\text{mol K}}\right)(297.2 \text{ K})}$$

$d = 1.28$ g/L for O_2

10.77 In general $PV = nRT$, where $n = \dfrac{PV}{RT}$

$$n = \frac{720 \text{ torr}\left(\frac{1 \text{ atm}}{760 \text{ torr}}\right)(30.0 \text{ L})}{\left(0.0821 \text{ L atm mol}^{-1} \text{ K}^{-1}\right)(298 \text{ K})} = 1.16 \text{ mol}$$

This is the total number of moles of products. The ratio of total moles of products (2 mole + 4 mol + 3 mol + 2 mol = 11 mol) to moles of reactants is
11 mol to 1 mol

$$1.16 \text{ mol gases product}\left(\frac{1 \text{ mol reactant}}{11 \text{ mole product}}\right) = 0.105 \text{ mol reactant}$$

$$\text{mass of reactant} = 0.105 \text{ mol } C(CH_2ONO_2)_4\left(\frac{316.17 \text{ g } C(CH_2ONO_2)_4}{\text{mol } C(CH_2ONO_2)_4}\right) = 33.2 \text{ g } C(CH_2ONO_2)_4$$

10.79 $\text{mol } C_3H_6 = (18.0 \text{ g } C_3H_6)\left(\dfrac{1 \text{ mol } C_3H_6}{42.08 \text{ g } C_3H_6}\right) = 0.428 \text{ mol } C_3H_6$

$$\text{mol } H_2 = (0.428 \text{ mol } C_3H_6)\left(\frac{1 \text{ mol } H_2}{1 \text{ mol } C_3H_6}\right) = 0.428 \text{ mol } H_2$$

$$V = \frac{nRT}{P} = \frac{(0.428 \text{ mol } H_2)\left(0.0821 \frac{\text{L atm}}{\text{mol K}}\right)(297.2 \text{ K})}{(740 \text{ torr})\left(\frac{1 \text{ atm}}{760 \text{ torr}}\right)} = 10.7 \text{ L } H_2$$

10.81 $P_{\text{Total}} = P_{N_2} + P_{O_2} + P_{\text{Ar}}$

$P_{\text{Total}} = 315 \text{ torr} + 275 \text{ torr} + 285 \text{ torr} = 875 \text{ torr}$

$$X_{O_2} = \frac{P_{O_2}}{P_{\text{Total}}} = \frac{275 \text{ torr}}{875 \text{ torr}} = 0.314$$

10.83 $P_{\text{Total}} = P_{N_2} + P_{O_2} + P_{\text{He}}$

$$P_{\text{Total}} = 20 \text{ cm Hg}\left(\frac{10 \text{ mm Hg}}{1 \text{ cm Hg}}\right)\left(\frac{1 \text{ torr}}{1 \text{ mm Hg}}\right) + 155 \text{ torr} + 0.450 \text{ atm}\left(\frac{760 \text{ torr}}{1 \text{ atm}}\right) =$$

$P_{\text{Total}} = 200 \text{ torr} + 155 \text{ torr} + 342 \text{ torr} = 697 \text{ torr}$

10.85 The mole fraction is defined as

$$X_{O_2} = \frac{P_{O_2}}{P_{\text{total}}} = \frac{116 \text{ torr}}{788 \text{ torr}} = 0.147 \text{ or } 14.7\%$$

10.87 Assume all gases behave ideally and recall that 1 mole of an ideal gas at 0 °C and 1 atm occupies a volume of 22.4 L. Therefore, the moles of gas equal the pressure of gas in atm: (RT/V = 1.000 atm mol^{-1})

$P_{N_2} = 0.30$ atm

$P_{O_2} = 0.20$ atm

$P_{He} = 0.40$ atm

$P_{CO_2} = 0.10$ atm

$$P_{N_2} = 0.30 \text{ atm}\left(\frac{760 \text{ torr}}{1 \text{ atm}}\right) = 228 \text{ torr} \qquad P_{N_2} = 0.30 \text{ atm}\left(\frac{1 \text{ bar}}{0.9868 \text{ atm}}\right) = 0.304 \text{ bar}$$

$$P_{O_2} = 0.20 \text{ atm}\left(\frac{760 \text{ torr}}{1 \text{ atm}}\right) = 152 \text{ torr} \qquad P_{O_2} = 0.20 \text{ atm}\left(\frac{1 \text{ bar}}{0.9868 \text{ atm}}\right) = 0.203 \text{ bar}$$

$$P_{He} = 0.40 \text{ atm}\left(\frac{760 \text{ torr}}{1 \text{ atm}}\right) = 304 \text{ torr} \qquad P_{He} = 0.40 \text{ atm}\left(\frac{1 \text{ bar}}{0.9868 \text{ atm}}\right) = 0.405 \text{ bar}$$

$$P_{CO_2} = 0.10 \text{ atm}\left(\frac{760 \text{ torr}}{1 \text{ atm}}\right) = 76 \text{ torr} \qquad P_{CO_2} = 0.10 \text{ atm}\left(\frac{1 \text{ bar}}{0.9868 \text{ atm}}\right) = 0.101 \text{ bar}$$

10.89 P_{CO_2} = 845 torr – 322 torr = 523 torr

$$n_{CO_2} = 0.200 \text{ mol}\left(\frac{523 \text{ torr}}{845 \text{ torr}}\right) = 0.124 \text{ moles}$$

10.91 From Table 10.2, the vapor pressure of water at 20 °C is 17.54 torr. Thus only (742 – 17.54) = 724 torr is due to "dry" methane. In other words, the fraction of the wet methane sample that is dry methane is 724/742 = 0.976. The question can now be phrased: What volume of wet methane, when multiplied by 0.976, equals 244 mL?

Volume "wet" methane × 0.976 = 244 mL

Volume "wet" methane = 244 mL/0.976 = 250 mL

In other words, one must collect 250 total mL of "wet methane" gas in order to have collected the equivalent of 244 mL of dry methane.

10.93 Effusion rates for gases are inversely proportional to the square root of the gas density, and the gas with the lower density ought to effuse more rapidly. Nitrogen in this problem has the higher effusion rate because it has the lower density:

$$\frac{\text{rate}(N_2)}{\text{rate}(CO_2)} = \sqrt{\frac{1.96 \text{ g L}^{-1}}{1.25 \text{ g L}^{-1}}} = 1.25$$

10.95 Ethylene, C_2H_4, the lightest of these three, diffuses the most rapidly, and Cl_2, the heaviest, will diffuse the slowest.

$Cl_2 < SO_2 < C_2H_4$

10.97
$$\frac{\text{effusion rate } X}{\text{effusion rate } C_3H_8} = \sqrt{\frac{M_{C_3H_8}}{M_X}}$$

$$M_X = M_{C_3H_8}\left(\frac{\text{effusion rate } C_3H_8}{\text{effusion rate } X}\right)^2 = 44.1 \text{ g/mol}\left(\frac{1}{1.65}\right)^2 = 16.2 \text{ g/mol}$$

Chapter 11
Intermolecular Attractions and the Properties of Liquids and Solids

Practice Exercises

11.1

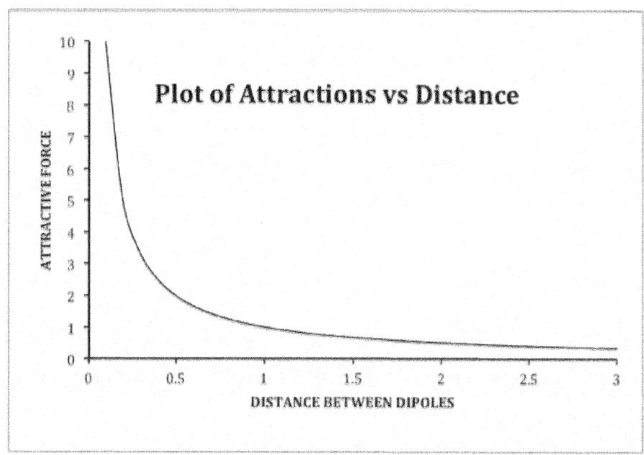

11.2 (a) $CH_3CH_2CH_2CH_2CH_3 < CH_3CH_2OH < KBr$

(b) $CH_3CH_2OCH_2CH_3 < CH_3CH_2NH_2 < HOCH_2CH_2CH_2CH_2OH$

11.3 Propylamine would have a substantially higher boiling point because of its ability to form hydrogen bonds (there are N—H bonds in propylamine, but not in trimethylamine.)

11.4 Evaporative cooling works in dry regions because as the water molecules evaporate, they require heat from the surroundings, and the surroundings include the air. This causes the air to become cooler. In climates that are more humid, there is already water in the air, and less water will evaporate, so the cooling effect that occurs by evaporation will be less pronounced.

11.5 Molecules in the gas phase will condense when they collide with a liquid phase because the kinetic energy that the gas phase molecules have is transferred to the liquid, and the gas molecule does not have enough kinetic energy to remain in the gas phase.

11.6 The piston should be pushed in. This will decrease the volume and increase the pressure, and when equilibrium is re–established, there will be fewer molecules in the gas phase.

11.7 The number of molecules in the vapor will decrease, and the number of molecules in the liquid will increase, but the sum of the molecules in the vapor and the liquid remains the same.

11.8 The boiling point is most likely (a) less than 10 °C above 100 °C.

11.9 We use the curve for water, and find that at 330 torr, the boiling point is approximately 75 °C.

11.10 You will need to calculate the joules of heat lost for the following:
(a) benzene vapor at 105.0 °C cools to it boiling point, 80.1 °C
(b) benzene vapor, at 80.1 °C condenses
(c) benzene liquid cools from 80.1 °C to 25.0 °C

$$q = [55 \text{ g} \times 1.92 \text{ J g}^{-1} \text{ °C}^{-1} \times (105.0 - 80.1 \text{ °C})] + [55 \text{ g} \times (1 \text{ mol}/78.11 \text{ g mol}^{-1}) \times 30{,}770 \text{ J/mol}] + [55 \text{ g} \times 1.8 \text{ J g}^{-1} \text{ °C}^{-1} \times (80.1 \text{ °C} - 25.0 \text{ °C})]$$

$$= 3.0 \times 10^4 \text{ J or } 3.0 \times 10^1 \text{ kJ}$$

11.11 The heat released when 10 g of water vapor condenses is:

$q = [10 \text{ g} \times (1 \text{ mol}/18.01 \text{ g}) \times 43{,}900 \text{ J mol}^{-1}] + [10 \text{ g} \times 4.184 \text{ J g}^{-1} \text{ °C}^{-1} \times (100 \text{ °C} - 37 \text{ °C})]$
$= 2.7 \times 10^4 \text{ J or } 2.7 \times 10^1 \text{ kJ}$

The heat content of 10 g of water at 100 °C cooling to 37 °C is

$q = 10 \text{ g} \times 4.184 \text{ J g}^{-1} \text{ °C}^{-1} \times (100 \text{ °C} - 37 \text{ °C}) = 2.6 \times 10^3 \text{ J or } 2.6 \text{ kJ}$

11.12 The line from the triple point to the critical point is the vapor pressure curve, see Figures 11.23 and 11.28.

11.13 Refer to the phase diagram for water, Figure 11.28. We "move" along a horizontal line marked for a pressure of 2.15 torr. At –20 °C, the sample is a solid. If we bring the temperature from –20 °C to 50 °C, keeping the pressure constant at 2.15 torr, the sample becomes a gas. The process is solid becoming a gas, i.e. sublimation.

11.14 As diagramed in Figure 11.28, this falls in the liquid region.

11.15 Adding heat will shift the equilibrium to the right, producing more vapor. This increase in the amount of vapor causes a corresponding increase in the pressure, such that the vapor pressure generally increases with increasing temperature.

11.16

Boiling	Endothermic
Melting	Endothermic
Condensing	Exothermic
Subliming	Endothermic
Freezing	Exothermic

No, each physical change is always exothermic, or always endothermic as shown.

11.17 $\ln\frac{P_1}{P_2} = \frac{\Delta H_{vap}}{R}\left(\frac{1}{T_2} - \frac{1}{T_1}\right)$

$\ln\frac{45.37 \text{ mm}}{P_2} = \frac{30{,}100 \text{ J mol}^{-1}}{8.314 \text{ J mol}^{-1} \text{ K}^{-1}}\left(\frac{1}{335.4 \text{ K}} - \frac{1}{273.2 \text{ K}}\right) = -2.458$

$\ln 45.37 - \ln P_2 = -2.458$

$-\ln P_2 = -2.458 - 3.815 = -6.273$

$P_2 = 530 \text{ mm Hg}$

11.18 $\ln\frac{P_1}{P_2} = \frac{\Delta H_{vap}}{R}\left(\frac{1}{T_2} - \frac{1}{T_1}\right)$

$\ln\frac{0.0992}{1.00} = \frac{40{,}500 \text{ J mol}^{-1}}{8.314 \text{ J mol}^{-1} \text{ K}^{-1}}\left(\frac{1}{T_2} - \frac{1}{300.5 \text{ K}}\right)$

$-2.311 = 4.871.3\left(\frac{1}{T_2} - 3.338 \times 10^{-3}\right)$

$T_2 = 350.5 \text{ K or } 77.4 \text{ °C}$

11.19 Since chromium crystallizes in a body centered cubic structure there are Cr atoms located at the corners of the cube and in the center of the cube. At the corner 1/8 of a Cr atom occupies the cube. There are eight corners of 8 × (1/8 Cr) = 1 Cr atom. In addition, there is a Cr atom in the center of the cube. Thus, there are a total of two Cr atoms per unit cell for a bcc structure.

11.20 For cesium:

8 corners × 1/8 Cs^+ per corner = 1 Cs^+

For chloride:
1 Cl^- in center, Total: 1 Cl^-
Thus, the ratio is 1 to 1. This matches the observed composition of CsCl.

11.21 Since the unit cell of polonium is a simple cubic structure, see Figure 11.36, the side of the cell is $2r$, if r is the radius of the polonium atom. If the side is 335 pm, then

$$r = \frac{335 \text{ pm}}{2} = 168 \text{ pm}$$

11.22 The density of polonium is the mass of the polonium in a unit cell divided by the volume of the unit cell.

To determine the mass of the unit cell, we need to determine how many atoms are in the unit cell:

$$8 \text{ corners} \times \frac{1}{8} \text{Po per corner} = 1 \text{ Po}$$

$$\text{mass of 1 atom Po} = \left(\frac{209 \text{ g Po}}{1 \text{ mol Po}}\right)\left(\frac{1 \text{ mol Po}}{6.022 \times 10^{23} \text{ atoms Po}}\right) = 3.47 \times 10^{-22} \text{ g Po/atom Po}$$

$$\text{Volume of the unit cell} = l^3 = (335 \text{ pm})^3 = 3.76 \times 10^7 \text{ pm}^3$$

$$3.76 \times 10^7 \text{ pm}^3\left(\frac{1 \text{ m}}{10^{12} \text{ pm}}\right)^3\left(\frac{100 \text{ cm}}{1 \text{ m}}\right)^3 = 3.76 \times 10^{-23} \text{ cm}^3$$

$$d = \frac{\text{mass}}{\text{volume}} = \frac{3.47 \times 10^{-22} \text{ g Po}}{3.76 \times 10^{-22} \text{ cm}} = 9.23 \text{ g cm}^{-3}$$

We can use the density of polonium to see if our density agrees with the density calculated for a given crystal structure. If it matches, then we can be relatively certain of that the crystal structure, as long as it is not one of the close-packed structures.

11.23 The compound is an organic molecule and the solid is held together by dipole–dipole attractions and London forces. It is also a soft solid with a low melting point, so it is a molecular crystal.

11.24 Because this is a high melting, hard material, it must be a covalent or network solid. Covalent bonds link the various atoms of the crystal.

11.25 Since the melt does not conduct electricity, it is not an ionic substance. The softness and the low melting point suggest that this is a molecular solid, and indeed the formula is properly written S_8.

Review Problems

11.91 London forces are possible in them all. Where another intermolecular force can operate, it is generally stronger than London forces, and this other type of interaction overshadows the importance of the London force. The substances in the list that can have dipole–dipole attractions are those with permanent dipole moments: (a), (b), and (d). SF_6, (c), is a non–polar molecular substance. HF, (a), has hydrogen bonding.

11.93 Diethyl ether should have a higher vapor pressure since it has weaker intermolecular forces.

Butanol has a higher boiling point since it has stronger intermolecular forces of attraction.

11.95 Chloroform would be expected to display larger dipole-dipole attractions because it has a larger dipole moment than bromoform. (Chlorine has a higher electronegativity which results in each C–Cl bond having a larger dipole than each C–Br bond.) On the other hand, bromoform would be expected to show stronger London forces due to having larger electron clouds which are more polarizable than those of chlorine.
Since bromoform in fact has a higher boiling point that chloroform, we must conclude that it experiences stronger intermolecular attractions than chloroform, which can only be due to London forces. Therefore, London forces are more important in determining the boiling points of these two compounds.

11.97 London forces are higher in chains than in branched isomers. Therefore, *n*-octane, the elongated structure, has higher London forces between molecules than 2,2,3,3-tetramethylbutane so *n*-octane would be more viscose.

11.99 Ethanol, because it has H-bonding.

11.101 diethyl ether < acetone < benzene < water < acetic acid

11.103 diethyl ether < acetone < benzene < water < acetic acid

11.105

	Compound	Intermolecular Forces Broken
(a)	C_2H_5OH	London, dipole-dipole, hydrogen bonding
(b)	H_3CCN	London, dipole-dipole
(c)	PCl_5	London
(c)	NaCl	London, ionic bonds

11.107 $\text{kJ} = (125\text{ g H}_2\text{O})\left(\dfrac{1\text{ mol H}_2\text{O}}{18.015\text{ g H}_2\text{O}}\right)\left(\dfrac{43.9\text{ kJ}}{1\text{ mol H}_2\text{O}}\right) = 305\text{ kJ}$

11.109 We can approach this problem by first asking either of two equivalent questions about the system: how much heat energy (q) is needed in order to melt the entire sample of solid water (105 g), or how much energy is lost when the liquid water (45.0 g) is cooled to the freezing point? Regardless, there is only one final temperature for the combined (150.0 g) sample, and we need to know if this temperature is at the melting point (0 °C, at which temperature some solid water remains in equilibrium with a certain amount of liquid water) or above the melting point (at which temperature all of the solid water will have melted).
Heat flow supposing that all of the solid water is melted:

$$q = 6.01\text{ kJ mol}^{-1} \times 105\text{ g}\left(\frac{1\text{ mol H}_2\text{O}}{18.0\text{ g H}_2\text{O}}\right) = 35.1\text{ kJ}$$

Heat flow on cooling the liquid water to the freezing point:
$q = 45.0\text{ g} \times 4.18\text{ J g}^{-1}\,°\text{C}^{-1} \times 85\,°\text{C} = 1.60 \times 10^4\text{ J} = 16.0\text{ kJ}$
The lesser of these two values is the correct one, and we conclude that 16.0 kJ of heat energy will be transferred from the liquid to the solid, and that the final temperature of the mixture will be 0 °C. The system will be an equilibrium mixture weighing 150 g and having some solid and some liquid in equilibrium with one another. The amount of solid that must melt in order to decrease the temperature of 45.0 g of water from 85 °C to 0 °C is:

$$\frac{16.0 \text{ kJ}}{6.01 \text{ kJ mol}^{-1}} = 2.66 \text{ mol of solid water}$$

$$2.66 \text{ mol} \times \left(\frac{18.0 \text{ g } H_2O}{1 \text{ mol } H_2O}\right) = 48.0 \text{ g of water must melt}$$

(a) The final temperature will be 0 °C.
(b) 47.9 g of solid water must melt.

11.111

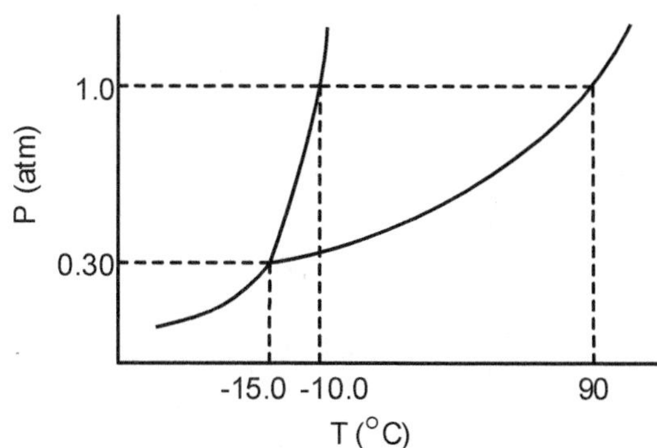

11.113 Sublimation is possible only below a pressure of 0.30 atm, as marked on the phase diagram. The density of the solid is higher than that of the liquid. Notice that the line separating the solid from the liquid slopes to the right, in contrast to the diagram for water, Figure 11.28 of the text.

11.115 The solid–liquid line slants toward the right.

11.117 $\ln\frac{P_1}{P_2} = \frac{\Delta H_{vap}}{R}\left(\frac{1}{T_2} - \frac{1}{T_1}\right)$

$$\ln\frac{1 \text{ atm}}{P_2} = \frac{59{,}200 \text{ J mol}^{-1}}{8.314 \text{ J mol}^{-1} \text{ K}^{-1}}\left(\frac{1}{298.2 \text{ K}} - \frac{1}{629.9 \text{ K}}\right)$$

$\ln 1 - \ln P_2 = 12.574$
$-\ln P_2 = 12.574$
$P_2 = 3.46 \times 10^{-6}$ atm

11.119 $\ln\frac{P_1}{P_2} = \frac{\Delta H_{vap}}{R}\left(\frac{1}{T_2} - \frac{1}{T_1}\right)$

$$\ln\frac{425 \text{ mm Hg}}{1.0 \text{ mm Hg}} = \frac{\Delta H_{vap}}{8.314 \text{ J mol}^{-1} \text{ K}^{-1}}\left(\frac{1}{(273.2 - 74.3) \text{ K}} - \frac{1}{(273.2 + 18.7) \text{ K}}\right)$$

$6.052 = \Delta H_{vap}(1.927 \times 10^{-4} \text{ J}^{-1} \text{ mol})$
$\Delta H_{vap} = 31{,}400 \text{ J mol}^{-1}$

11.121 For zinc:
4 surrounding center $= 4\ Zn^{2+}$
For sulfide:

8 corners $\times \frac{1}{8} S^{2-}$ per corner $= 1\ S^{2-}$

6 faces $\times \frac{1}{2} S^{2-}$ per face $= 3\ S^{2-}$

Total $= 4\ S^{2-}$

11.123 From Figure 11.38, we can see that the length of the face diagonal of the cell = $4r$, where r = radius of the atom. According to the Pythagorean theorem,

$a^2 + b^2 = c^2$

for a right triangle. Since $a = b$ here, we may re–write this as

$2l^2 = c^2$,

where l = length of the edge of the unit cell. The diagonal of the unit cell = $4r$, so we may say that

$2l^2 = (4r)^2$

$l^2 = (4r)^2/2$

$l^2 = 16r^2/2$

$l^2 = 8r^2$

$l = \sqrt{8r^2}$

Finally, substituting the value provided for r in the problem, $l = \sqrt{8(1.24\ \text{Å})^2} = 3.51$ Å. Using the conversion factor 1 pm = 100 Å, this is 351 pm.

11.125 Each edge is composed of 2 × radius of the cation plus 2 × radius of the anion. The edge is (2 × 133 pm) + (2 × 195 pm) = 656 pm.

11.127 Using the Bragg equation (eqn. 11.5), $n\lambda = 2d\sin\theta$

(a) $n(229\text{ pm}) = 2(1{,}000)\sin\theta$

$0.1145n = \sin\theta$ $\theta = 6.57°$

(b) $n(229\text{pm}) = 2(250)\sin\theta$

$0.458n = \sin\theta$ $\theta = 27.3°$

11.129 According to the Pythagorean theorem,

$a^2 + b^2 = c^2$

for a right triangle. First, we need to find the length of a diagonal on a face of the unit cell. Since $a = b$ here, we may re-write this as

$2l^2 = c^2$,

where l = length of the edge of the unit cell and c = the diagonal length. Using the given 412.3 pm as the length of the edge, c = 583.1 pm. The diagonal length inside the cell from corner to opposite corner may now be found by the same theorem:

$a^2 + b^2 = c^2$

$(412.3)^2 + (583.1)^2 = c^2$ $c = 714.1$ pm

This diagonal length inside the cell from corner to opposite corner is due to 1 Cs^+ ion and 1 Cl^- ion (see Figure 11.41). Therefore:

$2r_{Cs+} + 2r_{Cl-} = 714.1\text{pm}$

$2r_{Cs+} + 2(181\text{pm}) = 714.1\text{pm}$

$2r_{Cs+} = 352$ pm $r_{Cs+} = 176$ pm

11.131 This must be a molecular solid, because if it were ionic it would be high-melting, and the melt would conduct.

11.133 This is a metallic solid.

11.135
(a) molecular
(b) ionic
(c) ionic
(d) metallic
(e) covalent
(f) molecular
(g) ionic

Chapter Twelve
Mixtures at the Molecular Level: Properties of Solutions

Practice Exercises

12.1 a, b, and d

12.2 a, c, and d

12.3

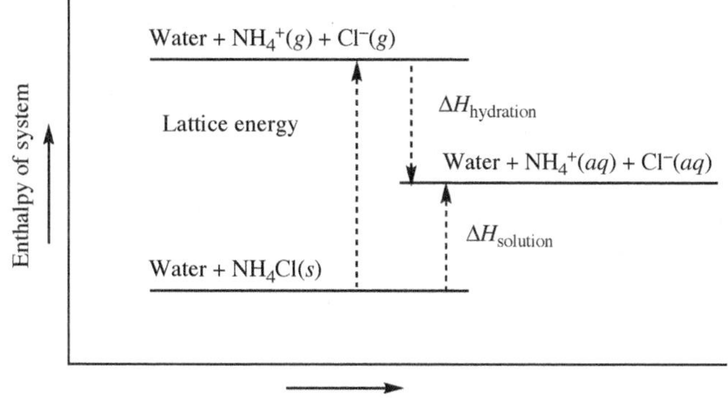

12.4 The dissolution of KOH in water evolves heat. In terms of the enthalpy, this means that more heat is released when the ions are solvated than the amount of heat required to form the gaseous ions.

12.5 $NH_4NO_3(s) + \text{heat} \longrightarrow NH_4^+(aq) + NO_3^-(aq)$
As heat is added, the reaction shifts towards the products, and a solution will form.

12.6 $NaOH(s) \longrightarrow Na^+(aq) + OH^-(aq) + \text{heat}$
The reaction is exothermic. As the reaction cools, the solubility of NaOH decreases, and NaOH will precipitate.

12.7 $C_{H2S} = k_H P_{H2S}$

$P_{H2S} = 1.0 \text{ atm} \quad C_{H2S} = \left(\frac{0.11 \text{ mol } H_2S}{L}\right)\left(\frac{34.08 \text{g } H_2S}{1 \text{ mol } H_2S}\right) = 3.7 \text{ g L}^{-1}$

$3.7 \text{ g L}^{-1} \; H_2S = k_H \; (1.0 \text{ atm } H_sS)$
$k_H = 3.7 \text{ g L}^{-1}$
Hydrogen sulfide is more soluble in water than nitrogen and oxygen. Hydrogen sulfide reacts with the water to form hydronium ions and HS^-.

12.8 With one atmosphere of air, the concentrations of the gases in the water depend on the partial pressures of the gases.

Oxygen: $C_{O2} = \left(\frac{0.00430 \text{ g } O_2}{100 \text{ mL } H_2O}\right)\left(\frac{159 \text{ mm Hg}}{760 \text{ mm Hg}}\right)\left(\frac{1000 \text{ mg } O_2}{1 \text{ g } O_2}\right) = 0.899 \text{ mg } O_2/ \text{ 100 mL}$

$\text{g } O_2$ in 125 g of water $= \left(\frac{0.899 \text{ mg } O_2}{100 \text{ mL } H_2O}\right) \times 125 \text{ mL} = 1.12 \text{ mg}$

Nitrogen $C_{N2} = \left(\frac{0.00190 \text{ g } N_2}{100 \text{ mL } H_2O}\right)\left(\frac{593 \text{ mm Hg}}{760 \text{ mm Hg}}\right) = 1.48 \text{ mg } N_2/\ 100 \text{ mL}$

g N_2 in 125 mL of water $= \left(\frac{1.48 \text{ mg } N_2}{100 \text{ mL } H_2O}\right) \times 125 \text{ mL} = 1.85 \text{ mg}$

12.9 A 10% w/w solution of sucrose will need 10 grams of sucrose for each 100 g of solution. For a solution with 45.0 g of sucrose:

$10\% \text{ solution} = \frac{45.0 \text{ g sucrose}}{x \text{ g solution}}$

$x = 450$ g solution
g water = 450 g solution – 45.0 g sucrose
g water = 405 g water

mL water $= 405 \text{ g water}\left(\frac{1 \text{ cm}^3}{0.9982 \text{ g}}\right) = 406 \text{ mL water}$

12.10 The total mass of the solution is to be 25.0 g.
If the solution is to be 1.00 % (w/w) NaBr, then the mass of NaBr will be:

$25.0 \text{ g} \times \frac{1.00 \text{ g NaBr}}{100 \text{ g solution}} = 0.250 \text{ g NaBr.}$

We therefore need 0.250 g of NaBr and (25.0 – 0.250) = 24.75 g H_2O.

The volume of water that is needed is: $24.75 \text{ g}\left(\frac{1 \text{ mL } H_2O}{0.998 \text{ g } H_2O}\right) = 24.8 \text{ mL } H_2O.$

12.11 We need to know the number of moles of Na_2SO_4 and the number of kg of water.

$44.00 \text{ g } Na_2SO_4\left(\frac{1 \text{ mol } Na_2SO_4}{142.0 \text{ g } Na_2SO_4}\right) = 0.3099 \text{ mol } Na_2SO_4$

$250 \text{ g } H_2O\left(\frac{1 \text{ kg } H_2O}{1000 \text{ g } H_2O}\right) = 0.250 \text{ kg } H_2O$

The molality is thus given by:

$m = \frac{0.3099 \text{ mol } Na_2SO_4}{0.25 \text{ kg } H_2O} = 1.239 \text{ mol } Na_2SO_4 / \text{kg } H_2O = 1.239\ m$

Molarity is moles solute per liters of solution. The moles of solute is the same for molarity and molality but the volume of solution would be larger than the kilograms of solvent so $M < m$.

12.12 g CH_3OH for 0.050 m = 0.200 kg $H_2O\left(\frac{0.050 \text{ mol } CH_3OH}{\text{kg } H_2O}\right)\left(\frac{32.0 \text{ g } CH_3OH}{1 \text{ mol } CH_3OH}\right) = 0.320 \text{ g } CH_3OH$

g CH_3OH for 0.100 m = 0.200 kg $H_2O\left(\frac{0.100 \text{ mol } CH_3OH}{\text{kg } H_2O}\right)\left(\frac{32.0 \text{ g } CH_3OH}{1 \text{ mol } CH_3OH}\right) = 0.640 \text{ g } CH_3OH$

g CH_3OH for 0.150 m = 0.200 kg $H_2O\left(\frac{0.150 \text{ mol } CH_3OH}{\text{kg } H_2O}\right)\left(\frac{32.0 \text{ g } CH_3OH}{1 \text{ mol } CH_3OH}\right) = 0.960 \text{ g } CH_3OH$

g CH_3OH for 0.200 m = 0.200 kg $H_2O\left(\frac{0.200 \text{ mol } CH_3OH}{\text{kg } H_2O}\right)\left(\frac{32.0 \text{ g } CH_3OH}{1 \text{ mol } CH_3OH}\right) = 1.28 \text{ g } CH_3OH$

$$\text{g CH}_3\text{OH for } 0.250\ m = 0.200 \text{ kg H}_2\text{O}\left(\frac{0.250 \text{ mol CH}_3\text{OH}}{\text{kg H}_2\text{O}}\right)\left(\frac{32.0 \text{ g CH}_3\text{OH}}{1 \text{ mol CH}_3\text{OH}}\right) = 1.60 \text{ g CH}_3\text{OH}$$

12.13 If a solution is 52% NaOH, then it has 52 g of NaOH for each 100 g of solution. The mass of water is 48 g of water for 52 g of NaOH. To calculate the molality of the solution, we need to find the moles of NaOH for each kilogram of water.

$$m = \left(\frac{52 \text{ g NaOH}}{48 \text{ g H}_2\text{O}}\right)\left(\frac{1 \text{ mol NaOH}}{40.0 \text{ g NaOH}}\right)\left(\frac{1000 \text{ g H}_2\text{O}}{1 \text{ kg H}_2\text{O}}\right) = 27 \text{ mol NaOH/kg H}_2\text{O} = 27\ m \text{ NaOH}$$

12.14 If a solution is 37.0% (w/w) HCl, then 37.0% of the mass of any sample of such a solution is HCl and (100.0 g solution – 37.0 g HCl) = 63.0 g H_2O of the mass is water. In order to determine the molality of the solution, we can conveniently choose 100.0 g of the solution as a starting point. Then 37.0 g of this solution are HCl and 63.0 g are H_2O. For molality, we need to know the number of moles of HCl and the mass in kg of the solvent:

$$37.0 \text{ g HCl}\left(\frac{1 \text{ mol HCl}}{36.46 \text{ g HCl}}\right) = 1.01 \text{ mol HCl}$$

$$63.0 \text{ g H}_2\text{O} \times \left(\frac{1 \text{ kg H}_2\text{O}}{1000 \text{ g H}_2\text{O}}\right) = 0.0630 \text{ kg H}_2\text{O}$$

$$\text{molality} = \frac{\text{mol HCl}}{\text{kg H}_2\text{O}} = \frac{1.01 \text{ mol HCl}}{0.630 \text{ kg H}_2\text{O}} = 16.1\ m$$

12.15 A solution of 40.0% HBr contains 40 g of HBr per 100 g of solution. Since molarity is defined as moles of solute per liter of solution we need to determine the volume of solution. The density of the solution allows us to determine the volume of the solution

$$\left(\frac{40 \text{ g HBr}}{100 \text{ g solution}}\right)\left(\frac{1 \text{ mol HBr}}{80.91 \text{ g HBr}}\right)\left(\frac{1.38 \text{ g solution}}{\text{mL solution}}\right)\left(\frac{1000 \text{ mL solution}}{\text{L solution}}\right) = 6.82\ M$$

12.16 First determine the number of moles of $Al(NO_3)_3$ dissolved in the liter of water.

$$\text{mol Al(NO}_3)_3 = (1.00 \text{ g Al(NO}_3)_3)\left(\frac{1 \text{ mol Al(NO}_3)_3}{212.996 \text{ g Al(NO}_3)_3}\right) = 0.00469 \text{ mol Al(NO}_3)_3$$

Next find the mass of the water:

$$\text{g H}_2\text{O} = 1.00 \text{ L H}_2\text{O}\left(\frac{1000 \text{ mL H}_2\text{O}}{1 \text{ L H}_2\text{O}}\right)\left(\frac{0.9982 \text{ g H}_2\text{O}}{1 \text{ mL H}_2\text{O}}\right) = 998.2 \text{ g H}_2\text{O}$$

To find the molarity of the solution, first we need to find the mass of the solution, and then the volume of the solution:

$$\text{g solution} = 998.2 \text{ g H}_2\text{O} + 1.00 \text{ g Al(NO}_3)_3 = 999.2 \text{ g solution}$$

$$\text{L solution} = (999.2 \text{ solution})\left(\frac{1 \text{ mL solution}}{0.9989 \text{ g solution}}\right)\left(\frac{1 \text{ L solution}}{1000 \text{ mL solution}}\right) = 1.0003 \text{ L}$$

$$M \text{ of solution} = \frac{0.00469 \text{ mol Al(NO}_3)_3}{1.0003 \text{ L solution}} = 0.00469\ M \text{ Al(NO}_3)_3$$

The molality of the solution can also be determined

$$m \text{ of solution} = \frac{0.00469 \text{ mol Al(NO}_3)_3}{0.9982 \text{ kg H}_2\text{O}} = 0.00470\ m \text{ Al(NO}_3)_3$$

12.17 First determine the number of moles of each component of the solution:

For $C_{16}H_{22}O_4$, 20.0 g$\left(\frac{1 \text{ mol } C_{16}H_{22}O_4}{278 \text{ g } C_{16}H_{22}O_4}\right) = 0.0719$ mol

For C_5H_{12}, 50.0 g$\left(\frac{1 \text{ mol } C_5H_{12}}{72.2 \text{ g } C_5H_{12}}\right) = 0.692$ mol

The mole fraction of solvent is:

$$\frac{0.692 \text{ mol } C_5H_{12}}{0.692 \text{ mol } C_5H_{12} + 0.0719 \text{ mol } C_{16}H_{22}O_4} = 0.906$$

Using Raoult's Law, we next find the vapor pressure to expect for the solution, which arises only from the solvent (since the solute is known to be nonvolatile):

$P_{\text{solvent}} = X_{\text{solvent}} \times P°_{\text{solvent}} = 0.906 \times 541 \text{ torr} = 4.90 \times 10^2$ torr

12.18 $P_{\text{acetone}} = X_{\text{acetone}} \times P°_{\text{acetone}}$

mol acetone = (156 g acetone)$\left(\frac{1 \text{ mol acetone}}{58.0 \text{ g acetone}}\right) = 2.690$ mol acetone

Do not round your answers until the end. The moles of stearic acid are small compared to the moles of acetone and rounding error may give you too high of a mass of stearic acid.

$$155 \text{ torr} = \left(\frac{2.690 \text{ mol acetone}}{2.690 \text{ mol acetone} + x \text{ mol stearic acid}}\right) \times 162 \text{ torr}$$

$$0.957 = \left(\frac{2.690 \text{ mol acetone}}{2.690 \text{ mol acetone} + x \text{ mol stearic acid}}\right)$$

2.574 mol acetone + $0.957x$ mol stearic acid = 2.690 mol acetone

$0.957x$ mol stearic acid = 0.116

x mol stearic acid = 0.121 mol stearic acid

Finally solve to find the number of grams of stearic acid

g stearic acid = (0.121 mol stearic acid)$\left(\frac{284.5 \text{ g stearic acid}}{1 \text{ mol stearic acid}}\right) = 34.5$ g stearic acid

12.19 $P_{\text{cyclohexane}} = X_{\text{cyclohexane}} \times P°_{\text{cyclohexane}} = 0.750 \times 66.9 \text{ torr} = 50.2$ torr

$P_{\text{toluene}} = X_{\text{toluene}} \times P°_{\text{toluene}} = 0.250 \times 21.1 \text{ torr} = 5.28$ torr

$P_{\text{total}} = P_{\text{cyclohexane}} + P_{\text{toluene}} = 50.2 \text{ torr} + 5.28 \text{ torr} = 55.4$ torr

12.20 First we need to find the moles of the cyclohexane and the moles of toluene.

mol cyclohexane: = (122 g cyclohexane)$\left(\frac{1 \text{ mol cyclohexane}}{84.15 \text{ g cyclohexane}}\right) = 1.450$ mol cyclohexane

mol toluene = (122 g toluene)$\left(\frac{1 \text{ mol toluene}}{92.14 \text{ g toluene}}\right) = 1.324$ mol toluene

Now, find the $X_{\text{cyclohexane}}$ and the X_{toluene}

$$X_{\text{cyclohexane}} = \frac{1.450 \text{ mol cyclohexane}}{1.450 \text{ mol cyclohexane} + 1.324 \text{ mol toluene}} = 0.523$$

$X_{\text{toluene}} = 1 - X_{\text{cyclohexane}} = 1 - 0.523 = 0.477$

$P_{\text{cyclohexane}} = X_{\text{cyclohexane}} \times P°_{\text{cyclohexane}} = 0.523 \times 66.9 \text{ torr} = 35.0$ torr

$P_{\text{toluene}} = X_{\text{toluene}} \times P°_{\text{toluene}} = 0.477 \times 21.1 \text{ torr} = 10.1$ torr

$P_{\text{total}} = P_{\text{cyclohexane}} + P_{\text{toluene}} = 35.0 \text{ torr} + 10.1 \text{ torr} = 45.1$ torr

12.21 First convert °F to °C. 235 °F = 112.78 °C and 240 °F = 115.56 °C
The corresponding boiling point elevations are then 12.78 °C and 15.56 °C respectively.
The molality of the two solutions is given as:

$$m = \frac{\Delta T_b}{k}$$

$$m = \frac{12.78\ °\text{C}}{0.51\ °\text{C}\ m^{-1}} = 25.06\ m$$

$$m = \frac{15.56\ °\text{C}}{0.51\ °\text{C}\ m^{-1}} = 30.51\ m$$

The solution that boils at 235 °F has the following mass percent of sugar:

$$\left(\frac{25.06\ \text{mol}\ C_{12}H_{22}O_{11}}{1000\ \text{g}\ H_2O}\right)\left(\frac{342\ \text{g}\ C_{12}H_{22}O_{11}}{1\ \text{mol}\ C_{12}H_{22}O_{11}}\right) = \frac{8.57\ \text{g}\ C_{12}H_{22}O_{11}}{\text{g}\ H_2O}$$

Total mass of the solution = 8.57 g $C_{12}H_{22}O_{11}$ + 1.00 g H_2O = 9.57 g

$$\text{Percent sugar} = \frac{8.57\ \text{g sugar}}{9.57\ \text{g solution}} \times 100\% = 89.5\%$$

For the solution that boils at 240 °F

$$\left(\frac{30.51\ \text{mol}\ C_{12}H_{22}O_{11}}{1000\ \text{g}\ H_2O}\right)\left(\frac{342\ \text{g}\ C_{12}H_{22}O_{11}}{\text{mol}\ C_{12}H_{22}O_{11}}\right) = \frac{10.43\ \text{g}\ C_{12}H_{22}O_{11}}{\text{g}\ H_2O}$$

$$\text{Percent sugar} = \frac{10.43\ \text{g sugar}}{11.43\ \text{g solution}} \times 100\% = 91.2\%$$

The mass percent range for the solutions is 89.5% to 91.2%

12.22 $\Delta T_b = K_b \times m = 0.51\ °\text{C}\ m^{-1} \times x\ m = 2.36\ °\text{C}$
$x\ m = 4.627\ m$
To find the number of grams of glucose, first we need to find the number of moles of glucose.
mol glucose = (m solution)(kg solvent)
mol glucose = (4.627 m)(0.255 kg H_2O) = 1.18 mol glucose

$$\text{g glucose} = (1.18\ \text{mol glucose})\left(\frac{180.9\ \text{g glucose}}{1\ \text{mol glucose}}\right) = 213\ \text{g glucose}$$

12.23 It is first necessary to obtain the values of the freezing point of pure benzene and the value of K_f for benzene from Table 12.4 of the text. We proceed to determine the number of moles of solute that are present and that have caused this depression in the freezing point: $\Delta T = K_f m$

$$\therefore m = \frac{\Delta T}{K_f} = \frac{5.45\ °\text{C} - 4.13\ °\text{C}}{5.07\ °\text{C kg mol}^{-1}} = 0.260\ m$$

Next, use this molality to determine the number of moles of solute that must be present:
0.260 mol solute/kg solvent × 0.0850 kg solvent = 0.0221 mol solute
Last, determine the formula mass of the solute:

$$\frac{3.46\ \text{g}}{0.221\ \text{mol}} = 157\ \text{g mol}^{-1}$$

12.24 To find the molar mass of the substance, first, we need to find the molality of the solution from the freezing point depression, and then using the 5.0% (wt./wt.) amount, determine the moles of the solute.

$$m = \frac{\Delta T}{K_f} = \frac{80.2\ °C - 77.3\ °C}{6.9\ °C\ m^{-1}} = 0.420\ m$$

Assume there is 100 g of solution:

$$5.0\%\ (\text{wt./wt.}) = \frac{5\ \text{g unknown substance}}{5\ \text{g unknown substance} + 95\ \text{g naphthalene}}$$

We have 95 g of naphthalene, or 0.095 kg naphthalene and 5 g of the unknown.
Using the equation for molality, we can determine the number of moles of the unknown
mol unknown = (m solution)(kg solvent) = (0.420 m)(0.095 kg naphthalene) = 0.0399 mol unknown

$$\text{molar mass} = \frac{5.0\ \text{g unknown}}{0.0399\ \text{mol unknown}} = 125\ \text{g mol}^{-1}$$

12.25 Use the equation $\Pi = MRT$

$$M = \frac{(5\ \text{g protein})\left(\frac{1\ \text{mol protein}}{235{,}000\ \text{g protein}}\right)}{0.1000\ \text{L solution}} = 2.13 \times 10^{-4}\ M\ \text{solution}$$

$R = 0.0821$ L atm/K mol

$T = 4.0 + 273.2 = 277.2$ K

$\Pi = (2.13 \times 10^{-4}\ M\ \text{solution})(0.0821\ \text{L atm K}^{-1}\ \text{mol}^{-1})(277.2\ \text{K}) = 4.84 \times 10^{-3}$ atm

$$\text{mm Hg} = 4.84 \times 10^{-3}\ \text{atm}\left(\frac{760\ \text{mm Hg}}{1\ \text{atm}}\right) = 3.68\ \text{mm Hg}$$

$$\text{mm } H_2O = 3.68\ \text{mm Hg}\left(\frac{13.6\ \text{mm } H_2O}{1\ \text{mm Hg}}\right) = 50.0\ \text{mm } H_2O$$

12.26 We can use the equation $\Pi = MRT$

$\Pi = (0.0115\ M)(0.0821\ \text{L atm K}^{-1}\ \text{mol}^{-1})(310\ \text{K})$

$\Pi = 0.293$ atm

$$\Pi = 0.293\ \text{atm}\left(\frac{760\ \text{torr}}{1\ \text{atm}}\right) = 222\ \text{torr}$$

To determine the boiling and freezing temperatures of the solution we can assume that the molality is equal to the molarity. At low concentrations the two values are nearly identical.

$T_f = 0\ °C - m\ K_f = -0.0115\ m \times 1.86\ °C\ m^{-1} = -0.021\ °C$

$T_{bp} = 100\ °C + m\ K_{bp} = 100 + (0.0115\ m \times 0.51\ °C\ m^{-1}) = 100.006\ °C$

Note that significant figures rules were not used for the boiling point answer.

12.27 $\Pi = MRT$

$$\Pi = 6.45\ \text{cm water}\left(\frac{10\ \text{mm } H_2O}{1\ \text{cm } H_2O}\right)\left(\frac{1.00\ \text{g mL}^{-1}}{13.6\ \text{g mL}^{-1}}\right)\left(\frac{1\ \text{atm}}{760\ \text{mm Hg}}\right) = 6.24 \times 10^{-3}\ \text{atm}$$

$R = 0.0821$ L atm mol^{-1} K^{-1}

$T = 277$ K

$$M = \frac{\Pi}{RT} = \frac{(6.24 \times 10^{-3}\ \text{atm})}{(0.0821\ \text{L atm mol}^{-1}\ \text{K}^{-1})(277\ \text{K})} = 2.74 \times 10^{-4}\ \text{mol L}^{-1}$$

mol protein = $(2.74 \times 10^{-3}\ \text{mol L}^{-1})(0.1000\ \text{L}) = 2.74 \times 10^{-5}$ mol

$$\text{molar mass} = \frac{0.1372 \text{ g protein}}{2.74 \times 10^{-5} \text{ mol protein}} = 5.00 \times 10^3 \text{ g mol}^{-1}$$

12.28 We can use the equation $\Pi = MRT$, remembering to convert pressure to atm:

$$\text{atm} = (25.0 \text{ torr})\left(\frac{1 \text{ atm}}{760 \text{ torr}}\right) = 0.0329 \text{ atm}$$

$\Pi = 0.0329$ atm $= M \times (0.0821$ L atm K^{-1} mol^{-1})(298 K)
$M = 1.34 \times 10^{-3}$ mol L^{-1}
mol $= 1.34 \times 10^{-3}$ mol L^{-1} $\times$ 0.100 L $= 1.34 \times 10^{-4}$ mol

$$\text{formula mass} = \frac{72.4 \times 10^{-3} \text{ g}}{1.34 \times 10^{-4} \text{ mol}} = 5.38 \times 10^2 \text{ g mol}^{-1}$$

12.29 For the solution as if the solute were 100% dissociated:
$\Delta T = (1.86\ °\text{C}\ m^{-1})(2 \times 0.237\ m) = 0.882\ °\text{C}$ and the freezing point should be –0.882 °C.
For the solution as if the solute were 0% dissociated:
$\Delta T = (1.86\ °\text{C}\ m^{-1})(1 \times 0.237\ m) = 0.441\ °\text{C}$ and the freezing point should be –0.441 °C.

12.30 Use the freezing point depression equation:
$\Delta T = K_f m$
Remember that there are two moles of ions for each mole of $MgSO_4$.
K_f water = 1.86 °C m^{-1}

(a) For 0.1 m $MgSO_4$ $\quad m = 0.2\ m$
$\Delta T = (1.86\ °\text{C}\ m^{-1})(0.2\ m) = 0.372\ °\text{C}$ thus, $T_f = -0.372\ °\text{C}$

(b) For 0.01 m $MgSO_4$ $\quad m = 0.02\ m$
$\Delta T = (1.86\ °\text{C}\ m^{-1})(0.02\ m) = 0.0372\ °\text{C}$ thus, $T_f = -0.0372\ °\text{C}$

(c) For 0.001 m $MgSO_4$ $\quad m = 0.002\ m$
$\Delta T = (1.86\ °\text{C}\ m^{-1})(0.002\ m) = 0.00372\ °\text{C}$ thus, $T_f = -0.00372\ °\text{C}$

The first freezing point depression could be measured using a laboratory thermometer that can measure
1 °C increments.

12.31
(a) colloid dispersion
(b) colloid dispersion
(c) suspension
(d) colloid dispersion

12.32
(a) colloid dispersion, oil in aqueous egg whites
(b) colloid dispersion, oil in solid whey
(c) colloid dispersion, air in sugar
(d) suspension, carbon particles and drops of water suspended in air
(e) suspension, air in aqueous soap

Review Problems

12.49 $\Delta H_{\text{soln}} = \Delta H_{\text{lattice energy}} + \Delta H_{\text{hydration}}$
$\Delta H_{\textit{lattice}\text{ energy}} = \Delta H_{\text{soln}} - \Delta H_{\text{hydration}}$
$\Delta H_{\text{lattice energy}} = -56$ kJ mol^{-1} – (–894 kJ mol^{-1}) = 838 kJ mol^{-1}

12.51 $\frac{C_1}{P_1} = \frac{C_2}{P_2}$

$$C_2 = \frac{C_1 \times P_2}{P_1} = \frac{(0.025 \text{ g L}^{-1}) \times (1.5 \text{ atm})}{(1.0 \text{ atm})} = 0.038 \text{ g/L}.$$

12.53 We can compare the solubility that is actually observed with the predicted solubility based on Henry's Law. If the actual and the predicted solubilities are the same, we conclude that the gas obeys Henry's Law.

$\frac{C_1}{P_1} = \frac{C_2}{P_2}$

$$C_2 = \frac{C_1 \times P_2}{P_1} = \frac{(0.018 \text{ g L}^{-1}) \times (620 \text{ torr})}{(740 \text{ torr})} = 0.015 \text{ g/L}.$$

The calculated value of C_2 is the same as the observed value, and we conclude that over this pressure range, nitrogen does obey Henry's Law.

12.55 $C_{gas} = k_H \times P_{gas}$

$$C_{O_2} = \frac{0.0039 \text{ g } O_2}{100 \text{ mL solution}} = 3.9 \times 10^{-5} \text{ g } O_2 \text{ mL}^{-1} \quad P_{O_2} = 1.0 \text{ atm}$$

$$k_H = \frac{3.9 \times 10^{-5} \text{ g mL}^{-1}}{1.0 \text{ atm}} = 3.9 \times 10^{-5} \text{ g mL}^{-1} \text{ atm}^{-1}$$

12.57 An HCl solution that is 30 % (w/w) has 3 grams of HCl for every 1.0×10^2 grams of solution.

$$\text{g solution} = 7.5 \text{ g HCl}\left(\frac{1.0 \times 10^2 \text{ g solution}}{30 \text{ g HCl}}\right) = 2.5 \times 10^1 \text{ g solution}$$

12.59 First we need to find the number of grams of $Fe(NO_3)_3$ for each kg of solvent.
For 0.853 *m* $Fe(NO_3)_3$

$$\text{mol } Fe(NO_3)_3 = 0.853 \text{ mol } Fe(NO_3)_3\left(\frac{241.86 \text{ g } Fe(NO_3)_3}{1 \text{ mol } Fe(NO_3)_3}\right) = 206.3 \text{ g } Fe(NO_3)_3$$

Then we need to find how the ratio of the moles of $Fe(NO_3)_3$ to the mass of the solution:

$$\text{ratio} = \frac{0.853 \text{ mol } Fe(NO_3)_3}{1000 \text{ g } H_2O + 206.3 \text{ g } Fe(NO_3)_3} = 7.072 \times 10^{-4} \text{ mol } Fe(NO_3)_3 \text{ / g solution}$$

(a) $$\text{g solution} = (0.0200 \text{ mol } Fe(NO_3)_3)\left(\frac{1 \text{ g solvent}}{7.07 \times 10^{-4} \text{ mol } Fe(NO_3)_3}\right) = 28.3 \text{ g solution}$$

(b) $$\text{g sol'n} = (0.0500 \text{ mol } Fe^{3+})\left(\frac{1 \text{mol } Fe(NO_3)_3}{1 \text{ mol } Fe^{3+}}\right)\left(\frac{1 \text{ g solvent}}{7.07 \times 10^{-4} \text{ mol } Fe(NO_3)_3}\right) = 70.7 \text{ g sol'n}$$

(c) $$\text{g sol'n} = (0.00300 \text{ mol } NO_3^-)\left(\frac{1 \text{mol } Fe(NO_3)_3}{3 \text{ mol } NO_3^-}\right)\left(\frac{1 \text{ g solvent}}{7.07 \times 10^{-4} \text{ mol } Fe(NO_3)_3}\right) = 1.41 \text{ g sol'n}$$

12.61 One liter of solution has a mass of:

$$\text{g solution} = 1\ \text{L solution}\left(\frac{1000\ \text{mL solution}}{1\ \text{L solution}}\right)\left(\frac{1.07\ \text{g solution}}{1\ \text{mL solution}}\right) = 1{,}070\ \text{g}$$

According to the given molarity, it contains 3.000 mol NaCl. This has a mass of:

$$\text{g NaCl} = 3.000\ \text{mol NaCl}\left(\frac{58.45\ \text{g NaCl}}{1\ \text{mol NaCl}}\right) = 175.4\ \text{g NaCl}$$

Thus, the mass of water in 1 L solution must be:
1,070 g – 175.4 g = 895 g water

$$m = \left(\frac{3.000\ \text{mol NaCl}}{0.895\ \text{kg solvent}}\right) = 3.35\ m$$

12.63 Concentration of neon in air in ppm

$$\text{ppm} = \left(\frac{2.6\times10^{-5}\ \text{mol Ne}}{1\ \text{mol air}}\right)\left(\frac{1\ \text{mol air}}{28.96\ \text{g air}}\right)\left(\frac{20.18\ \text{g Ne}}{1\ \text{mol Ne}}\right)\left(\frac{10^{6}\ \mu\text{g Ne}}{1\ \text{g Ne}}\right) = 18\ \mu\text{g Ne/ g air} = 18\ \text{ppm}$$

12.65 (a) $24.0\ \text{g glucose} \times \left(\frac{1\ \text{mol glucose}}{180\ \text{g glucose}}\right) = 0.133\ \text{mol glucose}$

molality = 0.133 mol glucose/1.00 kg solvent = 0.133 m

(b) $\text{mole fraction} = \frac{\text{moles glucose}}{\text{total moles}}$

moles glucose = 0.133

$$\text{moles } H_2O = (1.00 \times 10^{3}\ \text{g } H_2O)\left(\frac{1\ \text{mole } H_2O}{18.01\ \text{g } H_2O}\right) = 55.5\ \text{mol } H_2O$$

$$X_{\text{glucose}} = \frac{0.133}{55.5 + 0.133} = 2.39 \times 10^{-3}$$

(c) $\text{mass \%} = \frac{24.0\ \text{g glucose}}{1000\ \text{g } H_2O + 24\ \text{g glucose}} \times 100\% = 2.34\%$

(d) $\text{molarity} = \frac{\text{moles glucose}}{\text{volume of solution}}$

$$\text{volume of solution} = (1000\ \text{g } H_2O + 24.0\ \text{g glucose})\left(\frac{1\ \text{mL}}{1.0078\ \text{g}}\right)\left(\frac{1\ \text{L}}{1000\ \text{mL}}\right) = 1.016\ \text{L}$$

$$\text{molarity} = \frac{0.133\ \text{mol glucose}}{1.016\ \text{L}} = 0.131\ M$$

12.67 We need to know the mole amounts of both components of the mixture. It is convenient to work from an amount of solution that contains 1.25 mol of ethyl alcohol and, therefore, 1.00 kg of solvent. Convert the number of moles into mass amounts as follows:

For CH_3CH_2OH:

$$\frac{\text{g ethanol}}{\text{g solution}} = \left(\frac{1.25\ \text{mol ethanol}}{1\ \text{kg water}}\right)\left(\frac{46.08\ \text{g ethanol}}{1\ \text{mol ethanol}}\right)\left(\frac{1\ \text{kg water}}{1000\ \text{g water}}\right) = 57.6\ \text{g ethanol/1000 g water}$$

Mass % ethanol = (mass ethanol/(total solution mass) × 100%

$$\text{Mass \% ethanol} = \frac{57.6\ \text{g ethanol}}{1000\ \text{g water} + 57.6\ \text{g ethanol}} \times 100\% = 5.45\%$$

12.69 If we assume 100 g of solution we have 7.50 g NH_3 and 92.50 g H_2O.

$$\text{mol } NH_3 = (7.50 \text{ g } NH_3)\left(\frac{1 \text{ mole } NH_3}{17.03 \text{ g } NH_3}\right) = 0.440 \text{ mol } NH_3$$

$$\text{kg } H_2O = (92.5 \text{ g } H_2O)\left(\frac{1 \text{ kg}}{1000 \text{ g}}\right) = 0.0925 \text{ kg } H_2O$$

$$m = \frac{0.440 \text{ moles } NH_3}{0.0925 \text{ kg } H_2O} = 4.76\ m$$

$$\text{mol } H_2O = (92.5 \text{ g } H_2O)\left(\frac{1 \text{ mole } H_2O}{18.02 \text{ g } H_2O}\right) = 5.13 \text{ mol } H_2O$$

$$\text{mole percent} = \frac{0.440 \text{ mol } NH_3}{(0.440 \text{ mol } NH_3 + 5.13 \text{ mol } H_2O)} \times 100\% = 7.90\%$$

12.71 If we choose, for convenience, an amount of solution that contains 1 kg of solvent, then it also contains 0.363 moles of $NaNO_3$. The number of moles of solvent is:

$$\text{mol } H_2O = (1000 \text{ g})\left(\frac{1 \text{ mole } H_2O}{18.02 \text{ g } H_2O}\right) = 55.5 \text{ mol } H_2O$$

Now, convert the number of moles to a number of grams: for $NaNO_3$,
0.363 mol $NaNO_3$ × 85.0 g $NaNO_3$/mol $NaNO_3$ = 30.9 g $NaNO_3$;
for H_2O, 1000 g was assumed and the percent (w/w) values are:

$$\% \ NaNO_3 = \left(\frac{30.9 \text{ g } NaNO_3}{1030.9 \text{ g total}}\right) \times 100\% = 3.00\%$$

$$\% \ H_2O = \left(\frac{1000 \text{ g } H_2O}{1030.9 \text{ g total}}\right) \times 100\% = 97.0\%$$

To determine the molar concentration of $NaNO_3$ assume 1 kg of solvent which would then contain 0.363 mole of $NaNO_3$ or 30.9 g $NaNO_3$. The total mass of the solution would be 1000 g + 30.9 g = 1031 g of solution. Now, the ratio of moles of solute to grams of solution is 0.363 mol $NaNO_3$/1031 g solution. From this calculate the molarity of the solution

$$M \text{ of solution} = \left(\frac{0.363 \text{ mol } NaNO_3}{1031 \text{ g solution}}\right)\left(\frac{1.0185 \text{ g soln}}{1 \text{ mL soln}}\right)\left(\frac{1000 \text{ mL soln}}{1 \text{ L soln}}\right) = 0.359\ M\ NaNO_3$$

$$X_{NaNO3} = \frac{0.363 \text{ mol } NaNO_3}{55.5 \text{ mol } H_2O + 0.363 \text{ mol } NaNO_3} = 6.50 \times 10^{-3}$$

12.73 $P_{solution} = P°_{solvent} \times X_{solvent}$

We need to determine $X_{solvent}$:

$$\text{mol } C_6H_{12}O_6 = (65.0 \text{ g } C_6H_{12}O_6)\left(\frac{1 \text{ mol}}{180.2 \text{ g}}\right) = 0.361 \text{ mol } C_6H_{12}O_6$$

$$\text{mol } H_2O = (150 \text{ g } H_2O)\left(\frac{1 \text{ mol } H_2O}{18.02 \text{ g } H_2O}\right) = 8.32 \text{ mol } H_2O$$

The total number of moles is thus: 8.32 mol + 0.361 mol = 8.69 mol and the mole fraction of the solvent is: $X_{solvent} = \left(\frac{8.32 \text{ mol solvent}}{8.69 \text{ mol solution}}\right) = 0.957$. Therefore,

$P_{solution}$ = 23.8 torr × 0.957 = 22.8 torr

12.75 $P_{benzene} = X_{benzene} \times P°_{benzene}$

$P_{toluene} = X_{toluene} \times P°_{toluene}$

$P_{total} = P_{benzene} + P_{toluene}$

$$\text{mol benzene} = (35.0 \text{ g})\left(\frac{1 \text{ mol}}{78.11 \text{ g}}\right) = 0.448 \text{ mol benzene}$$

$$\text{mol toluene} = (65.0 \text{ g})\left(\frac{1 \text{ mol}}{92.14 \text{ g}}\right) = 0.705 \text{ mol toluene}$$

$$X_{benzene} = \frac{0.448}{0.448 + 0.705} = 0.389$$

$$X_{toluene} = \frac{0.705}{0.448 + 0.705} = 0.611$$

$P_{benzene} = (0.389)(93.4 \text{ torr}) = 36.3 \text{ torr}$

$P_{toluene} = (0.611)(26.9 \text{ torr}) = 16.4 \text{ torr}$

$P_{total} = 36.3 \text{ torr} + 16.4 \text{ torr} = 52.7 \text{ torr}$

12.77 The following relationships are to be established:

$P_{Total} = 96 \text{ torr} = (P°_{benzene} \times X_{benzene}) + (P°_{toluene} \times X_{toluene})$

The relationship between the two mole fractions is:

$X_{benzene} = 1 - X_{toluene}$, since the sum of the two mole fractions is one. Substituting this expression for $\chi_{benzene}$ into the first equation gives:

$96 \text{ torr} = [P°_{benzene} \times (1 - X_{toluene})] + [P°_{Toluene} \times X_{toluene}]$

$96 \text{ torr} = [180 \text{ torr} \times (1 - X_{toluene})] + [60 \text{ torr} \times X_{toluene}]$

Solving for $X_{toluene}$ we get: $120 \times X_{toluene} = 84$

$X_{toluene} = 0.70$ and $X_{benzene} = 0.30$. The mole % values are to be 70 mol% toluene and 30 mol% benzene.

12.79 (a) $X_{solvent} = \dfrac{P}{P°} = \dfrac{511 \text{ torr}}{526 \text{ torr}} = 0.971$

$X_{solute} = 1 - X_{solvent} = 0.029$

(b) We know $0.971 = \dfrac{1 \text{ mol}}{1 \text{ mol} + x \text{ mol}}$ $\quad x = 2.99 \times 10^{-2}$ moles

(c) $\text{molar mass} = \dfrac{8.3 \text{ g}}{2.99 \times 10^{-2} \text{ mol}} = 278 \text{ g/mol}$

12.81 $\Delta T_f = K_f m$

$$m = \frac{\Delta T_f}{K_f} = \frac{3.00 \text{ °C}}{1.86 \text{ °C kg mol}^{-1}} = 1.61 \text{ mol/kg}$$

$$\text{kg} = (125 \text{ g})\left(\frac{1 \text{ kg}}{1000 \text{ g}}\right) = 0.125 \text{ kg}$$

$$\text{mol} = \left(\frac{1.61 \text{ mol}}{1 \text{ kg}}\right)(0.125 \text{ kg}) = 0.201 \text{ mol}$$

$$\text{g} = (0.201 \text{ mol})\left(\frac{342.3 \text{ g}}{1 \text{ mol}}\right) = 68.9 \text{ g}$$

12.83 $\Delta T = (5.45 - 3.45) = 2.00\ °C = K_f \times m = 5.07\ °C\ kg\ mol^{-1} \times m$

$m = 0.394$ mol solute/kg solvent

0.394 mol/kg benzene × 0.200 kg benzene = 0.0788 mol solute

The molecular mass is: 12.00 g/0.0788 mol = 152 g/mol

12.85 $\Delta T_f = K_f m$

$m = \Delta T_f / K_f = 0.307\ °C / 5.07\ °C\ kg/mol = 0.0606\ mol/kg$

$$\text{mol} = \left(\frac{0.0606\ \text{mol}}{1\ \text{kg}}\right)(0.5\ \text{kg}) = 0.0303\ \text{mol}$$

$$\text{molar mass} = \frac{3.84\ \text{g}}{0.0303\ \text{mol}} = 127\ \text{g/mol}$$

The empirical formula has a mass of 64.1 g/mol. So the molecular formula is $C_8H_4N_2$.

12.87 (a) If the equation is correct, the units on both sides of the equation should be g/mol. The units on the right side of this equation are:

$$\frac{(\text{g}) \times (\text{L atm mol}^{-1}\ \text{K}^{-1}) \times (\text{K})}{\text{L} \times \text{atm}} = \text{g/mol}$$

which is correct.

(b) $\Pi = MRT = (n/V)RT, \quad n = \Pi V/RT$

This means that we can calculate the number of moles of solute in one L of solution, as follows:

$$n = \frac{(0.021\ \text{torr})\left(^{1\ \text{atm}}/_{760\ \text{torr}}\right)(1.0\ \text{L})}{(0.0821\ \text{L atm mol}^{-1}\ \text{K}^{-1})(298\ \text{K})} = 1.1 \times 10^{-6}\ \text{mol}$$

The molecular mass is the mass in 1 L divided by the number of moles in 1 L:

$2.0\ \text{g}/1.1 \times 10^{-6}\ \text{mol} = 1.8 \times 10^{6}\ \text{g/mol}$

12.89 The equation for the vapor pressure is:

$P_{solution} = P°_{H2O} \times X_{H2O}$

Where $P°_{H2O}$ is 17.5 torr. To calculate the vapor pressure we need to find the mole fraction of water first.

$$X_{H2O} = \frac{\text{mol H}_2\text{O}}{\text{mol H}_2\text{O} + \text{mol NaCl}}$$

Calculate the moles of NaCl in 23.0 g

$$\text{mol NaCl} = (23.0\ \text{g NaCl})\left(\frac{1\ \text{mol NaCl}}{58.44\ \text{g NaCl}}\right) = 0.394\ \text{moles NaCl}$$

When NaCl dissolves in water, Na^+ and Cl^- are formed. So, for every mole of NaCl that dissolves, two moles of ions are formed. For this solution, the number of moles of ions is 0.788. The number of moles of solvent (water) is:

$$\text{mol H}_2\text{O} = (100\ \text{g H}_2\text{O})\left(\frac{1\ \text{mol H}_2\text{O}}{18.02\ \text{g H}_2\text{O}}\right) = 5.55\ \text{moles H}_2\text{O}$$

Calculate the mole fraction as

$$X_{H_2O} = \frac{\text{moles H}_2\text{O}}{\text{moles H}_2\text{O} + \text{moles NaCl}} = \frac{5.55\ \text{mol}}{5.55\ \text{mol} + 0.788\ \text{mol}} = 0.876$$

The vapor pressure is then $P_{solution} = P°_{H2O} \times X_{H2O} = 17.5\ \text{torr} \times 0.876 = 15.3\ \text{torr}$

12.91 Assume 100 mL of solution, that is, 2.0 g NaCl and 0.100 L of solution:

$\Pi = MRT$

$$M = \frac{(2.0 \text{ g NaCl})\left(\dfrac{1 \text{ mol NaCl}}{58.44 \text{ g NaCl}}\right)}{0.100 \text{ L}} = 0.34\, M$$

For every NaCl there are two ions produces so $M = 0.68\ M$

$$\Pi = (0.68\, M)(0.0821 \text{ L atm/mol K})(298 \text{ K})\left(\frac{760 \text{ torr}}{1 \text{ atm}}\right) = 1.3 \times 10^4 \text{ torr}$$

12.93 $CaCl_2 \longrightarrow Ca^{2+} + 2Cl^-$; van't Hoff factor, i = 3

$\Delta T_f = i \times K_f \times m = (3)(1.86\ °C\ m^{-1})(0.20\ m) = 1.1\ °C$

The freezing point is –1.1 °C.

12.95 Any electrolyte such as $NiSO_4$, that dissociated to give 2 ions, if fully dissociated should have a van't Hoff factor of 2.

Chapter 13
Chemical Kinetics

Practice Exercises

13.1 Using the coefficients of the reaction, we find that the ratio of iodide production to sulfite disappearance in 1:3, and the ratio of sulfate production to sulfite disappearance is 3:3.

$$\text{Rate of production of } I^- = (2.4 \times 10^{-4}\text{ mol L}^{-1}\text{ s}^{-1})\left(\frac{1\text{ mol }I^-}{3\text{ mol }SO_3^{2-}}\right) = 8.0 \times 10^{-5}\text{ mol L}^{-1}\text{ s}^{-1}$$

$$\text{Rate of production of } SO_4^{2-} = (2.4 \times 10^{-4}\text{ mol L}^{-1}\text{ s}^{-1})\left(\frac{3\text{ mol }SO_4^{2-}}{3\text{ mol }SO_3^{2-}}\right) = 2.4 \times 10^{-4}\text{ mol L}^{-1}\text{ s}^{-1}$$

13.2 From the coefficients in the balanced equation we see that, for every two moles of SO_2 that is produced, 2 moles of H_2S are consumed, three moles of O_2 are consumed, and two moles of H_2O are produced.

$$\text{Rate of disappearance of } O_2 = \left(\frac{3\text{ mol }O_2}{2\text{ mol }SO_2}\right)\left(\frac{0.30\text{ mol}}{\text{L s}}\right) = 0.45\text{ mol L}^{-1}\text{ s}^{-1}$$

$$\text{Rate of disappearance of } H_2S = \left(\frac{2\text{ mol }H_2S}{2\text{ mol }SO_2}\right)\left(\frac{0.30\text{ mol}}{\text{L s}}\right) = 0.30\text{ mol L}^{-1}\text{ s}^{-1}$$

13.3 The rate of the reaction at 2.00 minutes (120 s) is equal to the slope of the tangent to the curve at 120 s. After drawing the tangent, the slope can be estimated as follows:

$$\text{rate}_{\text{with respect to HI}} = \left(\frac{[\text{HI}]_{\text{final}} - [\text{HI}]_{\text{initial}}}{t_{\text{final}} - t_{\text{initail}}}\right) = \left(\frac{0\text{ mol L}^{-1} - 0.075\text{ mol L}^{-1}}{360\text{ s} - 0\text{ s}}\right) = 2.1 \times 10^{-4}\text{ mol L}^{-1}\text{ s}^{-1}$$

13.4 The rate of the reaction after 250 seconds have elapsed is equal to the slope of the tangent to the curve at 250 seconds. First draw the tangent, and then estimate its slope as follows, where A is taken to represent one point on the tangent, and B is taken to represent another point on the tangent:

$$\text{rate} = \left(\frac{\text{A (mol/L)} - \text{B (mol/L)}}{\text{A (s)} - \text{B (s)}}\right) = \frac{\text{change in concentration}}{\text{change in time}} = \frac{0.02\text{ mol L}^{-1} - 0.058\text{ mol L}^{-1}}{400\text{ s} - 0\text{ s}}$$

$$= 9.5 \times 10^{-5}\text{ mol L}^{-1}\text{ s}^{-1}$$

A value near 1×10^{-4} mol L^{-1} s^{-1} is correct.

13.5 (a) Using the rate law:

$\text{Rate} = k[\text{NO}]^2[H_2]$

Substitute in the concentration and the rate and solve for the rate constant:

$7.86 \times 10^{-3}\text{ mol L}^{-1}\text{ s}^{-1} = k(2 \times 10^{-6}\text{ mol L}^{-1})^2(2 \times 10^{-6}\text{ mol L}^{-1})$

$k = 9.8 \times 10^{14}\text{ L}^2\text{ mol}^{-2}\text{ s}^{-1}$

(b) The units can be derived from the equation:

$\text{mol L}^{-1}\text{ s}^{-1} = k(\text{mol L}^{-1})^2(\text{mol L}^{-1})$

$\text{mol L}^{-1}\text{ s}^{-1} = k\text{ (mol}^3\text{ L}^{-3})$

$$k = \frac{\text{mol L}^{-1}\text{ s}^{-1}}{\text{mol}^3\text{ L}^{-3}} = \text{L}^2\text{mol}^{-2}\text{ s}^{-1}$$

13.6 (a) First use the given data in the rate law:

Rate = $k[\mathrm{HI}]^2$

2.5×10^{-4} mol L^{-1} s^{-1} = $k[5.58 \times 10^{-2}$ mol/L$]^2$

$k = 8.0 \times 10^{-2}$ L mol^{-1} s^{-1}

(b) L mol^{-1} s^{-1}

13.7 The order of the reaction with respect to a given substance is the exponent to which that substance is raised in the rate law:

order of the reaction with respect to $[BrO_3^-] = 1$

order of the reaction with respect to $[SO_3^{2-}] = 1$

overall order of the reaction = 1 + 1 = 2

13.8 The rate law is second order with respect to Cl_2 and first order with respect to NO. Therefore the exponent for the Cl_2 is two and the exponent for NO is one:

Rate = $k[Cl_2]^2[NO]$

13.9 Since the rate law is Rate = $k[Br_2]$ the order with respect to Br_2 is 1 and the order with respect to HCO_2H is zero. The overall order of the reaction is 1.

13.10 In each case, k = rate/[A][B]^2, and the units of k are L^2 mol^{-2} s^{-1}.

Each calculation is performed as follows, using the second data set as the example:

$$k = \frac{0.40 \text{ mol L}^{-1} \text{ s}^{-1}}{\left(0.20 \text{ mol L}^{-1}\right)\left(0.10 \text{ mol L}^{-1}\right)^2} = 2.0 \times 10^2 \text{ L}^2 \text{ mol}^{-2} \text{ s}^{-1}$$

Each of the other data sets also gives the same value:

$k = 2.0 \times 10^2$ L^2 mol^{-2} s^{-1}

13.11 The rate law is: rate = $k[A][B]^2$

(a) If the concentration of B is tripled, then the rate will increase nine–fold,

rate = $k[A][3B]^2$

rate = $9k[A][B]^2$

(b) If the concentration of A is tripled, then the rate will increase three–fold,

rate = $k[3A][B]^2$

rate = $3k[A][B]^2$

(c) If the concentration of A is tripled, and the concentration of B is halved, then the rate will decrease to three fourths, or seventy five percent

$$\text{rate} = k[3A][\frac{1}{2}B]^2$$

$$\text{rate} = \frac{3}{4}k[A][B]^2$$

13.12 $\text{rate} = k\left[\text{NO}\right]^n\left[\text{H}_2\right]^m$

(a) To find the rate law, take two reactions in which the concentration of one of the reactants is held constant, compare the two reactions and solve for the exponent:

For NO, use the first two reactions:

$$\frac{k\left[\text{H}_2\right]_1^n\left[\text{NO}\right]_1^m}{k\left[\text{H}_2\right]_2^n\left[\text{NO}\right]_2^m} = \frac{\text{rate}_1}{\text{rate}_2}$$

$$\frac{k\left[0.40\times10^{-4}\text{ mol L}^{-1}\right]^n\left[0.30\times10^{-4}\text{ mol L}^{-1}\right]^m}{k\left[0.80\times10^{-4}\text{ mol L}^{-1}\right]^n\left[0.30\times10^{-4}\text{ mol L}^{-1}\right]^m}=\frac{1.0\times10^{-8}\text{ mol L}^{-1}\text{ s}^{-1}}{4.0\times10^{-8}\text{ mol L}^{-1}\text{ s}^{-1}}$$

$$\frac{\left[0.40\times10^{-4}\text{ mol L}^{-1}\right]^n}{\left[0.80\times10^{-4}\text{ mol L}^{-1}\right]^n}=\frac{1.0\times10^{-8}\text{ mol L}^{-1}\text{ s}^{-1}}{4.0\times10^{-8}\text{ mol L}^{-1}\text{ s}^{-1}}$$

$$\left(\frac{1}{2}\right)^n=\frac{1}{4}\qquad n=2$$

For H_2, use the second two reactions:

$$\frac{k\left[H_2\right]_2^n\left[NO\right]_2^m}{k\left[H_2\right]_3^n\left[NO\right]_3^m}=\frac{\text{rate}_2}{\text{rate}_3}$$

$$\frac{k\left[0.80\times10^{-4}\text{ mol L}^{-1}\right]^n\left[0.30\times10^{-4}\text{ mol L}^{-1}\right]^m}{k\left[0.80\times10^{-4}\text{ mol L}^{-1}\right]^n\left[0.60\times10^{-4}\text{ mol L}^{-1}\right]^m}=\frac{4.0\times10^{-8}\text{ mol L}^{-1}\text{ s}^{-1}}{8.0\times10^{-8}\text{ mol L}^{-1}\text{ s}^{-1}}$$

$$\frac{\left[0.30\times10^{-4}\text{ mol L}^{-1}\right]^m}{\left[0.60\times10^{-4}\text{ mol L}^{-1}\right]^m}=\frac{4.0\times10^{-8}\text{ mol L}^{-1}\text{ s}^{-1}}{8.0\times10^{-8}\text{ mol L}^{-1}\text{ s}^{-1}}$$

$$\left(\frac{1}{2}\right)^m=\frac{1}{2}\qquad m=1$$

$$\text{rate}=k\left[H_2\right]^2\left[NO\right]^1$$

(b) To find the value for the rate constant, choose one of the reactions and use the values for the concentrations and rate and solve for the rate constant:

$$\text{rate}=k\left[H_2\right]^2\left[NO\right]^1$$

$$1.0\times10^{-8}\text{ mol L}^{-1}\text{ s}^{-1}=k\left[0.40\times10^{-4}\text{ mol L}^{-1}\right]^2\left[0.30\times10^{-4}\text{ mol L}^{-1}\right]^1$$

$k = 2.1\times10^5\text{ L}^2\text{ mol}^{-2}\text{ s}^{-1}$

(c) The units for the rate constant can be determined from the rate law and cancelling the units:

$$\text{mol L}^{-1}\text{ s}^{-1}=k\left[\text{mol L}^{-1}\right]^2\left[\text{mol L}^{-1}\right]^1$$

$$k=\frac{\left[\text{mol L}^{-1}\text{ s}^{-1}\right]}{\left[\text{mol L}^{-1}\right]^2\left[\text{mol L}^{-1}\right]^1}=\text{L}^2\text{ mol}^{-2}\text{ s}^{-1}$$

13.13 (a) The rate law will likely take the form rate = $k[A]^n$, where n is the order of the reaction with respect to A.

$$\frac{\text{rate}_2}{\text{rate}_1} = \frac{k[A]_2^n}{k[A]_1^n} = \frac{k(0.36)^n}{k(0.10)^n} = \frac{2.22 \times 10^{-4}}{6.17 \times 10^{-5}}$$

$3.6^n = 3.6$ Therefore, $n = 1$

$$\frac{\text{rate}_3}{\text{rate}_1} = \frac{k[A]_3^n}{k[A]_1^n} = \frac{k(0.58)^n}{k(0.10)^n} = \frac{3.58 \times 10^{-4}}{6.17 \times 10^{-5}}$$

$5.8^n = 5.8$ Therefore, $n = 1$

$$\frac{\text{rate}_3}{\text{rate}_1} = \frac{k[A]_3^n}{k[A]_2^n} = \frac{k(0.58)^n}{k(0.36)^n} = \frac{3.58 \times 10^{-4}}{2.22 \times 10^{-4}}$$

$1.6^n = 1.6$ Therefore, $n = 1$

(b) rate = k[sucrose]

$(6.17 \times 10^{-5}\ \text{mol L}^{-1}\ \text{s}^{-1}) = k(0.10\ \text{mol L}^{-1})$

$k = 6.17 \times 10^{-4}\ \text{s}^{-1}$

The rate constant is $6.17 \times 10^{-4}\ \text{s}^{-1}$

The other two data sets give the same value for k.

This reaction is actually a pseudo-first order reaction. We cannot determine whether H^+ or water are part of the rate expression. The experiment did not vary the acid concentration so we do not know how the rate might change with respect to acid. We cannot determine the effect changing water's concentration would have on the rate as the concentration of water is so large that its concentration would not change significantly during the reaction.

13.14 (a) The rate law will likely take the form rate = $k[A]^n[B]^m$, where n and m are the order of the reaction with respect to A and B, respectively.

$$\frac{\text{rate1}}{\text{rate}_2} = \frac{k[A]_2^n[B]_2^m}{k[A]_3^n[B]_3^m} = \frac{k(0.40)^n\ (0.30)^m}{k(0.60)^n\ (0.30)^m} = \frac{1.00 \times 10^{-4}}{2.25 \times 10^{-4}}$$

$0.6667^n = 0.44444$

$n = 2$

On comparing the second and third lines of data, neither the concentration of A nor the concentration of B are held constant, but we know that the reaction is second-order with respect to A and we can solve the two equations for the order with respect to B:

$$\frac{\text{rate}_2}{\text{rate}_3} = \frac{k[A]_2^2[B]_2^m}{k[A]_3^2[B]_3^m}$$

$$\frac{2.25 \times 10^{-4}\ \text{mol L}^{-1}\ \text{s}^{-1}}{1.60 \times 10^{-3}\ \text{mol L}^{-1}\ \text{s}^{-1}} = \frac{k(0.60\ M)^2(0.30\ M)^m}{k(0.80\ M)^2(0.60\ M)^m}$$

$$\frac{2.25 \times 10^{-4}}{1.60 \times 10^{-3}} = \frac{(0.36)(0.30\ M)^m}{(0.64)(0.60\ M)^m}$$

$$0.141 = 0.563\left(\frac{1}{2}\right)^m$$

$$0.25 = \left(\frac{1}{2}\right)^m$$

$$\frac{1}{4} = \left(\frac{1}{2}\right)^m$$

$m = 2$

Rate = $k[A]^2[B]^2$

(b) For the rate constant, use one of the experiments and insert the values and solve for k

Rate = $k[A]^2[B]^2$

1.00×10^{-4} mol L^{-1} s^{-1} = k[0.40 mol L^{-1}]^2[0.30 mol L^{-1}]^2

$k = 6.9 \times 10^{-3}$ L^3 mol^{-3} s^{-1}

(c) The units for the rate constant are

$$\frac{\text{mol L}^{-1}\text{ s}^{-1}}{\left(\text{mol L}^{-1}\right)^2\left(\text{mol L}^{-1}\right)^2} = \text{L}^3\text{ mol}^{-3}\text{ s}^{-1}$$

(d) The overall order for this reaction is 2 + 2 = 4

13.15 Since this is a first order reaction, then we can use the integrated rate law for a first order reaction:

$$\ln\frac{[A]_0}{[A]_t} = kt$$

If only 5% of the active ingredient can decompose in two years, then 95% must remain, therefore, $[A]_0 = 100$, $[A]_t = 95$, and t = 2 yr.

$$\ln\frac{[A]_0}{[A]_t} = kt$$

$$\ln\frac{100}{95} = k(2\text{ yr})$$

$k = 2.56 \times 10^{-2}$ yr.$^{-1}$

13.16 (a) In order to find the concentration at specific time for a first order reaction, we substitute into equation 14.5, first converting the time to seconds:

t = 2 hr. × 3600 s/hr. = 7200 s

$$\ln\frac{[A]_0}{[A]_t} = kt$$

$$\text{antiln}\left[\ln\frac{[A]_0}{[A]_t}\right] = \text{antiln}\left[kt\right]$$

$$\frac{[A]_0}{[A]_t} = \text{antiln}\left[kt\right]$$

$$\frac{[A]_0}{[A]_t} = \text{antiln}\left[\left(6.17 \times 10^{-4}\text{ s}^{-1}\right)\left(7200\text{ s}\right)\right]$$

$$\frac{0.40\ M}{[A]_t} = 84.98$$

$$[A]_t = \frac{0.40\ M}{84.98}$$

$= 4.71 \times 10^{-3}\ M$

(b) Again, we use equation 13.5, this time solving for time:

$$\ln\frac{[A]_0}{[A]_t} = kt$$

$$t = \frac{1}{k} \times \ln\frac{[A]_0}{[A]_t} = \frac{1}{6.17 \times 10^{-4}\ s^{-1}} \times \ln\frac{0.40\ M}{0.30\ M} = 466\ s$$

4.66×10^2 s × 1 min/60 s = 7.8 min

13.17 For a first–order reaction:

$$t_{1/2} = \frac{0.693}{k} = \frac{0.693}{6.17 \times 10^{-4}\ s^{-1}} = 1.12 \times 10^3\ s$$

$$t_{1/2} = 1.12 \times 10^3\ s \times \frac{1\ min}{60\ s}$$

= 18.7 min

If we refer to the chart given in the text in example 13.8, we see that two half lives will have passed if there is to be only one quarter of the original amount of material remaining. This corresponds to:

18.7 min per half-life × 2 half lives = 37.4 min

13.18 From practice exercise 13.15, the rate constant is: $k = 2.56 \times 10^{-2}$ yr.$^{-1}$, and use the half-life of a first order reaction equation:

$$t_{1/2} = \frac{\ln 2}{k} = \frac{\ln 2}{2.56 \times 10^{-2}\ yr^{-1}} = 27.1\ yr.$$

13.19 Recall that for a first order process

$$k = \frac{0.693}{t_{1/2}}$$

So $k = \dfrac{0.693}{14.26\ days} = 4.86 \times 10^{-2}$/days. Also,

$$\ln\frac{[A]_0}{[A]_t} = kt$$

Assume we started with 100 g ^{32}P

$\ln[A]_t = \ln[A]_0 - kt = \ln 100 - (4.86 \times 10^{-2})(60\ days)$

$\ln[A]_0 = 1.689$

$[A]_0 = 5.41$ g or 5.41%

Alternative method

$$\ln\frac{100}{A_t} = 4.86 \times 10^{-2}\ day^{-1}(60\ day)$$

$$\ln\frac{100}{A_t} = 2.916$$

$$\frac{100}{A_t} = e^{2.916}$$

$$\frac{100}{A_t} = 18.47$$

$A_t = 5.41$ g

$$\left[\frac{1}{2}\right]^{\frac{t}{t_{1/2}}} = [0.5]^{\frac{60\text{ days}}{14.26\text{ days}}} = 0.0541\text{ g}$$

Thus, 0.0541 g would be left if you started with 1.00 g, or 5.41% was left.

13.20 Recall that for a first order process

$$k = \frac{0.693}{t_{1/2}}$$

So $k = \dfrac{0.693}{5730\text{ yr}} = 1.21 \times 10^{-4}$/yr. Also,

$$\ln\frac{[A]_0}{[A]_t} = kt$$

$$t = \frac{1}{k}\ln\frac{[A]_0}{[A]_t} = \frac{1}{1.21 \times 10^{-4}\text{ /yr}}\ln\frac{10}{1} = 1.90 \times 10^4\text{ yrs}$$

13.21 Use the value of the rate constant for C-14 from the previous practice exercise: 1.21×10^{-4} y^{-1}. To find the upper and lower limits of dates before present, set the concentration of the C-14 to 5% and 95%, respectively, and then solve for the time:

For samples with less than 5% of the C-14 remaining:

$$\ln\frac{[A]_0}{[A]_t} = kt$$

$$t = \frac{1}{k}\ln\frac{[A]_0}{[A]_t} = \frac{1}{1.21 \times 10^{-4}\text{ yr}^{-1}}\ln\frac{100}{5} = 2.48 \times 10^4\text{ yr.}$$

Therefore the upper limit of dates is 25,000 years before present.

For samples with more than 95% of the C-14 remaining:

$$\ln\frac{[A]_0}{[A]_t} = kt$$

$$t = \frac{1}{k}\ln\frac{[A]_0}{[A]_t} = \frac{1}{1.21 \times 10^{-4}\text{ yr}^{-1}}\ln\frac{100}{95} = 4.24 \times 10^2\text{ yr.}$$

Therefore the lower limit of dates is 420 years before present.

13.22 This is a second-order reaction, and we use equation 13.10:

$$\frac{1}{[\text{NOCl}]_t} - \frac{1}{[\text{NOCl}]_0} = kt$$

$$\frac{1}{[0.010\ M]} - \frac{1}{[0.040\ M]} = \left(0.020\text{ L mol}^{-1}\text{ s}^{-1}\right) \times t$$

$t = 3.8 \times 10^3$ s

$$t = 3.8 \times 10^3\text{ s} \times \frac{1\text{ min}}{60\text{ s}} = 63\text{ min}$$

13.23 This is the same reaction as in the previous practice exercise, so it is a second-order reaction, and we use equation 13.10 and $k = 0.020 \text{ L mol}^{-1} \text{ s}^{-1}$:

$$\frac{1}{[\text{NOCl}]_t} - \frac{1}{[\text{NOCl}]_0} = kt$$

$$\frac{1}{[0.00035\ M]} - \frac{1}{[x\ M]} = (0.020 \text{ L mol}^{-1} \text{ s}^{-1}) \times t$$

We need to find the time in seconds:

From 10:35 to 3:15, 4 hours and 40 minutes has elapsed, or 280 minutes

$$t = (280 \text{ min})\left(\frac{60 \text{ s}}{1 \text{ min}}\right) = 1.68 \times 10^4 \text{ s}$$

$$\frac{1}{[0.00035\ M]} - \frac{1}{[x\ M]} = (0.020 \text{ L mol}^{-1} \text{ s}^{-1}) \times (1.68 \times 10^4 \text{ s})$$

$$-\frac{1}{[x\ M]} = (3.36 \times 10^2 \text{ L mol}^{-1}) - (2.86 \times 10^3 \text{ L mol}^{-1})$$

$x\ M = 4.0 \times 10^{-4}\ M$

13.24 Rate = $k[NO_2]^2$

To find the rate constant solve the rate law with the given data:

$4.42 \times 10^{-7} \text{ mol L}^{-1}\text{s}^{-1} = k(6.54 \times 10^{-4} \text{ mol L}^{-1})^2$

$$k = \frac{4.42 \times 10^{-7} \text{ mol L}^{-1} \text{ s}^{-1}}{(6.54 \times 10^{-4} \text{ mol L}^{-1})^2} = 1.03 \text{ L mol}^{-1} \text{ s}^{-1}$$

The half-life of the system is found using the half-life equation for a second-order reaction

$$t_{1/2} = \frac{1}{k \times (\text{initial concentration of reactant})}$$

$$t_{1/2} = \frac{1}{(1.03 \text{ L mol}^{-1} \text{ s}^{-1})(6.54 \times 10^{-4} \text{ mol L}^{-1})} = 1.48 \times 10^3 \text{ s}$$

13.25 The reaction is first-order. A second-order reaction should have a half-life that depends on the initial concentration according to equation 13.11.

13.26 (a) Use equation 13.16:

$$\ln \frac{k_2}{k_1} = \frac{-E_a}{R}\left[\frac{1}{T_2} - \frac{1}{T_1}\right]$$

$$\ln \left[\frac{23 \text{ L mol}^{-1} \text{ s}^{-1}}{3.2 \text{ L mol}^{-1} \text{ s}^{-1}}\right] = \frac{-E_a}{8.314 \text{ J mol}^{-1} \text{ K}^{-1}}\left[\frac{1}{678 \text{ K}} - \frac{1}{628 \text{ K}}\right]$$

Solving for E_a gives 1.4×10^5 J/mol = 1.4×10^2 kJ/mol

(b) We again use equation 13.16, substituting the values:

$k_1 = 3.2 \text{ L mol}^{-1} \text{ s}^{-1}$ at $T_1 = 628$ K

$k_2 = ?$ at $T_2 = 583$ K

$$\ln \frac{k_2}{k_1} = \frac{-E_a}{R}\left[\frac{1}{T_2} - \frac{1}{T_1}\right]$$

$$\ln\left[\frac{k_2}{3.2\ \text{L mol}^{-1}\ \text{s}^{-1}}\right] = \frac{-1.4 \times 10^5\ \text{J mol}^{-1}}{8.314\ \text{J mol}^{-1}\ \text{K}^{-1}}\left[\frac{1}{583\ \text{K}} - \frac{1}{628\ \text{K}}\right]$$

$$\ln\left[\frac{k_2}{3.2\ \text{L mol}^{-1}\ \text{s}^{-1}}\right] = -2.0697$$

$k_2 = (e^{-2.0697}) \times 3.2$

Solving for k_2 gives 0.40 L mol^{-1} s^{-1}

13.27 $$\ln\frac{k_2}{k_1} = \frac{-E_a}{R}\left[\frac{1}{T_2} - \frac{1}{T_1}\right]$$

$$\ln\left[\frac{3.37 \times 10^4\ M^{-1}\ \text{s}^{-1}}{3.37 \times 10^3\ M^{-1}\ \text{s}^{-1}}\right] = \frac{-9.31 \times 10^4\ \text{J mol}^{-1}}{8.314\ \text{J mol}^{-1}\ \text{K}^{-1}}\left[\frac{1}{T_2} - \frac{1}{600\ \text{K}}\right]$$

$$2.056 \times 10^{-4} = -\frac{1}{T_2} + \frac{1}{600}$$

$T_2 = 684$ K

13.28 We have the values for T_1 and E_a, but we need to think about the values for k_1 and k_2. We are given that the compound decomposes 5% over 2 years, or 104 weeks, and we want to see the same decomposition over 1 year. If we set the amount of decomposition over time, we get a rate.

k_1 = 5%/104 weeks and k_2 = 5%/1 week

Now use the values for k_1, k_2, T_1 = 298 K, and $E_a = 154 \times 10^3$ J mol^{-1}

$$\ln\frac{k_2}{k_1} = \frac{-E_a}{R}\left[\frac{1}{T_2} - \frac{1}{T_1}\right]$$

$$\ln\left[\frac{5\%\ /104\ \text{weeks}}{5\%\ /1\ \text{week}}\right] = \frac{-154 \times 10^3\ \text{J mol}^{-1}}{8.314\ \text{J mol}^{-1}\ \text{K}^{-1}}\left[\frac{1}{298\ \text{K}} - \frac{1}{T_2}\right]$$

$$\ln\left(\frac{1}{104}\right) = (-1.85 \times 10^4\ \text{K})\left[\left(3.36\times10^{-3}\ \text{K}^{-1}\right) - \frac{1}{T_2}\right]$$

$$2.51 \times 10^{-4}\ \text{K}^{-1} = \left[\left(3.36\times10^{-3}\ \text{K}^{-1}\right) - \frac{1}{T_2}\right]$$

$$-\frac{1}{T_2} = -3.11 \times 10^{-3}\ \text{K}^{-1}$$

T_2 = 322 K or 49 °C

Alternatively,

$$\ln\frac{104.4k}{k} = \frac{-154 \times 10^3\ \text{J mol}^{-1}}{8.314\ \text{J mol}^{-1}\ \text{K}^{-1}}\left[\frac{1}{T_2} - \frac{1}{298\ \text{K}}\right]$$

T_2 = 322 K

13.29 (a), (b), and (e) may be elementary processes.

Equations (c), (d), and (f) are not elementary processes because they have more than two molecules colliding at one time, and this is very unlikely.

13.30 If the reaction occurs in a single step, one molecule of each reactant must be involved, according to the balanced equation. Therefore, the rate law is expected to be: Rate = k[NO][O_3].

13.31 The slow step (second step) of the mechanism determines the rate law:

Rate = $k[NO_2Cl]^1[Cl]^1$

However, Cl is an intermediate and cannot be part of the rate law expression. We need to solve for the concentration of Cl by using the first step of the mechanism. Assuming that the first step is an equilibrium step, the rates of the forward and reverse reactions are equal:

Rate = $k_{forward}[NO_2Cl] = k_{reverse}[Cl][NO_2]$

Solving for [Cl] we get

$$[Cl] = \frac{k_f}{k_r}\frac{[NO_2Cl]}{[NO_2]}$$

Substituting into the rate law expression for the second step yields:

$$\text{Rate} = \frac{k[NO_2Cl]^2}{[NO_2]}$$, where all the constants have been combined into one new constant.

Review Problems

13.53 Since they are in a 1-to-1 mol ratio, the rate of formation of SO_2 is *equal* and *opposite* to the rate of consumption of SO_2Cl_2. This is equal to the slope of the curve at any point on the graph (see below). At 200 min, we obtain a value of about 1×10^{-4} *M*/min. At 600 minutes, this has decreased to about 7×10^{-5} *M*/min.

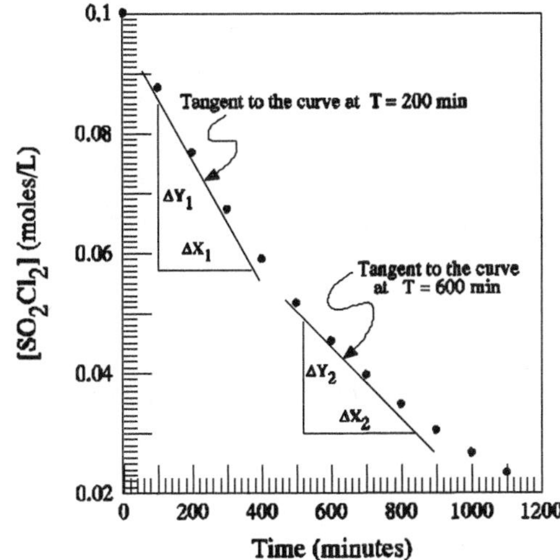

13.55 From the coefficients in the balanced equation we see that, for every mole of *B* that reacts, 2 mol of *A* are consumed, and three mol of *C* are produced. This means that *A* will be consumed twice as fast as *B*, and *C* will be produced three times faster than *B* is consumed.

rate of disappearance of $A = 0.30 \text{ mol } B \text{ L}^{-2} \text{ s}^{-1}\left(\frac{2 \text{ mol } A}{1 \text{ mol } B}\right) = 0.60 \text{ mol } A \text{ L}^{-1} \text{ s}^{-1}$

rate of appearance of $C = 0.30 \text{ mol } B \text{ L}^{-2} \text{ s}^{-1}\left(\frac{3 \text{ mol } C}{1 \text{ mol } B}\right) = 0.90 \text{ mol } C \text{ L}^{-1} \text{ s}^{-1}$

13.57 (a) rate for $O_2 = -1.20 \text{ mol } C_6H_{14} \text{ L}^{-1} \text{ s}^{-1} \times \left(\dfrac{19 \text{ mol } O_2}{2 \text{ mol } C_6H_{14}}\right) = -11.4 \text{ mol } O_2 \text{ L}^{-1} \text{ s}^{-1}$

By convention, this is reported as a positive number: $11.4 \text{ mol L}^{-1} \text{ s}^{-1}$

(b) rate for $CO_2 = +1.20 \text{ mol } C_6H_{14} \text{ L}^{-1} \text{ s}^{-1} \times \left(\dfrac{12 \text{ mol } CO_2}{2 \text{ mol } C_6H_{14}}\right) = 7.20 \text{ mol } CO_2 \text{ L}^{-1} \text{ s}^{-1}$

(c) rate for $H_2O = +1.20 \text{ mol } C_6H_{14} \text{ L}^{-1} \text{ s}^{-1} \times \left(\dfrac{14 \text{ mol } H_2O}{2 \text{ mol } C_6H_{14}}\right) = 8.40 \text{ mol } H_2O \text{ L}^{-1} \text{ s}^{-1}$

13.59 (a) $-\dfrac{\Delta[CH_3Cl]}{\Delta t} = -\dfrac{1}{3}\dfrac{\Delta[Cl_2]}{\Delta t} = \dfrac{\Delta[CCl_4]}{\Delta t} = \dfrac{1}{3}\dfrac{\Delta[HCl]}{\Delta t}$

(b) $\text{Rate} = \dfrac{1}{3}\left(0.029\ M \text{ s}^{-1}\right) = 9.7 \times 10^{-3}\ M \text{ s}^{-1}$

13.61 The rate can be found by simply inserting the given concentration values:
$\text{rate} = (5.0 \times 10^5 \text{ L}^5 \text{ mol}^{-5} \text{ s}^{-1})[H_2SeO_3][I^-]^3[H^+]^2$
$\text{rate} = (5.0 \times 10^5 \text{ L}^5 \text{ mol}^{-5} \text{ s}^{-1})(2.0 \times 10^{-2} \text{ mol L}^{-1})(2.0 \times 10^{-3} \text{ mol L}^{-1})^3(1.0 \times 10^{-3} \text{ mol L}^{-1})^2$
$\text{rate} = 8.0 \times 10^{-11} \text{ mol L}^{-1} \text{ s}^{-1}$

13.63 $\text{rate} = (7.1 \times 10^9 \text{ L}^2 \text{ mol}^{-2} \text{ s}^{-1})(1.0 \times 10^{-3} \text{ mol L}^{-1})^2(3.4 \times 10^{-2} \text{ mol L}^{-1})$
$\text{rate} = 2.4 \times 10^2 \text{ mol L}^{-1} \text{ s}^{-1}$

13.65 For a zero order reaction the rate is independent of concentration so rate = k
Therefore, the rate of the reaction is $6.4 \times 10^2\ M \text{ s}^{-1}$.

13.67 On comparing the data of the first and second experiments, we find that, the concentration of N is unchanged, and the concentration of M has been doubled, causing a doubling of the rate. This corresponds to the fourth case in Table 13.4, and we conclude that the order of the reaction with respect to M is 1. In the second and third experiments, we have a different result. When the concentration of M is held constant, the concentration of N is tripled, causing an increase in the rate by a factor of nine. This constitutes the eighth case in Table 13.4, and we conclude that the order of the reaction with respect to N is 2. This means that the overall rate expression is: rate = $k[M][N]^2$ and we can solve for the value of k by substituting the appropriate data:
$5.0 \times 10^{-3} \text{ mol L}^{-1} \text{ s}^{-1} = k \times [0.020 \text{ mol L}^{-1}][0.010 \text{ mol L}^{-1}]^2$
$k = 2.5 \times 10^3 \text{ L}^2 \text{ mol}^{-2} \text{ s}^{-1}$

13.69 Comparing the first two experiments:

$$\frac{\text{rate}_2}{\text{rate}_1} = \frac{k\left[OCl^-\right]_2^m\left[I^-\right]_2^n}{k\left[OCl^-\right]_1^m\left[I^-\right]_1^n}$$

$$\frac{8.31 \times 10^4}{1.75 \times 10^4} = \frac{k\left[7.6 \times 10^{-3}\right]^m\left[1.6 \times 10^{-3}\right]^n}{k\left[1.6 \times 10^{-3}\right]^m\left[1.6 \times 10^{-3}\right]^n}$$

$$4.75 = \left(\frac{7.6 \times 10^{-3}}{1.6 \times 10^{-3}}\right)^m = 4.75^m$$

log 4.75 = $m$log 4.75

$m = 1$

Compare the first and third experiments

$$\frac{\text{rate}_3}{\text{rate}_1} = \frac{k\left[\text{OCl}^-\right]_3^1\left[\text{I}^-\right]_3^n}{k\left[\text{OCl}^-\right]_1^1\left[\text{I}^-\right]_1^n}$$

$$\frac{1.05 \times 10^5}{1.75 \times 10^4} = \frac{k\left(1.6 \times 10^{-3}\right)^1\left(9.6 \times 10^{-3}\right)^n}{k\left(1.6 \times 10^{-3}\right)^1\left(1.6 \times 10^{-3}\right)^n}$$

$$6.00 = \left(\frac{9.6 \times 10^{-3}}{1.6 \times 10^{-3}}\right)^m = 6.00^n$$

$n = 1$

rate = $k[\text{OCl}^-][\text{I}^-]$

Using the last data set:

1.05×10^5 mol L^{-1} s^{-1} = k[1.6×10^{-3} mol L^{-1}][9.6×10^{-3} mol L^{-1}]

$k = 6.8 \times 10^9$ L mol^{-1} s^{-1}

13.71 Compare the first and second experiments.

$$\frac{\text{rate}_2}{\text{rate}_1} = \frac{k[\text{ICl}]_2^m[\text{H}_2]_2^n}{k[\text{ICl}]_1^m[\text{H}_2]_1^n}$$

$$\frac{0.0098}{0.0015} = \frac{k(0.78)^m(0.12)^n}{k(0.12)^m(0.12)^n}$$

$$6.5 = \left(\frac{0.78}{0.12}\right)^m = 6.5^m$$

$m = 1$

Compare the first and third experiments

$$\frac{\text{rate}_3}{\text{rate}_1} = \frac{k[\text{ICl}]_3^1[\text{H}_2]_3^n}{k[\text{ICl}]_1^1[\text{H}_2]_1^n}$$

$$\frac{0.0011}{0.0015} = \frac{k(0.12)^1(0.89)^n}{k(0.12)^1(0.12)^n}$$

$$0.733 = \left(\frac{0.089}{0.12}\right)^n = 0.74^n$$

log 0.73 = $n$log 0.74

$n = 1.04 = 1$
rate = k[ICl][H_2]
Using the data of the first experiment:
1.5×10^{-3} mol L^{-1} s^{-1} = k[0.12 mol L^{-1}][0.12 mol L^{-1}]
$k = 1.04 \times 10^{-1}$ L mol^{-1} s^{-1}

13.73 A graph of ln [SO_2Cl_2]$_t$ versus t will yield a straight line if the data obeys a first–order rate law.

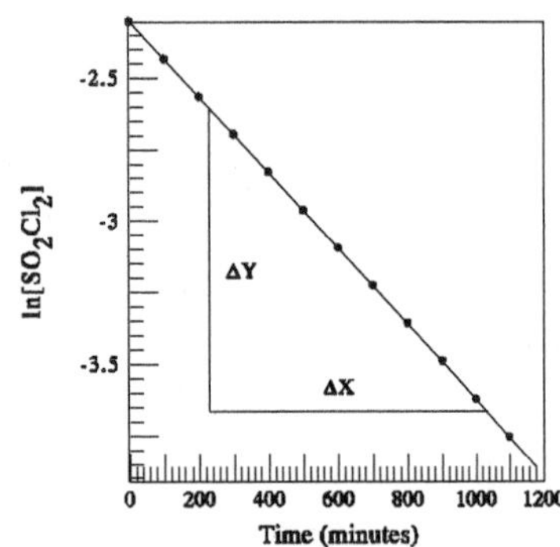

These data do yield a straight line when ln[SO_2Cl_2]$_t$ is plotted against the time, t. The slope of this line equals $-k$. Plotting the data provided and using linear regression to fit the data to a straight line yields a value of 1.32×10^{-3} min^{-1} for k.

13.75 (a) The time involved must be converted to a value in seconds:
1 hr. × 3600 s/hr. = 3.6×10^3 s, and then we make use of equation 13.5, where x is taken to represent the desired SO_2Cl_2 concentration:

$$\ln \frac{0.0040\ M}{x} = (2.2 \times 10^{-5}\ s^{-1})(3.6 \times 10^3\ s)$$

$x = 3.7 \times 10^{-3}\ M$

(b) The time is converted to a value having the units as seconds
24 hr. × 3600 s/hr. = 8.64×10^4 s, and then we use equation 13.5, where x is taken to represent the desired SO_2Cl_2 concentration:

$$\ln \frac{0.0040\ M}{x} = (2.2 \times 10^{-5}\ s^{-1})(8.64 \times 10^4\ s)$$

$x = 6.0 \times 10^{-4}\ M$

13.77 Use equation 13.5 and assume the initial concentration is 100 and the final concentration is, therefore, 25. The units are immaterial in this case since they will cancel out in the calculation:

$$\ln \frac{[A]_0}{[A]_t} = kt$$

$$\ln \frac{100}{25} = k(75.0\ \text{min})$$

Solving for k we get 1.85×10^{-2} min^{-1}
Alternatively,
75 min is equal to two half-lives

$$\frac{0.693}{37.5\ \text{min}} = 0.0185\ \text{min}^{-1}$$

13.79 Any consistent set of units for expressing concentration may be used in equation 13.5, where we let A represent the drug that is involved:

$$\ln \frac{[A]_0}{[A]_t} = kt$$

$$\ln \frac{25.0\ \text{mg/kg}}{15.0\ \text{mg/kg}} = k(120\ \text{min})$$

Solving for k we get $4.26 \times 10^{-3}\ \text{min}^{-1}$

13.81 We use the equation:

$$\frac{1}{[\text{HI}]_t} - \frac{1}{[\text{HI}]_0} = kt$$

$$\frac{1}{[8.0 \times 10^{-4}\ M]} - \frac{1}{[3.4 \times 10^{-2}\ M]} = (1.6 \times 10^{-3}\ \text{L mol}^{-1}\ \text{s}^{-1}) \times t$$

Solving for t gives:

$$t = 7.6 \times 10^5\ \text{s} \ \text{ or } \ t = (7.6 \times 10^5\ \text{s}) \times \frac{1\ \text{min}}{60\ \text{s}} = 1.3 \times 10^4\ \text{min}$$

13.83 Use the equation, taking time in minutes; 3 hr. = 180 min.

$$\ln \frac{x}{5.0\ \text{mg/kg}} = (4.26 \times 10^{-3}\ \text{min}^{-1})(180\ \text{min})$$

$x = 11$ mg/kg

13.85 $$\text{half lives} = (2.0\ \text{hrs})\left(\frac{60\ \text{min}}{1\ \text{hr}}\right)\left(\frac{1\ \text{half life}}{15\ \text{min}}\right) = 8.0\ \text{half lives}$$

Eight half lives correspond to the following fraction of original material remaining:

Number of half lives	Fraction remaining
1	1/2
2	1/4
3	1/8
4	1/16
5	1/32
6	1/64
7	1/128
8	1/256

Alternatively, since 8 half-lives have occurred
$0.5^8 = 0.0039$ or 1/256

13.87 It requires approximately 500 min (as determined from the graph) for the concentration of SO_2Cl_2 to decrease from 0.100 M to 0.050 M, i.e., to decrease to half its initial concentration. Likewise, in another 500 minutes, the concentration decreases by half again, i.e. from 0.050 M to 0.025 M. This means that the half-life of the reaction is independent of the initial concentration, and we conclude that the reaction is first–order in SO_2Cl_2.

13.89 $t_{\frac{1}{2}} = \dfrac{\ln 2}{k}$

$10 \text{ hr.} = \dfrac{\ln 2}{k}$

$k = 0.069 \text{ hr.}^{-1}$

$$\ln \frac{[A]_0}{[A]_t} = \left(0.069 \text{ hr}^{-1}\right)t$$

$$\ln \frac{1.0}{0.10} = \left(0.069 \text{ hr}^{-1}\right)t$$

$t = 30$ hr.

13.91 In order to solve this problem, it must be assumed that all of the argon–40 that is found in the rock must have come from the potassium–40, i.e., that the rock contains no other source of argon–40. If the above assumption is valid, then any argon–40 that is found in the rock represents an equivalent amount of potassium–40, since the stoichiometry is 1:1. Since equal amounts of potassium–40 and argon–40 have been found, this indicates that the amount of potassium–40 that remains is exactly half the amount that was present originally. In other words, the potassium–40 has undergone one half-life of decay by the time of the analysis. The rock is thus seen to be 1.3×10^9 years old.

13.93 Use equation 13.8 and 13.9.

$$\ln \frac{r_0}{r_t} = kt$$

$$\ln\left(\frac{1.2 \times 10^{-12}}{r_t}\right) = \left(1.21 \times 10^{-4} \text{ y}^{-1}\right)\left(9.0 \times 10^3 \text{ y}\right)$$

$$\left(\frac{1.2 \times 10^{-12}}{r_t}\right) = \exp\left[\left(1.21 \times 10^{-4} \text{ y}^{-1}\right)\left(9.0 \times 10^3 \text{ y}\right)\right]$$

$$\left(\frac{1.2 \times 10^{-12}}{2.97}\right) = \text{r}_\text{t} = 4.0 \times 10^{-13}$$

13.95 The graph is prepared exactly as in Example 13.12 of the text. The slope is found using linear regression, to be: -9.5×10^3 K. Thus $-9.5 \times 10^3 \text{ K} = -E_a/R$

$E_a = -(-9.5 \times 10^3 \text{ K})(8.314 \text{ J K}^{-1} \text{ mol}^{-1}) = 7.9 \times 10^4 \text{ J/mol} = 79 \text{ kJ/mol}$

Using the equation, we proceed as follows:

$$\ln \frac{k_2}{k_1} = \frac{-E_a}{R}\left[\frac{1}{T_2} - \frac{1}{T_1}\right]$$

$$\ln\left[\frac{1.94 \times 10^{-3} \text{ L mol}^{-1} \text{ s}^{-1}}{2.88 \times 10^{-4} \text{ L mol}^{-1} \text{ s}^{-1}}\right] = \frac{-E_a}{8.314 \text{ J mol}^{-1} \text{ K}^{-1}}\left[\frac{1}{673 \text{ K}} - \frac{1}{593 \text{ K}}\right]$$

$$1.907 = \frac{2.00 \times 10^{-4} \text{ K}^{-1}}{8.314 \text{ J mol}^{-1} \text{ K}^{-1}} \times E_a$$

$E_a = 7.93 \times 10^4 \text{ J/mol} = 79.3 \text{ kJ/mol}$

13.97 Using the equation we have:

$$\ln \frac{k_2}{k_1} = \frac{-E_a}{R}\left[\frac{1}{T_2} - \frac{1}{T_1}\right]$$

$$\ln\left[\frac{1.0 \times 10^{-3}\ \text{L mol}^{-1}\ \text{s}^{-1}}{9.3 \times 10^{-5}\ \text{L mol}^{-1}\ \text{s}^{-1}}\right] = \frac{-E_a}{8.314\ \text{J mol}^{-1}\ \text{K}^{-1}}\left[\frac{1}{403\ \text{K}} - \frac{1}{373\ \text{K}}\right]$$

$$2.37 = \frac{2.00 \times 10^{-4}\ \text{K}^{-1}}{8.314\ \text{J mol}^{-1}\ \text{K}^{-1}} \times E_a$$

$E_a = 9.89 \times 10^4$ J/mol = 99 kJ/mol

Equation states $k = A \exp\left(\frac{-E_a}{RT}\right)$

$$A = \frac{k}{\exp\left(\frac{-E_a}{RT}\right)}$$

$$A = \frac{9.3 \times 10^{-5}\ \text{L mol}^{-1}\ \text{s}^{-1}}{\exp\left(\frac{-9.89 \times 10^4\ \text{J/mol}}{(8.314\ \text{J/mol K})(373\ \text{K})}\right)}$$

$A = 6.6 \times 10^9$ L mol^{-1} s^{-1}

$$k = 6.6 \times 10^9\ \text{L mol}^{-1}\ \text{s}^{-1} \exp\left(\frac{-9.89 \times 10^4\ \text{J/mol}}{8.314\ \text{J mol}^{-1}\ \text{K}^{-1} \times 473\ \text{K}}\right)$$

$k = 7.9 \times 10^{-2}$ L mol^{-1} s^{-1}

13.99 Use equation 13.14:

(a) $k = A \exp\left(\frac{-E_a}{RT}\right)$

$$k = (4.3 \times 10^{13}\ \text{s}^{-1})\exp\left(\frac{-103 \times 10^3\ \text{J mol}^{-1}}{(8.314\ \text{J/mol K})(298\ \text{K})}\right)$$

$k = 3.8 \times 10^{-5}$ s^{-1}

(b) $k = A \exp\left(\frac{-E_a}{RT}\right)$

$$k = (4.3 \times 10^{13}\ \text{s}^{-1})\exp\left(\frac{-103 \times 10^3\ \text{J mol}^{-1}}{(8.314\ \text{J/mol K})(373\ \text{K})}\right)$$

$k = 1.6 \times 10^{-1}$ s^{-1}

13.101 Add all of the steps together:

$$CO + O_2 + NO \longrightarrow CO_2 + NO_2$$

13.103 For such a mechanism, the rate law should be:

rate = $k[NO_2][CO]$

Since this is not the same as the observed rate law, this is not a reasonable mechanism to propose.

13.105 If this reaction occurs in one step then the rate law is the product of the concentrations raised to the appropriate powers.
Rate = $k[AB][C]$

13.107 An intermediate is produced in one step and used up in another. The intermediate in this reaction is N_2O_2.
The overall reaction is the sum of the two steps

$2NO(g) + O_2(g) \longrightarrow 2NO_2(g)$

The rate law is determined from the slow step.
Rate = $k[N_2O_2][O_2]$
An intermediate cannot be part of the rate law expression. We can use the first step, which is an equilibrium step to solve for the concentration of N_2O_2.
Rate = $k_f[NO]^2 = k_r[N_2O_2]$
Solve for $[N_2O_2]$ to get

$$[N_2O_2] = \frac{k_f}{k_r}[NO]^2$$

Rate = $k[NO]^2[O_2]$ where all of the constants have been combined into a new constant

Chapter 14
Chemical Equilibrium

Practice Exercises

14.1 The chapter-opening photograph of the tight-rope walker represents a dynamic equilibrium because the tightrope walker is constantly adjusting his balance to keep himself upright. It differs from a dynamic equilibrium because two processes that are in equilibrium are not occurring at the same time.

14.2 (a) $$\frac{[H_2O]^2}{[H_2]^2[O_2]} = K_c$$

(b) $$\frac{[CO_2][H_2O]^2}{[CH_4][O_2]^2} = K_c$$

14.3 $2N_2O_3(g) + O_2(g) \longrightarrow 4NO_2(g)$

14.4 Since the starting equation has been reversed and divided by two, we must invert the equilibrium constant, and then take the square root: $K_c = 1.2 \times 10^{-13}$

14.5 If we divide both equations by 2 and reverse the second we get:

$CO(g) + 1/2O_2(g) \longrightarrow CO_2(g) \qquad K_c = 5.7 \times 10^{45}$

$H_2O(g) \longrightarrow H_2(g) + ½O_2(g) \quad K_c = 3.3 \times 10^{-41}$

Note that when we divide the equation by two, we need to take the square root of the rate constant. When we reverse the reaction, we need to take the inverse.

Adding these equations we get the desired equation so we need simply multiply the values for K_c in order to obtain the new value: $K_c = 1.9 \times 10^5$

14.6 $$K_P = \frac{(P_{N_2O})^2}{(P_{N_2})^2(P_{O_2})}$$

14.7 $$K_P = \frac{(P_{HI})^2}{(P_{H_2})(P_{I_2})}$$

14.8 Use the equation $K_p = K_c(RT)^{\Delta n_g}$. In this reaction, $\Delta n_g = 3 - 2 = 1$, so

$$K_p = K_c(RT)^{\Delta n_g} = (7.3 \times 10^{34})\left(\left(0.0821 \frac{\text{L atm}}{\text{mol K}}\right)(298\text{ K})\right)^1 = 1.8 \times 10^{36}$$

14.9 We would expect K_P to be smaller than K_c since Δn_g is negative.

Use the equation:

$$K_p = K_c(RT)^{\Delta n_g}$$

$$K_c = \frac{K_p}{(RT)^{\Delta n_g}}$$

In this case, $\Delta n_g = (1 - 3) = -2$, and we have:

$$K_c = \frac{K_p}{(RT)^{\Delta n_g}} = \frac{3.8 \times 10^{-2}}{\left(\left(0.0821 \frac{\text{L atm}}{\text{mol K}}\right)(473 \text{ K})\right)^{-2}} = 57$$

14.10 $K_c = \dfrac{1}{[NH_3(g)][HCl(g)]}$

14.11 (a) $K_c = \dfrac{1}{[Cl_2(g)]}$

(b) $K_c = [Ag^+(aq)]^2[CrO_4^{2-}(aq)]$

(c) $K_c = \dfrac{[Ca^{+2}(aq)][HCO_3^-(aq)]^2}{[CO_2(aq)]}$

14.12 To solve this problem, the first step is to find Q, the mass action expression, and then compare it to the value for K_c. If Q is larger than K_c, then there are more products than the equilibrium concentration and the reaction will move to reactants. If Q is smaller than K_c, then there are more reactants than the equilibrium concentrations and the reaction will move to products.
For this reaction, the concentrations of H_2, Br_2 and HBr are all equal. If we set them to x, then we see that they cancel to give $Q = 1$:

$$Q = \frac{[HBr]^2}{[H_2][Br_2]} = \frac{[x]^2}{[x][x]} = 1$$

Q is larger than K_c, therefore the reaction will move to reactants.

14.13 Reaction (a) will proceed the least amount towards completion followed by reaction (c) and finally, reaction (b) will proceed farthest to completion since it has the largest value for K_c.

14.14 Only the gases will be affected by the volume of the container. But the stoichiometric ratio of the reactants to products is 5:5, so as the volume is changes, the number of moles of gas will not change, therefore the reaction will not change, and there will be no change in the amount of H_3PO_4.

14.15 (a) The equilibrium will shift to the right, decreasing the concentration of Cl_2 at equilibrium, and consuming some of the added PCl_3. The value of K_p will be unchanged.

(b) The equilibrium will shift to the left, consuming some of the added PCl_5 and increasing the amount of Cl_2 at equilibrium. The value of K_p will be unchanged.

(c) For any exothermic equilibrium, an increase in temperature causes the equilibrium to shift to the left, in order to remove energy in response to the stress. This equilibrium is shifted to the left, making more Cl_2 and more PCl_3 at the new equilibrium. The value of K_p is given by the following:

$$K_p = \frac{P_{PCl_5}}{P_{PCl_3} \times P_{Cl_2}}$$

In this system, an increase in temperature (which causes an increase in the equilibrium concentrations of both PCl_3 and Cl_2 and a decrease in the equilibrium concentration of PCl_5) causes an increase in the denominator of the above expression as well as a decrease in the numerator of the above expression. Both of these changes serve to decrease the value of K_p.

(d) Decreasing the container volume for a gaseous system will produce an increase in partial pressures for all gaseous reactants and products. In order to lower the increase in partial pressures, the equilibrium will shift so as to favor the product side having the smaller number of gaseous molecules, in this case to the right. This shift will decrease the amount of Cl_2 and PCl_3 at equilibrium, and it will increase the amount of PCl_5 at equilibrium. While the change in volume will change the position of the equilibrium, it does not change the value for K_p.

14.16 $2CO(g) + O_2(g) \longrightarrow 2CO_2(g)$

Using the stoichiometry of the reaction we can see that for every mol of O_2 that is used, twice as much CO will react and twice as much CO_2 will be produced. Consequently, if the $[O_2]$ decreases by 0.030 mol/L, the [CO] decreases by 0.060 mol/L and $[CO_2]$ increases by 0.060 mol/L.

14.17 $K_c = \frac{[CO_2][H_2]}{[CO][H_2O]} = \frac{(0.150)(0.200)}{(0.180)(0.0411)} = 4.06$

14.18 (a) The initial concentrations were:

$[PCl_3] = 0.200 \text{ mol}/1.00 \text{ L} = 0.200\ M$

$[Cl_2] = 0.100 \text{ mol}/1.00 \text{ L} = 0.100\ M$

$[PCl_5] = 0.00 \text{ mol}/1.00 \text{ L} = 0.000\ M$

(b) The change in concentration of PCl_3 was $(0.200 - 0.120)\ M = 0.080$ mol/L. The other materials must have undergone changes in concentration that are dictated by the coefficients of the balanced chemical equation, namely: $PCl_3 + Cl_2 \longrightarrow PCl_5$ so both PCl_3 and Cl_2 have decreased by 0.080 M and PCl_5 has increased by 0.080 M.

(c) As stated in the problem, the equilibrium concentration of PCl_3 is 0.120 M. The equilibrium concentration of PCl_5 is 0.080 M since initially there was no PCl_5. The equilibrium concentration of Cl_2 equals the initial concentration minus the amount that reacted, $0.100\ M - 0.080\ M = 0.020\ M$.

(d) $K_c = \frac{[PCl_5]}{[PCl_3][PCl_2]} = \frac{(0.080)}{(0.120)(0.020)} = 33$

14.19 $K_c = \frac{[NO_2]^2}{[N_2O_4]}$

$$4.61 \times 10^{-3} = \frac{[NO_2]^2}{\left(\frac{0.0466}{2}\right)}$$

$[NO_2] = 1.03 \times 10^{-2}\ M$

14.20 $$K_c = \frac{[CH_3CO_2C_2H_5][H_2O]}{[CH_3CO_2H][C_2H_5OH]} = \frac{(0.910)(0.00850)}{(0.210)[C_2H_5OH]} = 4.10$$

$[C_2H_5OH] = 8.98 \times 10^{-3}\ M$

14.21 Initially we have $[H_2] = [I_2] = 0.200\ M$.

	$[H_2]$	$[I_2]$	$[HI]$
I	0.200	0.200	–
C	$-x$	$-x$	$+2x$
E	$0.200 - x$	$0.200 - x$	$+ 2x$

Substituting the above values for equilibrium concentrations into the mass action expression gives:

$$K_c = \frac{[HI]^2}{[H_2][I_2]} = \frac{(2x)^2}{(0.200-x)(0.200-x)} = 49.5$$

Take the square root of both sides of this equation to get; $\frac{2x}{(0.200-x)} = 7.04$. This equation is easily solved giving $x = 0.156$. The substances then have the following concentrations at equilibrium: $[H_2] = [I_2] = 0.200 - 0.156 = 0.044\ M$, $[HI] = 2(0.156) = 0.312\ M$.

14.22 Initially we have $[H_2] = 0.200\ M$, $[I_2] = 0.100\ M$.

	$[H_2]$	$[I_2]$	$[HI]$
I	0.200	0.100	–
C	$-x$	$-x$	$+2x$
E	$0.200 - x$	$0.100 - x$	$+ 2x$

Substituting the above values for equilibrium concentrations into the mass action expression gives:

$$K_c = \frac{[HI]^2}{[H_2][I_2]} = \frac{(2x)^2}{(0.200-x)(0.100-x)} = 49.5$$

$4x^2 = 49.5(0.0200 - 0.300x + x^2)$

$45.5x^2 - 14.9x + 0.990 = 0$

Solve the quadratic equation:

$$x = \frac{-b \pm \sqrt{b^2 - 4ac}}{2a} = \frac{-(-14.9) \pm \sqrt{(14.9)^2 - 4(45.5)(0.990)}}{2(45.5)} = 0.0.0934$$

The substances then have the following concentrations at equilibrium:

$[H_2] = 0.200 - 0.0934 = 0.107\ M$

$[I_2] = 0.100 - 0.0934 = 0.007\ M$

$[HI] = 2(0.0.0934) = 0.187\ M$

14.23 $2NH_3(g) \longrightarrow N_2(g) + 3H_2(g)$

	$[NH_3]$	$[N_2]$	$[H_2]$
I	0.041	–	–
C	$-2x$	$+x$	$+3x$
E	$0.041 - x$	$+ x$	$+ 3x$

Substitute the values for the equilibrium concentrations into the mass action expression:

$$K_c = \frac{[N_2][H_2]^3}{[NH_3]^2} = \frac{(x)(3x)^3}{(0.041-x)^2} = 2.3 \times 10^{-9}$$

We will assume that x is small compared to the concentration of NH_3, so the equation will simplify to:

$$K_c = \frac{27x^4}{(0.041)^2} = 2.3 \times 10^{-9}$$

$x^4 = 1.4 \times 10^{-13}$
$x = 6.2 \times 10^{-4}$
$[N_2] = 6.2 \times 10^{-4}$
$[H_2] = 1.9 \times 10^{-3}$

14.24 $N_2(g) + O_2(g) \rightleftharpoons 2NO(g)$

	$[N_2]$	$[O_2]$	$[NO]$
I	0.033	0.00810	–
C	$-x$	$-x$	$+2x$
E	$0.033 - x$	$0.00810 - x$	$+ 2x$

Substituting the above values for equilibrium concentrations into the mass action expression gives:

$$K_c = \frac{[NO]^2}{[N_2][O_2]} = \frac{(2x)^2}{(0.033-x)(0.00810-x)} = 4.8 \times 10^{-31}$$

If we assume that $x << 0.033$ and $x << 0.00810$, we can simplify this equation. (Because the value of K_c is so low, this assumption should be valid.) The equation simplifies as:

$$K_c = \frac{(2x)^2}{(0.033)(0.00810)} = 4.8 \times 10^{-31}$$

This equation is easily solved to give $x = 5.7 \times 10^{-18}$ *M*. The equilibrium concentration of NO is $2x$ according to the ICE table so, $[NO] = 1.1 \times 10^{-17}$ *M*.

Review Problems

14.27 (a) $K_c = \dfrac{[POCl_3]^2}{[PCl_3]^2[O_2]}$ (d) $K_c = \dfrac{[NO_2]^2[H_2O]^8}{[N_2H_4][H_2O_2]^6}$

(b) $K_c = \dfrac{[SO_2]^2[O_2]}{[SO_3]^2}$ (e) $K_c = \dfrac{[SO_2][HCl]^2}{[SOCl_2][H_2O]}$

(c) $K_c = \dfrac{[NO]^2[H_2O]^2}{[N_2H_4][O_2]^2}$

14.29 (a) $K_c = \dfrac{[Ag(NH_3)_2{}^+]}{[Ag^+][NH_3]^2}$ (b) $K_c = \dfrac{[Cd(SCN)_4{}^{2-}]}{[Cd^{2+}][SCN^-]^4}$

14.31 The first equation has been reversed in making the second equation. We therefore take the inverse of the value of the first equilibrium constant in order to determine a value for the second equilibrium constant: $K_c = 1 \times 10^{85}$

14.33 (a) $K_c = \dfrac{[HCl]^2}{[H_2][Cl_2]}$ (b) $K_c = \dfrac{[HCl]}{[H_2]^{1/2}[Cl_2]^{1/2}}$

K_c for reaction (b) is the square root of K_c for reaction (a).

14.35 (a) $K_p = \dfrac{\left(P_{POCl_3}\right)^2}{\left(P_{PCl_3}\right)^2\left(P_{O_2}\right)}$ (d) $K_p = \dfrac{\left(P_{NO_2}\right)^2\left(P_{H_2O}\right)^8}{\left(P_{N_2H_4}\right)\left(P_{H_2O_2}\right)^6}$

(b) $K_p = \dfrac{\left(P_{SO_2}\right)^2\left(P_{O_2}\right)}{\left(P_{SO_3}\right)^2}$ (e) $K_p = \dfrac{\left(P_{SO_2}\right)\left(P_{HCl}\right)^2}{\left(P_{SOCl_2}\right)\left(P_{H_2O}\right)}$

(c) $K_p = \dfrac{\left(P_{NO}\right)^2\left(P_{H_2O}\right)^2}{\left(P_{N_2H_4}\right)\left(P_{O_2}\right)^2}$

14.37 $M = P/RT$

$$M = \frac{(745\text{ torr})\left(\dfrac{1\text{ atm}}{760\text{ torr}}\right)}{\left(0.0821\ \frac{\text{L atm}}{\text{mol K}}\right)(318\text{ K})} = 0.0375\ M$$

14.39 a, since $\Delta n_g = 0$

14.41 $K_p = K_c \times (RT)^{\Delta n_g}$
$6.3 \times 10^{-3} = K_c \times [(0.0821\text{ L atm K}^{-1}\text{ mol}^{-1})(498\text{ K})]^{-2} = (5.98 \times 10^{-4}) \times K_c$
$K_c = 11$

14.43 $K_p = K_c \times (RT)^{\Delta n_g}$

$K_p = 4.2 \times 10^{-4} \times [(0.0821\text{ L atm K}^{-1}\text{ mol}^{-1})(773\text{ K})]^1 = 2.7 \times 10^{-2}$

14.45 $K_p = K_c \times (RT)^{\Delta n_g}$

$K_p = (0.40)[(0.0821\text{ L atm K}^{-1}\text{ mol}^{-1})(1046\text{ K})]^{-2} = 5.4 \times 10^{-5}$

14.47 In each case we get approximately 55.5 *M*:

(a) $$\text{mol H}_2\text{O} = (18.0\text{ mL H}_2\text{O})\left(\frac{1\text{ g}}{1\text{ mL}}\right)\left(\frac{1\text{ mol H}_2\text{O}}{18.02\text{ g H}_2\text{O}}\right) = 0.999\text{ mol H}_2\text{O}$$

$$M = \left(\frac{0.999\text{ mol H}_2\text{O}}{18.0\text{ mL H}_2\text{O}}\right)\left(\frac{1000\text{ mL}}{1\text{ L}}\right) = 55.5\ M$$

(b) $$\text{mol H}_2\text{O} = (100.0\text{ mL H}_2\text{O})\left(\frac{1\text{ g}}{1\text{ mL}}\right)\left(\frac{1\text{ mol H}_2\text{O}}{18.02\text{ g H}_2\text{O}}\right) = 5.549\text{ mol H}_2\text{O}$$

$$M = \left(\frac{5.549\text{ mol H}_2\text{O}}{100.0\text{ mL H}_2\text{O}}\right)\left(\frac{1000\text{ mL}}{1\text{ L}}\right) = 55.49\ M$$

(c) $$\text{mol H}_2\text{O} = (1.00\text{ L H}_2\text{O})\left(\frac{1000\text{ mL}}{1\text{ L}}\right)\left(\frac{1\text{ g}}{1\text{ mL}}\right)\left(\frac{1\text{ mol H}_2\text{O}}{18.02\text{ g H}_2\text{O}}\right) = 55.5\text{ mol H}_2\text{O}$$

$$M = \left(\frac{55.5\text{ mol H}_2\text{O}}{1.00\text{ L H}_2\text{O}}\right) = 55.5\ M$$

14.49 (a) $K_c = \dfrac{[CO]^2}{[O_2]}$

(b) $K_c = [H_2O][SO_2]$

(c) $K_c = \dfrac{[CH_4][CO_2]}{[H_2O]^2}$

(d) $K_c = \dfrac{[H_2O][CO_2]}{[HF]^2}$

(e) $K_c = [H_2O]^5$

14.51 $2HCl(g) + I_2(s) \rightleftharpoons 2HI(g) + Cl_2(g)$

	[HCl]	[HI]	[Cl$_2$]
I	0.100	–	–
C	$-2x$	$+2x$	$+x$
E	$0.100 - 2x$	$+ 2x$	$+ x$

Note: Since the $I_2(s)$ has a constant concentration, it may be neglected.

$$K_c = \frac{(2x)^2(x)}{(0.100-2x)^2} = 1.6 \times 10^{-34}$$

Because the value of K_c is so small, we make the simplifying assumption that $(0.100 - 2x) \approx 0.100$, and the above equation becomes:

$$K_c = \frac{[HI]^2[Cl_2]}{[HCl]^2} = 1.6 \times 10^{-34}$$

$$K_c = \frac{(2x)^2(x)}{(0.100)^2} = 1.6 \times 10^{-34}$$

$4x^3 = 1.6 \times 10^{-36}$

$\therefore\ x = 7.37 \times 10^{-13}$, and the above assumption is seen to have been valid.

$[HI] = 2x = 1.47 \times 10^{-12}\ M$

$[Cl_2] = x = 7.37 \times 10^{-13}\ M$

$[HCl] = (0.100 - 2x) \approx 0.100\ M$

14.53 (a) increase; we are adding a reactant
(b) decrease; we are removing a reactant
(c) increase; decreasing the volume favors the side with fewer gas molecules
(d) no change; a catalyst increases the rate but does not affect the concentration
(e) decrease; this is an exothermic reaction so heat may be considered a product

14.55 (a) right
(b) left
(c) left
(d) right
(e) no effect
(f) left

14.57 $K_c = \frac{[CH_3OH]}{[CO][H_2]^2} = \frac{(0.00261)}{(0.105)(0.250)^2} = 0.398$

14.59 $2HBr(g) \rightleftharpoons H_2(g) + Br_2(g)$

	[HBr]	[H_2]	[Br_2]
I	0.500	–	–
C	$-2x$	+x	$+x$
E	$0.500 - 2x$	$+x$	$+x$

The problem tells us that $[Br_2] = 0.0955\ M = x$ at equilibrium. Using the ICE table as a guide we see that the equilibrium concentrations are; $[H_2] = [Br_2] = 0.0955\ M$ and $[HBr] = 0.500-2(0.0955) = 0.309\ M$.

$K_c = \frac{[H_2][Br_2]}{[HBr]^2} = \frac{(0.0955)(0.0955)}{(0.309)^2} = 0.0955$

14.61 According to the problem, the concentration of NO_2 increases in the course of this reaction. This means our ICE table will look like the following:

$NO_2(g) + NO(g) \rightleftharpoons N_2O(g) + O_2(g)$

	[NO_2]	[NO]	[N_2O]	[O_2]
I	0.0560	0.294	0.184	0.377
C	$+x$	$+x$	$-x$	$-x$
E	$0.0560 + x$	$0.294 + x$	$0.184 - x$	$0.377 - x$

The problem tell us that $[NO_2] = 0.118\ M = 0.0560 + x$ at equilibrium. Solving we get; $x = 0.062\ M$. Using the ICE table as a guide we see that the equilibrium concentrations are; $[NO] = 0.356\ M$, $[N_2O] = 0.122\ M$ and $[O_2] = 0.315\ M$.

$$K_c = \frac{[N_2O][O_2]}{[NO_2][NO]} = \frac{(0.122)(0.315)}{(0.118)(0.356)} = 0.915$$

14.63 The mass action expression for this equilibrium is:

$$K_c = \frac{[PCl_5]}{[PCl_3][PCl_2]} = 0.18$$

(a) The value for the reaction quotient for this system is:

$$Q = \frac{(0.00500)}{(0.0420)(0.0240)} = 4.96$$

This is not the value of the equilibrium constant, and we conclude that the system is not at equilibrium.

(b) Since the value of the reaction quotient for this system is larger than that of the equilibrium constant, the system must shift to the left to reach equilibrium.

14.65 $$K_c = \frac{[CH_3OH]}{[CO][H_2]^2} = \frac{[CH_3OH]}{(0.180)(0.220)^2} = 0.500$$

$[CH_3OH] = 4.36 \times 10^{-3}$ *M*.

14.67 $2BrCl \rightleftharpoons Br_2 + Cl_2$

	[BrCl]	**[Br$_2$]**	**[Cl$_2$]**
I	0.050	–	–
C	$-2x$	$+x$	$+x$
E	$0.050 - 2x$	$+x$	$+x$

Substituting the above values for equilibrium concentrations into the mass action expression gives:

$$K_c = \frac{[Br_2][Cl_2]}{[BrCl]^2} = \frac{(x)(x)}{(0.050-2x)^2} = 0.145$$

Take the square root of both sides to get

$$K_c = \frac{x}{0.050-2x} = 0.381$$

Solving for x gives: $x = 0.012$ $M = [Br_2] = [Cl_2]$

14.69 The initial concentrations are each 0.240 mol/2.00 L = 0.120 *M*.

	[SO$_3$]	**[NO]**	**[NO$_2$]**	**[SO$_2$]**
I	0.120	0.120	–	–
C	$-x$	$-x$	$+x$	$+x$
E	$0.120 - x$	$0.120 - x$	$+x$	$+x$

Substituting the above values for equilibrium concentrations into the mass action expression gives:

$$K_c = \frac{[NO_2][SO_2]}{[SO_3][NO]} = \frac{(x)(x)}{(0.120 - x)(0.120 - x)} = 0.500$$

Taking the square root of both sides of this equation gives: $0.707 = \frac{x}{(0.120 - x)}$

Solving for x we have: $1.707(x) = 0.0848$
$x = 0.0497$ mol/L = $[NO_2] = [SO_2]$
$[NO] = [SO_3] = 0.120 - x = 0.0703$ mol/L

14.71 The initial concentrations are all 1.00 mol/100 L = 0.0100 M. Since the initial concentrations are all the same, the reaction quotient is equal to 1.0, and we conclude that the system must shift to the left to reach equilibrium since $Q > K_c$.

	[CO]	$[H_2O]$	$[CO_2]$	$[H_2]$
I	0.0100	0.0100	0.0100	0.0100
C	$+x$	$+x$	$-x$	$-x$
E	$0.0100 + x$	$0.0100 + x$	$0.0100 - x$	$0.0100 - x$

Substituting the above values for equilibrium concentrations into the mass action expression gives:

$$K_c = \frac{[CO_2][H_2]}{[CO][H_2O]} = \frac{(0.0100 - x)(0.0100 - x)}{(0.0100 + x)(0.0100 + x)} = 0.400$$

We take the square root of both sides of the above equation:

$$\frac{(0.0100 - x)}{(0.0100 + x)} = 0.632$$

and $(0.632)(0.0100 + x) = 0.0100 - x$
$(1.632)x = 3.68 \times 10^{-3}$
$x = 2.25 \times 10^{-3}$ mol/L
The equilibrium concentrations are then:
$[H_2] = [CO_2] = [0.0100 - (2.25 \times 10^{-3})] = 7.7 \times 10^{-3}\ M$
$[CO] = [H_2O] = [0.0100 + (2.25 \times 10^{-3})] = 0.0123\ M$

14.73

	[HCl]	$[H_2]$	$[Cl_2]$
I	0.0500	–	–
C	$-2x$	$+x$	$+x$
E	$0.0500 - 2x$	$+x$	$+x$

$$K_c = \frac{[H_2][Cl_2]}{[HCl]^2} = \frac{(x)(x)}{(0.0500 - 2x)^2} = 3.2 \times 10^{-34}$$

Because K_c is so exceedingly small, we can make the simplifying assumption that x is also small enough to make $(0.0500 - 2x) \approx 0.0500$. Thus we have: $3.2 \times 10^{-34} = (x)^2/(0.0500)^2$
Taking the square root of both sides, and solving for the value of x gives:
$x = 8.9 \times 10^{-19}\ M = [H_2] = [Cl_2]$
$[HCl] = (0.0500 - x) \approx 0.0500$ mol/L

14.75 $K_c = \frac{[CO]^2[O_2]}{[CO_2]^2} = 6.4 \times 10^{-7}$

	$[CO_2]$	**[CO]**	**$[O_2]$**
I	1.0×10^{-2}	–	–
C	$-2x$	$+2x$	$+x$
E	$1.0 \times 10^{-2} - 2x$	$+ 2x$	$+ x$

$$K_c = \frac{[2x]^2[x]}{[1.0 \times 10^{-2} - 2x]^2} = 6.4 \times 10^{-7}$$

Assume $x << 1.0 \times 10^{-2}$.

$\frac{4x^3}{(1.0 \times 10^{-2})^2} = 6.4 \times 10^{-7}$ $\qquad x = 2.5 \times 10^{-4}$

$[CO] = 2x = 5.0 \times 10^{-4}\ M$

14.77 We first approach the problem in the normal fashion with an initial concentration of $PCl_5 = 0.013\ M$.

	$[PCl_3]$	**$[Cl_2]$**	**$[PCl_5]$**
I	–	–	0.013
C	$+x$	$+x$	$-x$
E	$+x$	$+x$	$0.013 - x$

Substituting the above values for equilibrium concentrations into the mass action expression gives:

$$K_c = \frac{[PCl_5]}{[PCl_3][Cl_2]} = \frac{(0.013 - x)}{(x)(x)} = 0.18$$

rearranging $0.18x^2 + x - 0.013 = 0$

We next attempt to use the quadratic equation to solve for the value of x, setting $a = 0.18$; $b = 1$; $c = -0.013$.

However, we find that unless we carry one more significant figure than is allowed, the quadratic formula for this problem gives us a concentration of zero for PCl_5. A better solution is obtained by "allowing" the initial equilibrium to shift **completely** to the left, giving us a new initial situation from which to work:

	$[PCl_3]$	**$[Cl_2]$**	**$[PCl_5]$**
I	0.013	0.013	–
C	$-x$	$-x$	$+x$
E	$0.013 - x$	$0.013 - x$	$+ x$

Substituting the above values for equilibrium concentrations into the mass action expression gives:

$$K_c = \frac{[PCl_5]}{[PCl_3][Cl_2]} = \frac{(+x)}{(0.013-x)(0.013-x)} = 0.18$$

Now, we may assume that $x << 0.013$. The equation is simplified and we solve for x
$x = [PCl_5] = 3.0 \times 10^{-5}\ M$
$[PCl_3] = 1.297 \times 10^{-2}\ M = 1.3 \times 10^{-2}\ M$

14.79

	$[SO_3]$	[NO]	$[NO_2]$	$[SO_2]$
I	0.0500	0.100	–	–
C	$-x$	$-x$	$+x$	$+x$
E	$0.0500 - x$	$0.100 - x$	$+x$	$+x$

$$K_c = \frac{[NO_2][SO_2]}{[SO_3][NO]} = \frac{(x)(x)}{(0.0500-x)(0.100-x)} = 0.500$$

Since the equilibrium constant is not much larger than either of the values 0.0500 or 0.100, we cannot neglect the size of x in the above expression. A simplifying assumption is not therefore possible, and we must solve for the value of x using the quadratic equation. Multiplying out the above denominator, collecting like terms, and putting the result into the standard quadratic form gives:
$0.500x^2 + (7.50 \times 10^{-2})x - (2.50 \times 10^{-3}) = 0$

$$x = \frac{-7.50 \times 10^{-2} \pm \sqrt{(7.50 \times 10^{-2})^2 - 4(0.500)(-2.50 \times 10^{-3})}}{2(0.500)} = 0.0281\ M$$

using the (+) root. So, $[NO_2] = [SO_2] = 0.0281\ M$
$[SO_3] = 0.219\ M$ and $[NO] = 0.0719\ M$

14.81 $K_c = \dfrac{[CO][H_2O]}{[HCHO_2]} = 4.3 \times 10^5$

Since K_c is large, start by assuming all of the $HCHO_2$ decomposes to give CO and H_2O

	$[HCHO_2]$	[CO]	$[H_2O]$
I	–	0.200	0.200
C	$+x$	$-x$	$-x$
E	$+x$	$0.200 - x$	$0.200 - x$

$$K_c = \frac{[0.200][0.200]}{[x]} = 4.3 \times 10^5$$

$x = 9.3 \times 10^{-8}$
so, at equilibrium
$[CO] = [H_2O] = 0.200 - x = 0.200\ M$

Chapter Fifteen
Acids and Bases, A Molecular Look

Practice Exercises

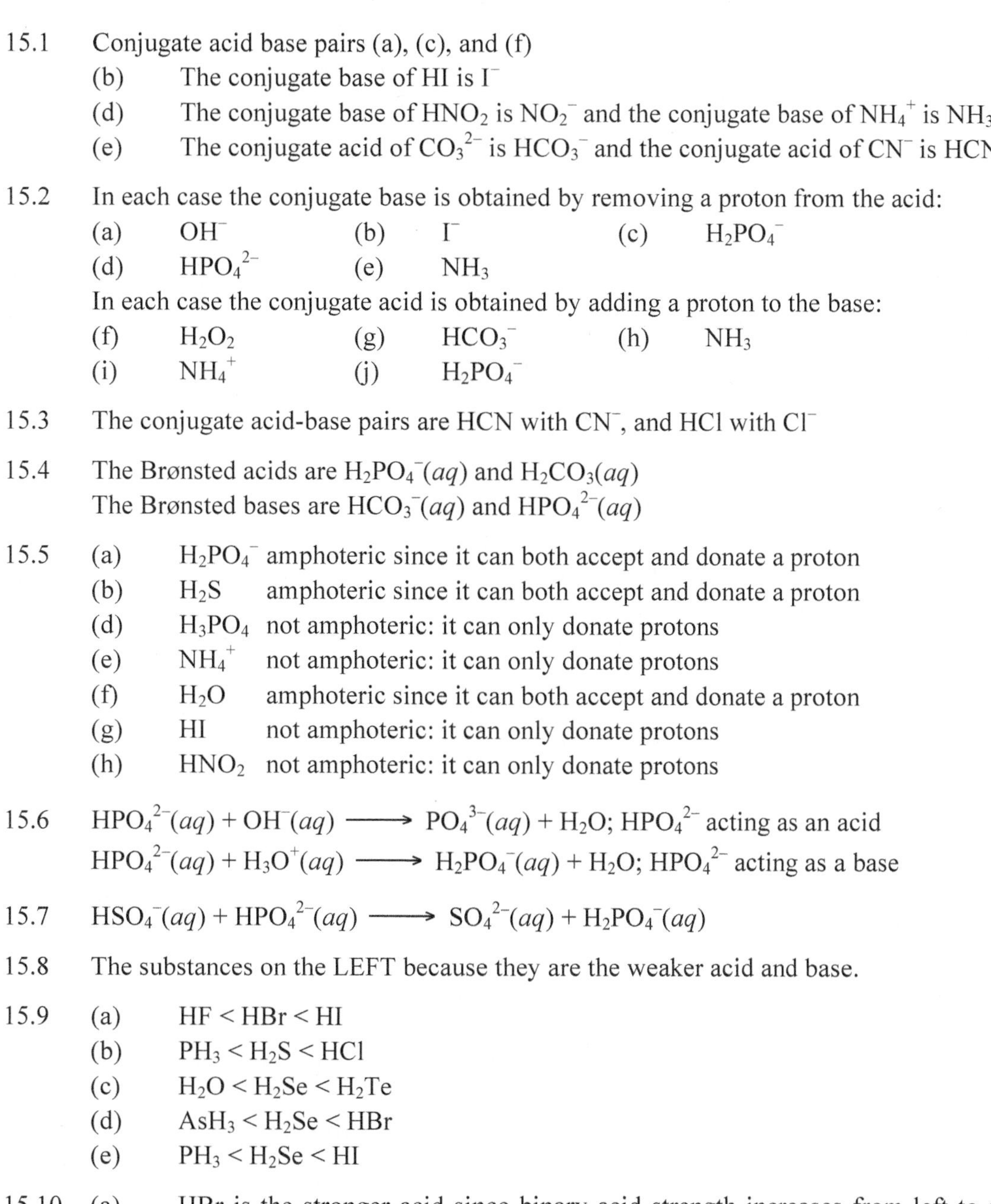

15.1 Conjugate acid base pairs (a), (c), and (f)
(b) The conjugate base of HI is I^-
(d) The conjugate base of HNO_2 is NO_2^- and the conjugate base of NH_4^+ is NH_3
(e) The conjugate acid of CO_3^{2-} is HCO_3^- and the conjugate acid of CN^- is HCN

15.2 In each case the conjugate base is obtained by removing a proton from the acid:
(a) OH^- (b) I^- (c) $H_2PO_4^-$
(d) HPO_4^{2-} (e) NH_3
In each case the conjugate acid is obtained by adding a proton to the base:
(f) H_2O_2 (g) HCO_3^- (h) NH_3
(i) NH_4^+ (j) $H_2PO_4^-$

15.3 The conjugate acid-base pairs are HCN with CN^-, and HCl with Cl^-

15.4 The Brønsted acids are $H_2PO_4^-(aq)$ and $H_2CO_3(aq)$
The Brønsted bases are $HCO_3^-(aq)$ and $HPO_4^{2-}(aq)$

15.5 (a) $H_2PO_4^-$ amphoteric since it can both accept and donate a proton
(b) H_2S amphoteric since it can both accept and donate a proton
(d) H_3PO_4 not amphoteric: it can only donate protons
(e) NH_4^+ not amphoteric: it can only donate protons
(f) H_2O amphoteric since it can both accept and donate a proton
(g) HI not amphoteric: it can only donate protons
(h) HNO_2 not amphoteric: it can only donate protons

15.6 $HPO_4^{2-}(aq) + OH^-(aq) \longrightarrow PO_4^{3-}(aq) + H_2O$; HPO_4^{2-} acting as an acid
$HPO_4^{2-}(aq) + H_3O^+(aq) \longrightarrow H_2PO_4^-(aq) + H_2O$; HPO_4^{2-} acting as a base

15.7 $HSO_4^-(aq) + HPO_4^{2-}(aq) \longrightarrow SO_4^{2-}(aq) + H_2PO_4^-(aq)$

15.8 The substances on the LEFT because they are the weaker acid and base.

15.9 (a) $HF < HBr < HI$
(b) $PH_3 < H_2S < HCl$
(c) $H_2O < H_2Se < H_2Te$
(d) $AsH_3 < H_2Se < HBr$
(e) $PH_3 < H_2Se < HI$

15.10 (a) HBr is the stronger acid since binary acid strength increases from left to right within a period.
(b) H_2Te is the stronger acid since binary acid strength increases from top to bottom within a group.
(c) H_2S since acid strength increases from top to bottom within a group.

15.11 (a) $HClO_3$ is the stronger acid because oxoacids strengths increase with the more electronegative central atom.

(b) H_2SO_4 is the stronger acid because oxoacids strengths increase with the more electronegative central atom.

15.12 (a) H_3AsO_4 is the weaker acid because oxoacid strength decreases as the electronegativity of the central atom decreases.

(b) H_2TeO_4 is the weaker acid because oxoacid strength decreases as the electronegativity of the central atom decreases.

15.13 (a) HIO_4 is the stronger acid because oxoacid strength increases as the number of oxygen atoms around the central atom increases.

(b) H_2TeO_4 is the stronger acid because oxoacid strength increases as the number of oxygen atoms around the central atom increases.

(c) H_3AsO_4 is the stronger acid because oxoacid strength increases as the number of oxygen atoms around the central atom increases.

15.14 (a) H_2SO_4 is the weaker acid because oxoacid strength decreases as the electronegativity of the central atom decreases.

(b) H_3AsO_4 is the weaker acid because oxoacid strength decreases as the electronegativity of the central atom decreases.

15.15 The acid strength decreases as follows:
$FCH_2COOH > ClCH_2COOH > BrCH_2COOH$

15.16 The acid strength increases as
$CH_3CO_2H < CH_2FCO_2H < CHF_2CO_2H < CF_3CO_2H$

15.17 (a) NH_3 is the Lewis base since it has an unshared pair of electrons.
H^+ is the Lewis acid since it can accept a pair of electrons

(b) Na_2O is the Lewis base (O^{2-} would also be a correct answer.) Na_2O cannot exist in water.
SeO_3 is the Lewis acid

(c) Ag^+ is the Lewis acid
NH_3 is the Lewis base

15.18 (a) Fluoride ions have a filled octet of electrons and are likely to behave as Lewis bases, i.e., electron pair donors.

(b) $BeCl_2$ is a likely Lewis acid since it has an incomplete shell. The Be atom has only two valence electrons and it can easily accept a pair of electrons.

(c) It could reasonably be considered a potential Lewis base since it contains three oxygens, each with lone pairs and partial negative charges. However, it is more effective as a Lewis acid, since the central sulfur bears a significant positive charge.

Review Problems

15.49 (a) HF (b) $C_5H_5NH^+$ (c) H_2CrO_4 (d) HCN

15.51 (a) NH_2O^- (b) CN^- (c) NO_2^- (d) PO_4^{3-}

15.53 (a)

conjugate pair (acid HNO_3 / base NO_3^-)

$$HNO_3 + N_2H_4 \longrightarrow NO_3^- + N_2H_5^+$$

acid base base acid

conjugate pair (base N_2H_4 / acid $N_2H_5^+$)

(b)

conjugate pair (base NH_3 / acid NH_4^+)

$$NH_3 + N_2H_5^+ \rightleftharpoons NH_4^+ + N_2H_4$$

base acid acid base

conjugate pair (acid $N_2H_5^+$ / base N_2H_4)

(c)

conjugate pair (acid $H_2PO_4^-$ / base HPO_4^{2-})

$$H_2PO_4^- + CO_3^{2-} \rightleftharpoons HPO_4^{2-} + HCO_3^-$$

acid base base acid

conjugate pair (base CO_3^{2-} / acid HCO_3^-)

(d)

conjugate pair (acid HIO_3 / base IO_3^-)

$$HIO_3 + HC_2O_4^- \longrightarrow IO_3^- + H_2C_2O_4$$

acid base base acid

conjugate pair (base $HC_2O_4^-$ / acid $H_2C_2O_4$)

15.55 (a) HBr, HBr bond is weaker
(b) HF, more electronegative F polarizes and weakens the bond
(c) HBr, larger Br forms a weaker bond with H

15.57 (a) $HClO_2$, because it has more oxygen atoms
(b) H_2SeO_4, because it has more lone oxygen atoms

15.59 (a) $HClO_3$, because Cl is more electronegative
(b) $HClO_3$, because the charge is more evenly distributed
(c) $HBrO_4$, because the negative charge is more evenly distributed

15.61

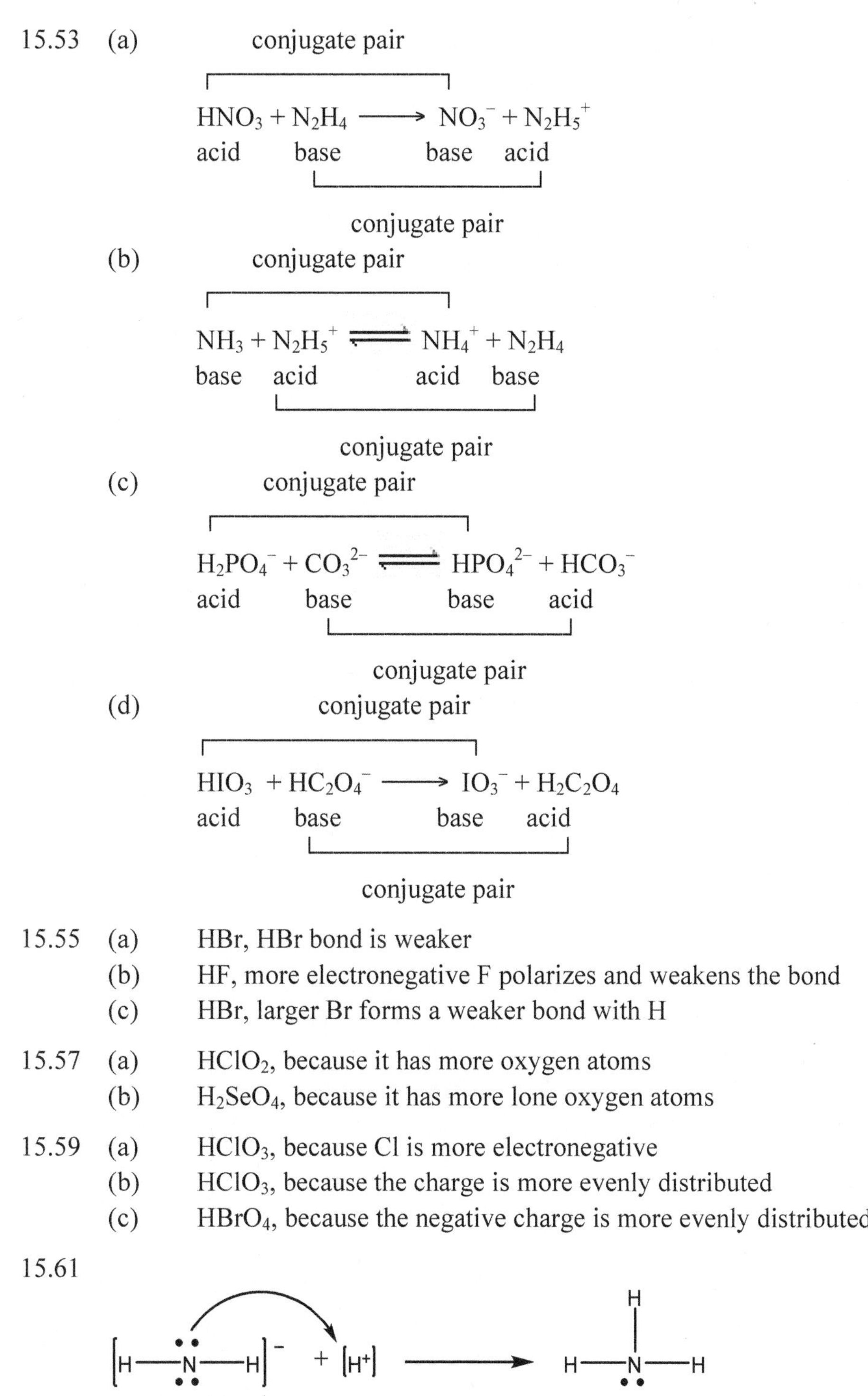

Lewis acid: H^+
Lewis base: NH_2^-

15.63

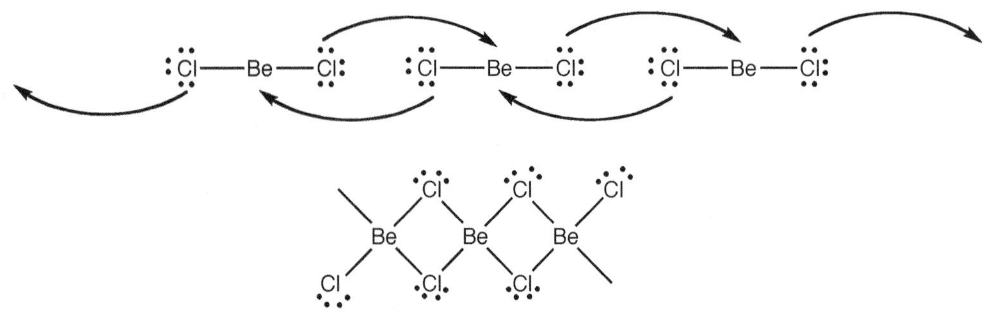

15.65 Lewis base   Lewis acid

15.67

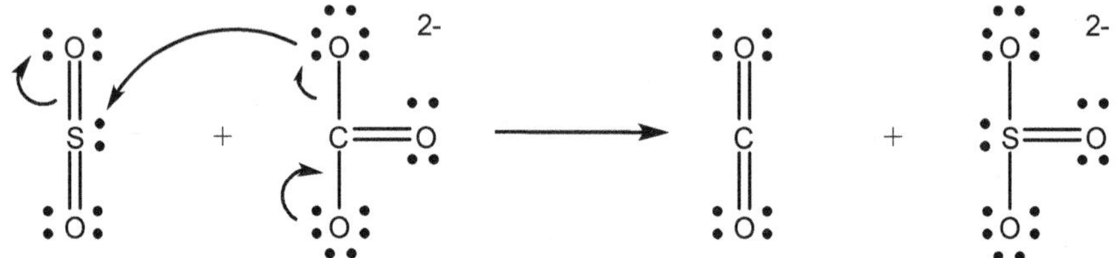

Lewis bases: an O in CO_3^{2-} and SO_3^{2-} Lewis acids: SO_2 and CO_2

15.69 $Cr(H_2O)_6^{3+}(aq) + H_2O \longrightarrow Cr(H_2O)_5OH^{2+}(aq) + H_3O^+(aq)$

15.71 $Mg(OH)_2$ is an ionic species consisting of Mg^{2+} ions and OH^- ions. When the compound dissociates in water it produces OH^- ions so the compound is a basic species.

$Si(OH)_4$ is a molecular compound with the OH groups attached to the silicon atom. This bonding weakens the OH bond so that a hydrogen ion will dissociate when the compound is dissolved in water. Thus, $Si(OH)_4$ is an acidic species.

Chapter Sixteen
Acid-Base Equilibria in Aqueous Solutions

Practice Exercises

16.1 $[H_3O^+] = 12\ M$
$10 \times 10^{-14} = [H_3O^+][OH^-] = (12\ M)[OH^-]$
$[OH^-] = 8.3 \times 10^{-16}\ M$

16.2 $1.00 \times 10^{-14} = [H_3O^+][OH^-]$

$$[H_3O^+] = \frac{1.00 \times 10^{-14}}{7.8 \times 10^{-6}\ M} = 1.3 \times 10^{-9}\ M.$$

The solution is basic.

16.3 $pOH = 14 - pH = 14 - 3.75 = 10.25$
$[H_3O^+] = 10^{-3.75} = 1.8 \times 10^{-4}\ M$
$[OH^-] = 10^{-10.25} = 5.6 \times 10^{-11}\ M$

16.4 $pH = -\log[H_3O^+] = -\log[3.2 \times 10^{-5}] = 4.49$
$pOH = 14.00 - pH = 14.00 - 4.49 = 9.51$
The solution is acidic.

16.5 In general, we have the following relationships between pH, $[H^+]$, and $[OH^-]$:
$[H_3O^+] = 10^{-pH}$
$[H_3O^+][OH^-] = 1.00 \times 10^{-14}$

(a) $[H_3O^+] = 10^{-2.90} = 1.3 \times 10^{-3}\ M$

$$[OH^-] = \frac{1.00 \times 10^{-14}}{1.3 \times 10^{-3}\ M} = 7.7 \times 10^{-12}\ M$$

The solution is acidic.

(b) $[H_3O^+] = 10^{-3.85} = 1.4 \times 10^{-4}\ M$

$$[OH^-] = \frac{1.00 \times 10^{-14}}{1.4 \times 10^{-4}\ M} = 7.1 \times 10^{-11}\ M$$

The solution is acidic.

(c) $[H_3O^+] = 10^{-10.81} = 1.54 \times 10^{-11}\ M = 1.5 \times 10^{-11} M$

$$[OH^-] = \frac{1.00 \times 10^{-14}}{1.5 \times 10^{-11}\ M} = 6.46 \times 10^{-4}\ M = 6.5 \times 10^{-4} M$$

The solution is basic.

(d) $[H_3O^+] = 10^{-4.11} = 7.8 \times 10^{-5}\ M$

$$[OH^-] = \frac{1.00 \times 10^{-14}}{7.8 \times 10^{-5}\ M} = 1.3 \times 10^{-10}\ M$$

The solution is acidic.

(e) $[H_3O^+] = 10^{-11.61} = 2.45 \times 10^{-12}\ M = 2.5 \times 10^{-12}\ M$

$$[OH^-] = \frac{1.00 \times 10^{-14}}{2.5 \times 10^{-12}\ M} = 4.1 \times 10^{-3}\ M$$

The solution is basic.

16.6 $[H_3O^+] = 0.0050\ M$
$pH = -\log[H^+] = -\log[0.0050] = 2.30$
$pOH = 14.0 - pH = 14.00 - 2.30 = 11.70$

16.7 First determine the number of moles of *K*OH and then the molarity of the solution

$$\text{mol } KOH = (1.20 \text{ g KOH})\left(\frac{1 \text{ mol KOH}}{56.11 \text{ g KOH}}\right) = 0.0214 \text{ mol KOH}$$

$$\text{molarity} = \frac{0.0214 \text{ mol KOH}}{0.250 \text{ L solution}} = 0.0855\ M\ KOH$$

$pOH = -\log 0.0855\ M = 1.07$
$pH = 14 - 1.068 = 12.93$
$[H_3O^+] = 1.2 \times 10^{-13}\ M$

16.8 (a) $HC_2H_3O_2 + H_2O \rightleftharpoons H_3O^+ + C_2H_3O_2^-$

$$K_a = \frac{[H_3O^+][C_2H_3O_2^-]}{[HC_2H_3O_2]}$$

(b) $(CH_3)_3NH^+ + H_2O \rightleftharpoons H_3O^+ + (CH_3)_3N$

$$K_a = \frac{[H_3O^+][(CH_3)_3N]}{[(CH_3)_3NH^+]}$$

(c) $H_3PO_4 + H_2O \rightleftharpoons H_3O^+ + H_2PO_4^-$

$$K_a = \frac{[H_3O^+][H_2PO_4^-]}{[H_3PO_4]}$$

16.9 (a) $HCHO_2 + H_2O \rightleftharpoons H_3O^+ + CHO_2^-$

$$K_a = \frac{[H_3O^+][CHO_2^-]}{[HCHO_2]}$$

(b) $(CH_3)_2NH_2^+ + H_2O \rightleftharpoons H_3O^+ + (CH_3)_2NH$

$$K_a = \frac{[H_3O^+][(CH_3)_2NH]}{[(CH_3)_2NH_2^+]}$$

(c) $H_2PO_4^- + H_2O \rightleftharpoons H_3O^+ + HPO_4^{2-}$

$$K_a = \frac{[H_3O^+][HPO_4^{2-}]}{[H_2PO_4^-]}$$

16.10 The smaller the value of pK_a, the stronger the acid. The acids stronger than acetic acid and weaker than formic acid from Table 16.2 are barbituric acid and hydrazoic acid.

16.11 The acid with the smaller pK_a (HA) is the stronger acid.
Since pK_a = – log K_a, $K_a = 10^{-pKa}$

For HA: $K_a = 10^{-3.16} = 6.9 \times 10^{-4}$

For HB: $K_a = 10^{-4.14} = 7.2 \times 10^{-5}$

16.12 (a) $(CH_3)_3N + H_2O \rightleftharpoons (CH_3)_3NH^+ + OH^-$

$$K_b = \frac{[(CH_3)_3NH^+][OH^-]}{[(CH_3)_3N]}$$

(b) $SO_3^{2-} + H_2O \rightleftharpoons HSO_3^- + OH^-$

$$K_b = \frac{[HSO_3^-][OH^-]}{[SO_3^{2-}]}$$

(c) $NH_2OH + H_2O \rightleftharpoons NH_3OH^+ + OH^-$

$$K_b = \frac{[NH_3OH^+][OH^-]}{[NH_2OH]}$$

16.13 (a) $HS^- + H_2O \rightleftharpoons H_2S + OH^-$

$$K_b = \frac{[H_2S][OH^-]}{[HS^-]}$$

(b) $H_2PO_4^- + H_2O \rightleftharpoons H_3PO_4 + OH^-$

$$K_b = \frac{[H_3PO_4][OH^-]}{[H_2PO_4^-]}$$

(c) $HPO_4^{2-} + H_2O \rightleftharpoons H_2PO_4^- + OH^-$

$$K_b = \frac{[H_2PO_4^-][OH^-]}{[HPO_4^{2-}]}$$

(d) $HCO_3^- + H_2O \rightleftharpoons H_2CO_3 + OH^-$

$$K_b = \frac{[H_2CO_3][OH^-]}{[HCO_3^-]}$$

(e) $HSO_3^- + H_2O \rightleftharpoons H_2SO_3 + OH^-$

$$K_b = \frac{[H_2SO_3][OH^-]}{[HSO_3^-]}$$

16.14 For conjugate acid base pairs, $K_a \times K_b = K_w$:

$$K_a = \frac{K_w}{K_b} = \frac{1.0 \times 10^{-14}}{4.5 \times 10^{-4}} = 2.2 \times 10^{-11}$$

16.15 For conjugate acid base pairs, $K_a \times K_b = K_w$:

$$K_b = \frac{K_w}{K_a} = \frac{1.0 \times 10^{-14}}{1.8 \times 10^{-4}} = 5.6 \times 10^{-11}$$

16.16 $HSal + H_2O \rightleftharpoons H_3O^+ + Sal^-$

$$K_a = \frac{[H_3O^+][Sal^-]}{[HSal]}$$

$[H_3O^+] = 10^{-pH} = 10^{-1.83} = 0.0148\ M$

	[HSal]	$[H_3O^+]$	$[Sal^-]$
I	0.200	–	–
C	$-x$	$+x$	$+x$
E	$0.200 - x$	$+x$	$+x$

We know the $[H^+]$ is 0.0148 M

$x = 0.0148$

[HSal] = 0.200 – 0.0148 = 0.185 M

$[Sal^-] = 0.0148\ M$

$$K_a = \frac{[H_3O^+][Sal^-]}{[HSal]} = \frac{(0.0148)(0.0148)}{(0.185)} = 1.2 \times 10^{-3}$$

$pK_a = -\log(K_a) = -\log(1.18 \times 10^{-3}) = 2.93$

16.17 $HBu + H_2O \rightleftharpoons H_3O^+ + Bu^-$

$$K_a = \frac{[H_3O^+][Bu^-]}{[HBu]}$$

	[HBu]	$[H_3O^+]$	$[Bu^-]$
I	0.01000	–	–
C	$-x$	$+x$	$+x$
E	$0.01000 - x$	$+x$	$+x$

We know that the acid is 3.8% ionized so $x = 0.01000\ M \times 0.038 = 0.00038\ M$. Therefore, our equilibrium concentrations are

$[H^+] = [Bu^-] = 0.00038\ M$, and

$[HBu] = 0.01000\ M - 0.00038\ M = 0.00962\ M$.

Substituting these values into the mass action expression gives:

$$K_a = \frac{(0.00038)(0.00038)}{0.00962} = 1.5 \times 10^{-5}$$

$pK_a = -\log(K_a) = -\log(1.5 \times 10^{-5}) = 4.82$

16.18 We will use the symbol Mor and $HMor^+$ for the base and its conjugate acid respectively:

$Mor + H_2O \rightleftharpoons HMor^+ + OH^-$

$$K_b = \frac{[HMor^+][OH^-]}{[Mor]}$$

	[Mor]	$[HMor^+]$	$[OH^-]$
I	0.010	–	–
C	$-x$	$+x$	$+x$
E	$0.010 - x$	$+x$	$+x$

At equilibrium, the pOH = 3.90.
The $[OH^-] = 10^{-pOH} = 10^{-3.90} = 1.26 \times 10^{-4}\ M = x$.

Substituting these values into the mass action expression gives:

$$K_b = \frac{(1.26 \times 10^{-4})(1.26 \times 10^{-4})}{0.010 - (1.26 \times 10^{-4})} = 1.6 \times 10^{-6}$$

$pK_b = -\log(K_b) = -\log(1.61 \times 10^{-6}) = 5.79$

16.19 $HC_3H_5O_2 + H_2O \rightleftharpoons H_3O^+ + C_3H_5O_2^-$

$$K_a = \frac{[C_3H_5O_2^-][H_3O^+]}{[HC_3H_5O_2]} = 1.3 \times 10^{-5}$$

	$HC_3H_5O_2$	$[H_3O^+]$	$C_3H_5O_2^-$
I	0.10	–	–
C	$-x$	$+x$	$+x$
E	$0.10 - x$	$+x$	$+x$

Assume that $x << 0.10$ and substitute the equilibrium values into the mass action expression to get:

$$K_a = \frac{[C_3H_5O_2^-][H_3O^+]}{[HC_3H_5O_2]} = 1.3 \times 10^{-5}$$

$$K_a = \frac{(x)(x)}{(0.10)} = 1.3 \times 10^{-5}$$

$x = 1.1 \times 10^{-3}$
$[H_3O^+] = x = 1.1 \times 10^{-3}\ M$
$pH = -\log[H_3O^+] = -\log(1.1 \times 10^{-3}) = 2.96$

16.20 $HC_6H_5CO_2 + H_2O \rightleftharpoons H_3O^+ + C_6H_5CO_2^-$

$$K_a = \frac{[C_6H_5CO_2^-][H_3O^+]}{[HC_6H_5CO_2]} = 6.3 \times 10^{-5}$$

	$HC_6H_5CO_2$	$[H_3O^+]$	$C_6H_5CO_2^-$
I	0.023	–	–
C	$-x$	$+x$	$+x$
E	$0.023 - x$	$+x$	$+x$

Assume that x << 0.023 and substitute the equilibrium values into the mass action expression to get:

$$K_a = \frac{[C_6H_5CO_2^-][H_3O^+]}{[HC_6H_5CO_2]} = 6.3 \times 10^{-5}$$

$$K_a = \frac{(x)(x)}{(0.023)} = 6.3 \times 10^{-5}$$

$x = 1.2 \times 10^{-3}$
$[H_3O^+] = x = 1.2 \times 10^{-3}\ M$
$pH = -\log[H_3O^+] = -\log(1.2 \times 10^{-3}) = 2.92$

16.21 $HC_6H_4NO_2 + H_2O \rightleftharpoons H_3O^+ + C_6H_4NO_2^-$

$$K_a = \frac{[H_3O^+][C_6NH_4O_2^-]}{[HC_6NH_4O_2]} = 1.4 \times 10^{-5}$$

	$[C_5NH_4COOH]$	$[H_3O^+]$	$[C_5NH_4COO^-]$
I	0.050	–	–
C	$-x$	$+x$	$+x$
E	$0.050 - x$	$+x$	$+x$

Assume that x << 0.050 and substitute the equilibrium values into the mass action expression to get:

$$K_a = \frac{[x][x]}{[0.050]} = 1.4 \times 10^{-5}$$

Solving for x we determine that
$x = 8.4 \times 10^{-4}\ M = [H^+]$
$pH = -\log[H^+] = -\log(8.4 \times 10^{-4}) = 3.08$

16.22 $C_6H_5NH_2 + H_2O \rightleftharpoons C_6H_5NH_3^+ + OH^-$

$$K_b = \frac{[C_6H_5NH_3^+][OH^-]}{[C_6H_5NH_2]} = 4.3 \times 10^{-10}$$

	$[C_6H_5NH_2]$	$[C_6H_5NH_3^+]$	$[OH^-]$
I	0.025	–	–
C	$-x$	$+x$	$+x$
E	$0.025 - x$	$+x$	$+x$

Assume that x << 0.025 and substitute the equilibrium values into the mass action expression to get:

$$K_b = \frac{(x)(x)}{(0.025)} = 4.3 \times 10^{-10}$$

$x = 3.27 \times 10^{-6}$
$[OH^-] = 3.3 \times 10^{-6}$
$pOH = -\log[OH^-] = -\log(3.3 \times 10^{-6}) = 5.48$

16.23 $C_5H_5N + H_2O \rightleftharpoons C_5H_5NH^+ + OH^-$

$$K_b = \frac{[C_5H_5NH^+][OH^-]}{[C_5H_5N]} = 1.8 \times 10^{-9}$$

	$[C_5H_5N]$	$[C_5H_5NH^+]$	$[OH^-]$
I	0.010	–	–
C	$-x$	$+x$	$+x$
E	$0.010 - x$	$+x$	$+x$

Assume that $x << 0.010$ and substitute the equilibrium values into the mass action expression to get:

$$K_b = \frac{[x][x]}{[0.010]} = 1.8 \times 10^{-9}$$

Solving for x we determine that

$x = 4.2 \times 10^{-6}\ M = [OH^-]$

$pOH = -\log[OH^-] = -\log(4.2 \times 10^{-6}) = 5.37$

$pH = 14.00 - pOH = 14.00 - 5.37 = 8.63$

16.24 We will use the notation phenol and phenolate$^-$ for the acid and its conjugate base, respectively:

phenol $\rightleftharpoons$ H^+ + phenolate$^-$

$$K_a = \frac{[H^+][\text{phenolate}^-]}{[\text{phenol}]} = 1.3 \times 10^{-10}$$

	[phenol]	$[H^+]$	[phenolate$^-$]
I	0.15	–	–
C	$-x$	$+x$	$+x$
E	$0.15 - x$	$+x$	$+x$

If we assume that $x << 0.15$, a good assumption based upon the size of K_a, we can substitute the equilibrium values in to the mass action expression to get:

$$K_a = \frac{(x)(x)}{0.15} = 1.3 \times 10^{-10}$$

Solving gives

$x = 4.42 \times 10^{-6}\ M = [H^+]$

$pH = -\log[H^+] = -\log(4.42 \times 10^{-6}) = 5.35$

16.25 Examine the ions in solution one at a time: The cation for acidity and the anion for basicity.

(a) $NaNO_2$

Na^+ is an ion of a Group 1 metal and is not acidic

NO_2^- is the conjugate base of HNO_2, a weak acid, therefore it is a weak base.

The solution should be basic.

(b) KCl

K^+ is an ion of a Group 1 metal and is not acidic

Cl^- is the conjugate base of HCl, a strong acid, therefore it is not a strong base.

The solution should be neutral.

(c) NH_4Br
NH_4^+ is the conjugate acid of NH_3, a weak base, therefore it is a weak acid
Br^- is the conjugate base of HBr, a strong acid, therefore it is not a strong base.
The solution should be acidic.

16.26 Examine the ions in solution one at a time: The cation for acidity and the anion for basicity.

(a) $NaNO_3$
Na^+ is an ion of a Group 1 metal and is not acidic
NO_3^- is the conjugate base of HNO_3, a strong acid, therefore it is not a strong base.
The solution should be neutral.

(b) KF
K^+ is an ion of a Group 1 metal and is not acidic
F^- is the conjugate base of HF, a weak acid, therefore it is a weak base.
The solution should be basic.

(c) NH_4NO_3
NH_4^+ is the conjugate acid of NH_3, a weak base, therefore it is a weak acid
NO_3^- is the conjugate base of HNO_3, a strong acid, therefore it is not a strong base.
The solution should be acidic.

16.27 The chloride ion is neutral since it is the conjugate base of a strong acid, HCl. The ion $CH_3NH_3^+$ forms from the reaction of CH_3NH_2 with water and will act as a weak acid.

$$CH_3NH_3^+ + H_2O \rightleftharpoons H_3O^+ + CH_3NH_2$$

$$K_a = \frac{K_w}{K_b} = \frac{1.0\times10^{-14}}{4.5\times10^{-4}} = 2.2 \times 10^{-11}$$

$$K_a = \frac{[CH_3NH_2][H_3O^+]}{[CH_3NH_3^+]}$$

The concentration of the salt, CH_3NH_3Cl, in water is

$$M\ CH_3NH_3C = (25.0\text{ g }CH_3NH_3Cl)\left(\frac{1\text{ mol }CH_3NH_3Cl}{67.53\text{ g }CH_3NH_3Cl}\right)\left(\frac{1}{0.500\text{ L}}\right) = 0.740\ M\ CH_3NH_3Cl$$

	$[CH_3NH_3^+]$	$[H_3O^+]$	$[CH_3NH_2]$
I	0.740	–	–
C	$-x$	$+x$	$+x$
E	$0.740 - x$	$+x$	$+x$

Assume that x is small relative to 0.740 M CH_3NH_3Cl.

$$K_a = \frac{(x)(x)}{(0.740)} = 2.2 \times 10^{-11}$$

$x = 4.03 \times 10^{-6} = [H^+]$

$pH = -\log[H^+] = -\log(4.03 \times 10^{-6}) = 5.39$

16.28 The sodium ion is neutral since it is the salt of the strong base, NaOH. The nitrite ion is basic since it is the salt of nitrous acid, HNO_2, a weak acid. The equilibrium we are interested in for this problem is:

$$NO_2^- + H_2O \rightleftharpoons HNO_2 + OH^-$$

$$K_b = \frac{[HNO_2][OH^-]}{[NO_2^-]}$$

In order to determine the value for K_b recall that $K_a \times K_b = K_w$. We can look for the value of K_a for HNO_2

$$K_b = \frac{K_w}{K_a} = \frac{1.0\times10^{-14}}{4.6\times10^{-4}} = 2.17 \times 10^{-11}$$

	$[NO_2^-]$	$[HNO_2]$	$[OH^-]$
I	0.10	–	–
C	$-x$	$+x$	$+x$
E	$0.10 - x$	$+x$	$+x$

Assume that $x << 0.10$ and substitute the equilibrium values into the mass action expression to get:

$$K_b = \frac{[x][x]}{[0.10]} = 2.17 \times 10^{-11}$$

Solving we determine that $x = 1.47 \times 10^{-6}\ M = [OH^-]$.
$pOH = -\log[OH^-] = -\log(1.2 \times 10^{-6}) = 5.83$
$pH = 14.00 - pOH = 14.00 - 5.83 = 8.17$

16.29 Upon mixing, the NH_3 will react with HBr to form NH_4Br. The initial concentration of the NH_4Br will be:

$$\text{mol } NH_3 = (500 \text{ mL solution})\left(\frac{0.20 \text{ mol } NH_3}{1000 \text{ mL solution}}\right) = 0.10 \text{ mol } NH_3$$

$$\text{mol HBr} = (500 \text{ mL solution})\left(\frac{0.20 \text{ mol HBr}}{1000 \text{ mL solution}}\right) = 0.10 \text{ mol HBr}$$

The NH_3 and HBr are in a 1:1 ratio, therefore the number of moles of NH_4Br is 0.10. The volume of the solution is 500 mL + 500 mL = 1000 mL = 1.0 L
The concentration of NH_4Br is:

$$M\ NH_4Br = \frac{0.10 \text{ mol } NH_4Br}{1.00 \text{ L solution}} = 0.10\ M\ NH_4Br$$

As previously determined, a solution of NH_4Br will be acidic since NH_4^+ is the salt of a weak base and Br^- is the salt of a strong acid. As in the previous Practice Exercise, we need to determine the value for the dissociation constant using the relationship $K_a \times K_b = K_w$ and the value of K_b for NH_3 as listed in Table 16.3.

$$K_a = \frac{K_w}{K_b} = \frac{1.0\times10^{-14}}{1.8\times10^{-5}} = 5.6 \times 10^{-10}$$

The equilibrium reaction is:
$NH_4^+ \rightleftharpoons NH_3 + H^+$

	$[NH_4^+]$	$[NH_3]$	$[H^+]$
I	0.10	–	–
C	$-x$	$+x$	$+x$
E	$0.10 - x$	$+x$	$+x$

Assume that $x << 0.10$ and substitute the equilibrium values into the mass action expression to get:

$$K_a = \frac{[x][x]}{[0.10]} = 5.6 \times 10^{-10}$$

Solving we determine that
$x = 7.5 \times 10^{-6}\ M = [H^+]$
$pH = -\log[H^+] = -\log(7.5 \times 10^{-6}) = 5.13$

16.30 NH_3 is a weak base and its conjugate, the ammonium ion, is a weak acid. The cyanide ion is the conjugate base of HCN, a weak acid. In order to determine if the solution is acidic or basic, we need to determine the relative strength of the two components. Use the relationship $K_a \times K_b = K_w$ in order to determine the dissociation constants for the cyanide ion and the ammonium ion.

$$K_a(NH_4^+) = \frac{K_w}{K_b} = \frac{1.0 \times 10^{-14}}{1.8 \times 10^{-5}} = 5.6 \times 10^{-10}$$

$$K_b(CN^-) = \frac{K_w}{K_a} = \frac{1.0 \times 10^{-14}}{4.9 \times 10^{-10}} = 2.0 \times 10^{-5}$$

Since the $K_b(CN^-)$ is larger than the $K_a(NH_4^+)$ the NH_4CN solution will be basic.

16.31 Using the same logic as the previous question:
NH_4^+ is the salt of a weak base, and is acidic
NO_2^- is the salt of a weak acid, and is basic
The relative strengths of the two components must be compared.

$$K_a(NH_4^+) = \frac{K_w}{K_b} = \frac{1.0 \times 10^{-14}}{1.8 \times 10^{-5}} = 5.6 \times 10^{-10}$$

$$K_b(NO_2^-) = \frac{K_w}{K_a} = \frac{1.0 \times 10^{-14}}{4.6 \times 10^{-4}} = 2.2 \times 10^{-11}$$

The $K_a\ (NH_4^+)$ is larger than the $K_b(NO_2^-)$, thus the solution will be acidic.

16.32 In the acetate buffer, there are $HC_2H_3O_2$ and $C_2H_3O_2^-$ present.
Upon addition of a strong acid, the concentration of $HC_2H_3O_2$ will increase:
$H^+ + C_2H_3O_2^- \longrightarrow HC_2H_3O_2$

When a strong base is added, it reacts with the acid to form more of the acetate ion; therefore the concentration of the acetic acid will decrease:
$HC_2H_3O_2 + OH^- \longrightarrow C_2H_3O_2^- + H_2O$

16.33 (a) $H^+ + NH_3 \longrightarrow NH_4^+$

(b) $OH^- + NH_4^+ \longrightarrow H_2O + NH_3$

16.34 The equation is: $C_2H_3O_2^- + H_2O \rightleftharpoons HC_2H_3O_2 + OH^-$
Start by determining K_b for acetate ion using $K_w = K_aK_b$

$$K_b = \frac{K_w}{K_b} = \frac{1.0 \times 10^{-14}}{1.8 \times 10^{-5}} = 5.56 \times 10^{-10}$$

$$K_b = \frac{[HC_2H_3O_2][OH^-]}{[C_2H_3O_2^-]}$$

	$[C_2H_3O_2^-]$	$[HC_2H_3O_2]$	$[OH^-]$
I	0.11	0.090	–
C	$-x$	$+x$	$+x$
E	$0.11 - x$	$0.090 + x$	$+x$

$$K_b = \frac{(x)(0.090+x)}{(0.11-x)} = 5.56 \times 10^{-10}$$

Assume $x << 0.090$ and solve for x

$x = [OH^-] = 6.79 \times 10^{-10}$

pOH = 9.17

pH = 14.00 – 9.17 = 4.83, the difference is due to rounding errors.

16.35 Find the concentrations of the acetic acid and the acetate ion:

$$M\ HC_2H_3O_2 = (100.0\text{ g }HC_2H_3O_2)\left(\frac{1\text{ mol }HC_2H_3O_2}{60.053\text{ g }HC_2H_3O_2}\right)\left(\frac{1}{1\text{ L solution}}\right) = 1.665\ M$$

$$M\ NaC_2H_3O_2 = (100.0\text{ g }HC_2H_3O_2)\left(\frac{1\text{ mol }HC_2H_3O_2}{82.035\text{ g }HC_2H_3O_2}\right)\left(\frac{1}{1\text{ L solution}}\right) = 1.219\ M$$

$$HC_2H_3O_2 \rightleftharpoons C_2H_3O_2^- + H^+$$

$$K_a = \frac{[C_2H_3O_2^-][H^+]}{[HC_2H_3O_2]} = 1.8 \times 10^{-5}$$

	$[HC_2H_3O_2]$	$[C_2H_3O_2^-]$	$[H^+]$
I	1.665	1.219	–
C	$-x$	$+x$	$+x$
E	$1.665 - x$	$1.219 + x$	$+x$

$$K_a = \frac{(1.219+x)(x)}{(1.665-x)} = 1.8 \times 10^{-5}$$

Assume that x is small and solve for x

$x = 2.5 \times 10^{-5}$

$pH = -\log[H^+] = -\log(2.5 \times 10^{-5}) = 4.61$

16.36 Use propanoic acid. It has a pK_a of 4.89 which is within the range of $pH = pK_a \pm 1$.

You may assume no volume change when adding the sodium salt of the acid to the solution.

$$pH = pK_a + \log\frac{[\text{salt}]}{[\text{acid}]}$$

[acid] = 0.200 M

$$5.25 = 4.89 + \log\frac{[\text{salt}]}{[0.200]}$$

[salt] = 0.458 M

$$\text{g salt} = (0.5000\text{ L})\left(\frac{0.458\text{ mol NaC}_3\text{H}_5\text{O}_2}{1\text{ L}}\right)\left(\frac{96.06\text{ g NaC}_3\text{H}_5\text{O}_2}{1\text{ mol NaC}_3\text{H}_5\text{O}_2}\right) = 22.0\text{ g NaC}_3\text{H}_5\text{O}_2$$

Propanoic acid buffer, 22.1 g $NaC_3H_5O_2$

Using the same calculations, the following buffers can also be used:
Hydrazoic acid buffer, 29.1 NaN_3,
Acetic acid buffer, 26.2 g $NaC_2H_3O_2$
Butanoic acid buffer, 29.6 g $NaC_4H_7O_2$,

16.37 Yes, formic acid and sodium formate would make a good buffer solution since pK_a = 3.74 and the desired pH is within one pH unit of this value.
Desired [H^+] = $10^{-3.90} = 1.259 \times 10^{-4}$ M

Using Equation 16.18 $\left[\text{H}^+\right] = K_a \times \dfrac{(\text{mol HCHO}_2)_{\text{initial}}}{(\text{mol CHO}_2^-)_{\text{initial}}}$

Rearranging and substituting the known values we get

$$\frac{\text{mol HCHO}_{2\,\text{initial}}}{\text{mol CHO}_2^-{}_{\text{initial}}} = \frac{\left[\text{H}^+\right]}{K_a} = \frac{1.259 \times 10^{-4}}{1.8 \times 10^{-4}} = 0.70\text{ mol}$$

mol HCHO = 0.70 × (mol CHO_2^-)

$$\text{mol CHO}_2^- = \frac{0.1\text{ mol HCHO}}{0.70} = 0.1428\text{ mol CHO}_2^-$$

(0.1428 mol $NaCHO_2$)(68.007 g/1 mol) = 9.7 g $NaCHO_2$.

16.38 The NaOH added to a buffer solution will react with the $HC_2H_3O_2$.

$NaOH + HC_2H_3O_2 \longrightarrow NaC_2H_3O_2 + H_2O$

First, calculate the amount of acid and base after the addition of the NaOH:
mol $NaC_2H_3O_2$ = 1.00 mol $NaC_2H_3O_2$ + 0.15 mol $NaC_2H_3O_2$ = 1.15 mol $NaC_2H_3O_2$
mol $HC_2H_3O_2$ = 1.00 mol $HC_2H_3O_2$ – 0.15 mol $NaC_2H_3O_2$ = 0.85 mol $HC_2H_3O_2$
Then calculate the molarity, assuming no change in the volume.

$$M\,\text{NaC}_2\text{H}_3\text{O}_2 = \frac{1.15\text{ mol NaC}_2\text{H}_3\text{O}_2}{1\text{ L solution}}$$

$$M\,\text{HC}_2\text{H}_3\text{O}_2 = \frac{0.85\text{ mol HC}_2\text{H}_3\text{O}_2}{1\text{ L solution}}$$

$$\text{pH} = \text{p}K_a + \log\frac{[\text{salt}]}{[\text{acid}]} = 4.74 + \log\frac{(1.15)}{(0.85)} = 4.87$$

The pH of the initial solution was 4.74 since the concentration of the acid equals the concentration of the salt.
The pH change:
4.74 – 4.87 = 0.13 pH units

16.39 Calculate the moles of the salt and the acid, then calculate the pH of the solution.

$$\text{mol NH}_3 = (50.0\text{ g NH}_3)\left(\frac{1\text{ mol NH}_3}{17.03\text{ g NH}_3}\right) = 2.94\text{ mol NH}_3$$

$$\text{mol NH}_4\text{Cl} = (50.0\text{ g NH}_4\text{Cl})\left(\frac{1\text{ mol NH}_4\text{Cl}}{53.49\text{ g NH}_4\text{Cl}}\right) = 0.935\text{ mol NH}_4\text{Cl}$$

$NH_4^+ \longrightarrow NH_3 + H^+$

$$K_a = \frac{K_w}{K_b} = \frac{1.0\times10^{-14}}{1.8\times10^{-5}} = 5.6 \times 10^{-10}$$

$pK_a = -\log(K_a) = -\log(5.6 \times 10^{-10}) = 9.25$

$$\text{pH} = pK_a + \log\frac{[\text{salt}]}{[\text{acid}]} = 9.25 + \log\frac{(2.94)}{(0.935)} = 9.75$$

If 5.00 g of HCl is add, the solution will become more acidic:

$NH_3 + H^+ \longrightarrow NH_4^+$

$$\text{mol HCl} = (5.00\text{ g HCl})\left(\frac{1\text{ mol HCl}}{36.46\text{ g HCl}}\right) = 0.137\text{ mol HCl}$$

The number of moles of NH_4^+ will increase by this amount and the number of moles of NH_3 will decrease by this amount.

mol NH_3 = 2.94 mol NH_3 – 0.137 mol = 2.80 mol NH_3

mol NH_4^+ = 0.935 mol + 0.137 mol = 1.07 mol NH_4^+

$$\text{pH} = pK_a + \log\frac{[\text{salt}]}{[\text{acid}]} = 9.25 + \log\frac{(2.80)}{(1.07)} = 9.67$$

16.40 $H_3PO_4 + H_2O \rightleftharpoons H_3O^+ + H_2PO_4^-$ $\qquad K_a = \frac{[H_3O^+][H_2PO_4^-]}{[H_3PO_4]}$

$H_2PO_4^- + H_2O \rightleftharpoons H_3O^+ + HPO_4^{2-}$ $\qquad K_a = \frac{[H_3O^+][HPO_4^{2-}]}{[H_2PO_4^-]}$

$HPO_4^{2-} + H_2O \rightleftharpoons H_3O^+ + PO_4^{3-}$ $\qquad K_a = \frac{[H_3O^+][PO_4^{3-}]}{[HPO_4^{2-}]}$

16.41 $[H^+]$ is determined by the first equilibrium:

$H_2C_6H_6O_6 \rightleftharpoons H^+ + HC_6H_6O_6^-$

The mass action expression is: $K_{a_1} = 9.1 \times 10^{-5} = \frac{x^2}{0.10}$

$x = [H^+] = 3.0 \times 10^{-3}\ M$

pH = –log(3.0 × 10^{-3}) = 2.52

The concentration of the anion, $[HC_6H_6O_6^-]$, is given almost entirely by the second ionization equilibrium: $HC_6H_6O_6^- \rightleftharpoons H^+ + C_6H_6O_6^{2-}$ for which the mass action expression is:

$$K_{a2} = \frac{[H^+][C_6H_6O_6^{2-}]}{[HC_6H_6O_6^-]} = 1.6 \times 10^{-12}$$

We have used the value for K_{a_2} from Table 16.4. Using the value of x from the first step above gives:

$$1.6 \times 10^{-12} = \frac{\left(3.0 \times 10^{-3}\right)\left[C_6H_6O_6^{2-}\right]}{\left(3.0 \times 10^{-3}\right)}$$

$[HC_6H_6O_6^-] = 1.6 \times 10^{-12}$

16.42 $CO_3^{2-}(aq) + H_2O \rightleftharpoons HCO_3^-(aq) + OH^-(aq)$

$$K_b = \frac{K_w}{K_{a_2}} = \frac{1.0 \times 10^{-14}}{5.6 \times 10^{-11}} = 1.8 \times 10^{-4}$$

$$K_b = 1.8 \times 10^{-4} = \frac{\left[HCO_3^-\right]\left[OH^-\right]}{\left[CO_3^{2-}\right]}$$

	$[CO_3^{2-}]$	$[HCO_3^-]$	$[OH^-]$
I	0.10	–	–
C	$-x$	$+x$	$+x$
E	$0.10-x$	$+x$	$+x$

$$K_b = 1.8 \times 10^{-4} = \frac{(x)(x)}{(0.10-x)}$$

$x = [OH^-] = 4.2 \times 10^{-3}$ M
$pOH = -\log(4.2 \times 10^{-3}) = 2.37$
$pH = 14.00 - 2.37 = 11.63$
It is too basic to be a substitute for $NaHCO_3$.

16.43 The equilibrium we are interested in for this problem is:
$SO_3^{2-}(aq) + H_2O \rightleftharpoons HSO_3^-(aq) + OH^-(aq)$

$$K_b = \frac{K_w}{K_{a_2}} = \frac{1.0 \times 10^{-14}}{6.3 \times 10^{-8}} = 1.6 \times 10^{-7}$$

$$K_b = 1.6 \times 10^{-7} = \frac{\left[HSO_3^-\right]\left[OH^-\right]}{\left[SO_3^{2-}\right]}$$

	$[SO_3^{2-}]$	$[HSO_3^-]$	$[OH^-]$
I	0.20	–	–
C	$-x$	$+x$	$+x$
E	$0.20-x$	$+x$	$+x$

Substituting these values into the mass action expression gives:

$$K_b = 1.6 \times 10^{-7} = \frac{(x)(x)}{0.20-x}$$

Assume that $x << 0.20$ and solving gives $x = 1.8 \times 10^{-4}$.
$x = [OH^-] = 1.8 \times 10^{-4}$ M
$pOH = -\log(1.8 \times 10^{-4}) = 3.75$
$pH = 14.00 - pOH = 14.00 - 3.75 = 10.25$

16.44 For weak polyprotic acids, the concentration of the polyvalent ions is equal to the value of K_{a_n} where n is the valency. By analogy, the concentration of H_2SO_3 in 0.010 M Na_2SO_3 will be equal to K_{b_2} for SO_3^{2-}, 6.7×10^{-13}.

16.45 (a) H_2O, K^+, $HC_2H_3O_2$, H^+, $C_2H_3O_2^-$, and OH^-
$[H_2O] > [K^+] > [C_2H_3O_2^-] > [OH^-] > [HC_2H_3O_2] > [H^+]$

(b) H_2O, $HC_2H_3O_2$, H^+, $C_2H_3O_2^-$, and OH^-
$[H_2O] > [HC_2H_3O_2] > [H^+] = [C_2H_3O_2^-] > [OH^-]$

(c) H_2O, K^+, $HC_2H_3O_2$, H^+, $C_2H_3O_2^-$, and OH^-
$[H_2O] > [K^+] > [OH^-] > [C_2H_3O_2^-] > [HC_2H_3O_2] > [H^+]$

(d) H_2O, K^+, $HC_2H_3O_2$, H^+, $C_2H_3O_2^-$, and OH^-
$[H_2O] > [HC_2H_3O_2] > [K^+] > [C_2H_3O_2^-] > [H^+] > [OH^-]$

16.46 $HCHO_2 + H_2O \rightleftharpoons H_3O^+ + CHO_2^-$

$$K_a = \frac{[H_3O^+][CHO_2^-]}{[HCHO_2]} = 1.8 \times 10^{-4}$$

(a)

	$[HCHO_2]$	$[H_3O^+]$	$[CHO_2^-]$
I	0.100	–	–
C	$-x$	$+x$	$+x$
E	$0.100 - x$	$+x$	$+x$

Assume $x \ll 0.100$. Solving we get $x = [H_3O^+] = 4.2 \times 10^{-3}$. The pH is 2.37.

(b) $[HCHO_2] = [CHO_2^-]$ so $[H_3O^+] = K_a = 1.8 \times 10^{-4}$ and the pH = 3.74.

(c) $$\text{mol base added} = (15.0\text{ mL})\left(\frac{0.100\text{ mol}}{1000\text{ mL}}\right) = 1.50 \times 10^{-3}\text{ mol}$$

$$\text{mol acid initially present} = (20.0\text{ mL})\left(\frac{0.10\text{ mol}}{1000\text{ mL}}\right) = 2.00 \times 10^{-3}\text{ mol}$$

$$\text{excess acid} = 2.00 \times 10^{-3} - 1.50 \times 10^{-3} = 0.50 \times 10^{-3}\text{ moles acid}$$

$$[HCHO_2] = [\text{acid}] = \frac{0.50 \times 10^{-3}\text{ moles}}{(35\text{ mL})\left(\frac{1\text{ L}}{1000\text{ mL}}\right)} = 1.43 \times 10^{-2}\ M$$

$$[CHO_2^-] = [\text{base}] = \frac{1.50 \times 10^{-3}\text{ moles}}{(35\text{ mL})\left(\frac{1\text{ L}}{1000\text{ mL}}\right)} = 4.29 \times 10^{-2}\ M$$

Substituting into the equilibrium expression and solving we get $[H_3O^+] = 6.00 \times 10^{-5}$ and the pH = 4.22.

(d) We now have a solution of formate ion with a concentration of 0.0500 M. We need K_b for formate ion: $K_b = K_w/K_a = 5.6 \times 10^{-11}$. If we set up the equilibrium problem and solve we get: $[OH^-] = 1.7 \times 10^{-6}$.
The pOH = 5.78 and the pH = 8.22.

16.47 mol base added = $(30.0 \text{ mL})\left(\frac{0.15 \text{ mol}}{1000 \text{ mL}}\right) = 4.50 \times 10^{-3} \text{ mol OH}^-$

mol acid initially present = $(50.0 \text{ mL})\left(\frac{0.20 \text{ mol}}{1000 \text{ mL}}\right) = 1.00 \times 10^{-2} \text{ mol HCHO}_2$

excess acid = $(1.00 \times 10^{-2}) - (4.50 \times 10^{-3}) = 5.50 \times 10^{-3}$ mol acid

$$[HCHO_2] = [\text{acid}] = \frac{5.50 \times 10^{-3} \text{ moles}}{(80 \text{ mL})\left(\frac{1 \text{ L}}{1000 \text{ mL}}\right)} = 6.88 \times 10^{-2}\ M$$

$$[CHO_2^-] = [\text{base}] = \frac{4.50 \times 10^{-3} \text{ moles}}{(80 \text{ mL})\left(\frac{1 \text{ L}}{1000 \text{ mL}}\right)} = 5.63 \times 10^{-2}\ M$$

$$HCHO_2 + H_2O \rightleftharpoons H_3O^+ + CHO_2^-$$

$$K_a = \frac{[H_3O^+][CHO_2^-]}{[HCHO_2]}$$

	$[HCHO_2]$	$[H_3O^+]$	$[CHO_2^-]$
I	6.88×10^{-2}	–	5.63×10^{-2}
C	$-x$	$+x$	$+x$
E	$6.88 \times 10^{-2} - x$	$+x$	$5.63 \times 10^{-2} + x$

Assume $x << 5.63 \times 10^{-2}$

$$K_a = \frac{[x][5.63 \times 10^{-2}]}{[6.88 \times 10^{-2}]} = 1.8 \times 10^{-4}$$

$x = 2.20 \times 10^{-4} = [H_3O^+]$

pH = 3.66

Review Problems

16.49 At 25 °C, $K_w = 1.0 \times 10^{-14} = [H^+] \times [OH^-]$. Let $x = [H^+]$, for each of the following:

(a) $x(0.0068) = 1.0 \times 10^{-14}$

$$[H^+] = \frac{1.0 \times 10^{-14}}{0.0068} = 1.5 \times 10^{-12}\ M$$

pH = $-\log[H^+] = -\log(1.5 \times 10^{-12}) = 11.83$

pOH = 14 – pH = 14 – 11.83 = 2.17

(b) $x(6.4 \times 10^{-5}) = 1.0 \times 10^{-14}$

$$[H^+] = \frac{1.0 \times 10^{-14}}{6.4 \times 10^{-5}} = 1.6 \times 10^{-10}\ M$$

pH = $-\log[H^+] = -\log(1.6 \times 10^{-10}) = 9.81$

pOH = 14 – pH = 14 – 9.81 = 4.19

(c) $x(1.6 \times 10^{-8}) = 1.0 \times 10^{-14}$

$$[H^+] = \frac{1.0 \times 10^{-14}}{1.6 \times 10^{-8}} = 6.3 \times 10^{-7}\ M$$

$pH = -\log[H^+] = -\log(6.3 \times 10^{-7}) = 6.20$

$pOH = 14 - pH = 14 - 6.20 = 7.80$

(d) $x(8.2 \times 10^{-12}) = 1.0 \times 10^{-14}$

$$[H^+] = \frac{1.0 \times 10^{-14}}{8.2 \times 10^{-12}} = 1.2 \times 10^{-3}\ M$$

$pH = -\log[H^+] = -\log(1.2 \times 10^{-3}) = 2.91$

$pOH = 14 - pH = 14 - 2.91 = 11.09$

16.51 $[H^+] = 10^{-pH}$ and $[OH^-] = 10^{-pOH}$

At 25 °C, pH + pOH = 14.00

(a) $[H^+] = 10^{-pH} = 10^{-8.14} = 7.2 \times 10^{-9}\ M$

$pOH = 14.00 - pH = 14.00 - 8.14 = 5.86$

$[OH^-] = 10^{-pOH} = 10^{-5.86} = 1.4 \times 10^{-6}\ M$

(b) $[H^+] = 10^{-pH} = 10^{-2.56} = 2.8 \times 10^{-3}\ M$

$pOH = 14.00 - pH = 14.00 - 2.56 = 11.44$

$[OH^-] = 10^{-pOH} = 10^{-11.44} = 3.6 \times 10^{-12}\ M$

(c) $[H^+] = 10^{-pH} = 10^{-11.25} = 5.6 \times 10^{-12}\ M$

$pOH = 14.00 - pH = 14.00 - 11.25 = 2.75$

$[OH^-] = 10^{-pOH} = 10^{-2.75} = 1.8 \times 10^{-3}\ M$

(d) $[H^+] = 10^{-pH} = 10^{-13.28} = 5.2 \times 10^{-14}\ M$

$pOH = 14.00 - pH = 14.00 - 13.28 = 0.72$

$[OH^-] = 10^{-pOH} = 10^{-0.76} = 1.9 \times 10^{-1}\ M$

(e) $[H^+] = 10^{-pH} = 10^{-6.70} = 2.0 \times 10^{-7}\ M$

$pOH = 14.00 - pH = 14.00 - 6.70 = 7.30$

$[OH^-] = 10^{-pOH} = 10^{-7.30} = 5.0 \times 10^{-8}\ M$

16.53 (a) $[OH^-] = 10^{-pOH} = 10^{-7.19} = 6.5 \times 10^{-8}\ M$

$pH = 14.00 - pOH = 14.00 - 7.19 = 6.81$

$[H^+] = 10^{-pH} = 10^{-6.81} = 1.5 \times 10^{-7}\ M$

(b) $[OH^-] = 10^{-pOH} = 10^{-1.26} = 5.5 \times 10^{-2}\ M$

$pH = 14.00 - pOH = 14.00 - 1.26 = 12.74$

$[H^+] = 10^{-pH} = 10^{-12.74} = 1.8 \times 10^{-13}\ M$

(c) $[OH^-] = 10^{-pOH} = 10^{-10.85} = 1.4 \times 10^{-11}\ M$

$pH = 14.00 - pOH = 14.00 - 10.85 = 3.15$

$[H^+] = 10^{-pH} = 10^{-3.15} = 7.1 \times 10^{-4}\ M$

(d) $[OH^-] = 10^{-pOH} = 10^{-13.15} = 7.1 \times 10^{-14}\ M$

$pH = 14.00 - pOH = 14.00 - 13.15 = 0.85$

$[H^+] = 10^{-pH} = 10^{-0.85} = 1.4 \times 10^{-1}\ M$

(e) $[OH^-] = 10^{-pOH} = 10^{-5.24} = 5.8 \times 10^{-6}\ M$

$pH = 14.00 - pOH = 14.00 - 5.24 = 8.76$

$[H^+] = 10^{-pH} = 10^{-8.76} = 1.7 \times 10^{-9}\ M$

16.55 $pH = -\log[H^+] = -\log(1.9 \times 10^{-5}) = 4.72$

16.57 $pH = 14 - pOH$
$pOH = -\log[OH^-] = -\log(6.3 \times 10^{-5}) = 4.20$
$pH = 14 - 4.20 = 9.80$

16.59 $D_2O \rightleftharpoons D^+ + OD^-$, $K_w = [D^+] \times [OD^-] = 8.9 \times 10^{-16}$
Since $[D^+] = [OD^-]$, we can rewrite the above expression to give:
$8.9 \times 10^{-16} = ([D^+])^2$
$[D^+] = 3.0 \times 10^{-8}\ M = [OD^-]$
$pD = -\log[D^+] = -\log(3.0 \times 10^{-8}) = 7.52$
$pOD = -\log[OD^-] = -\log(3.0 \times 10^{-8}) = 7.52$
neutral pD would be 7.52

16.61 $[H^+] = 10^{-pH} = 10^{-5.7} = 2 \times 10^{-6}\ M$
$pOH = 14.00 - pH = 14.00 - 5.7 = 8.3$
$[OH^-] = 10^{-pOH} = 10^{-8.3} = 5 \times 10^{-9}\ M$

16.63 HNO_3 is a strong acid so $[H^+] = [HNO_3] = 0.00065\ M$
$pH = -\log[H^+] = -\log(0.00065) = 3.19$
$pOH = 14.00 - pH = 14.00 - 3.19 = 10.81$
$[OH^-] = 10^{-pOH} = 10^{-10.81} = 1.55 \times 10^{-11}\ M$

16.65 $$M\,OH^- = \frac{\text{moles OH}^-}{\text{L solution}} = \left(\frac{6.0\text{ g NaOH}}{1.00\text{ L solution}}\right)\left(\frac{1\text{ mole NaOH}}{40.0\text{ g NaOH}}\right)\left(\frac{1\text{ mole OH}^-}{1\text{ mole NaOH}}\right) = 0.15\ M\,OH^-$$

$pOH = -\log[OH^-] = -\log(0.15) = 0.82$
$pH = 14.00 - pOH = 14.00 - 0.82 = 13.18$
$[H^+] = 10^{-pH} = 10^{-13.18} = 6.67 \times 10^{-14}\ M$

16.67 $pOH = 14.00 - pH = 14.00 - 11.60 = 2.40$
$[OH^-] = 10^{-pOH} = 10^{-2.40} = 4.0 \times 10^{-3}\ M$
$$[Ca(OH)_2] = \left(\frac{4.0 \times 10^{-3}\text{ mol OH}^-}{1\text{ L solution}}\right)\left(\frac{1\text{ mol Ca(OH)}_2}{2\text{ mol OH}^-}\right)$$
$= 2.0 \times 10^{-3}\ M\ Ca(OH)_2$
$pH = 10.60$
$pOH = 14 - 10.60 = 3.4$
$[OH^-] = 10^{-pOH} = 10^{-3.40} = 4.0 \times 10^{-4}\ M$
$$[Ca(OH)_2] = \left(\frac{4.0 \times 10^{-4}\text{ mol OH}^-}{1\text{ L solution}}\right)\left(\frac{1\text{ mol Ca(OH)}_2}{2\text{ mol OH}^-}\right) = 2.0 \times 10^{-4}\ M\ Ca(OH)_2$$

16.69 Since NaOH is a strong base, $[NaOH] = [OH^-] = 0.0020\ M\ OH^-$. (Next, we make the simplifying assumption that the amount of hydroxide ion formed from the dissociation of water is so small that we can neglect it in calculating pOH for the solution.)
$$[H^+] = \frac{1.0\times10^{-14}}{2.0\times10^{-3}} = 5.0 \times 10^{-12}\ M\ H^+$$
The only source of H^+ is the autoionization of water. Therefore, the molarity of OH^- from the ionization water is $5.0 \times 10^{-12}\ M$.

16.71 First calculate the molarity of OH^- produced from the $Ba(OH)_2$.

$$2.64\times10^{-6}\text{ g Ba(OH)}_2\left(\frac{1\text{ mol Ba(OH)}_2}{171.3\text{ g}}\right)\left(\frac{2\text{ mol OH}^-}{\text{mol Ba(OH)}_2}\right)\left(\frac{1}{1\text{ L}}\right)=3.08\times10^{-8}\,M\text{ OH}^-$$

$pOH = -\log(3.08\times10^{-8}) = 7.51$

$pH = 14 - 7.51 = 6.49$

Notice the pH calculated from the $Ba(OH)_2$ is less than 7. We would expect the solution to be basic but since the amount of base added to water is so small, the salt is not the major contributor to the pH of the solution. Thus, we cannot neglect the ionization of water in this case, since water's ionization is a significant contributor to the H^+ and OH^- concentrations in the solution. The problem is more complicated than when acid or base concentrations are larger than $10^{-7}M$. Let's analyze all of the reactions:

$Ba(OH)_2 \longrightarrow Ba^{2+} + 2OH^-$ complete ionization of a strong base

$2H_2O \rightleftharpoons H_3O^+ + OH^-$ $K_w = [H_3O^+][OH^-] = 1.00\times10^{-14}$

We know that the concentration of cations must equal the concentration of anions for the solution to be electrically neutral. Therefore,

$[H_3O^+] + 2[Ba^{2+}] = [OH^-]$ This is the charge balanced equation.

Remember, the Ba^{2+} ion's total charge is twice its concentration.

We also know that $[OH^-] = \dfrac{K_w}{[H_3O^+]}$. Thus,

$$[H_3O^+] + 2[Ba^{2+}] = \frac{K_w}{[H_3O^+]}$$

$[H_3O^+]^2 + 2[Ba^{2+}][H_3O^+] - K_w = 0$

$[Ba^{2+}] = (1/2)[OH^-] = (1/2)\,3.08\times10^{-8}\,M = 1.54\times10^{-8}\,M$

$[H_3O^+]^2 + 3.08\times10^{-8}[H_3O^+] - 1.00\times10^{-14} = 0$

Solve using the quadratic equation.

$[H_3O^+] = 8.58\times10^{-8}$

$pH = 7.07$ A basic solution, though just barely!

16.73 $pH = 5.5$

$[H^+] = 10^{-5.5} = 3.16\times10^{-6}\,M\,H^+$

16.75 At 25 °C, $K_a\times K_b = K_w$

$$K_b = K_w/K_a = \frac{1.0\times10^{-14}}{3.5\times10^{-4}} = 2.9\times10^{-11}$$

16.77 (a) The conjugate base is IO_3^-.

$pK_b = 14.00 - pK_a = 14.00 - 0.77 = 13.23$

$K_b = 10^{-pK_b} = 10^{-13.23} = 5.9\times10^{-14}$

(b) IO_3^- is a weaker base than an acetate anion, because its K_b value is smaller than that of an acetate anion.

16.79 $[H^+] = 10^{-pH} = 10^{-2.37} = 4.27\times10^{-3}\,M$

$$\text{Percentage ionization} = \frac{\text{moles ionized per liter}}{\text{moles available per liter}}\times100\%$$

$$\% \text{ ionization} = \frac{4.27 \times 10^{-3}}{1.0} \times 100\% = 0.43\%$$

16.81 pH = 3.22 $[H^+] = 10^{-3.22} = 6.03 \times 10^{-4}\ M$

$$\text{Percentage ionization} = \frac{\text{moles ionized per liter}}{\text{moles available per liter}} \times 100\%$$

$$\text{Percentage ionization} = \frac{6.03 \times 10^{-4}\ M}{0.20\ M} \times 100\% = 0.30\%$$

$HA \rightleftharpoons H^+ + A^-$

	[H*A*]	**[H⁺]**	**[*A*⁻]**
I	0.20	–	–
C	-6.03×10^{-4}	$+6.03 \times 10^{-4}$	$+6.03 \times 10^{-4}$
E	0.20	6.03×10^{-4}	6.03×10^{-4}

$$K_a = \frac{[H^+][A^-]}{[HA]} = \frac{[6.03 \times 10^{-4}][6.03 \times 10^{-4}]}{[0.20]} = 1.8 \times 10^{-6}$$

16.83 $\text{Percentage ionization} = \frac{\text{moles ionized per liter}}{\text{moles available per liter}} \times 100\%$

Let x = moles of OH^- in solution

$0.012\% = \frac{x\ M}{0.12\ M} \times 100\%$ $\qquad x = 1.4 \times 10^{-5}\ M\ OH^-$

$pOH = -\log[OH^-] = -\log[1.4 \times 10^{-5}] = 4.85$

$pH = 14 - pOH = 14 - 4.85 = 9.15$

$B + H_2O \rightleftharpoons HB^+ + OH^-$

	[*B*]	**[OH⁻]**	**[H*B*⁺]**
I	0.12	–	–
C	-1.4×10^{-5}	$+1.4 \times 10^{-5}$	$+1.4 \times 10^{-5}$
E	0.12	1.4×10^{-5}	1.4×10^{-5}

$$K_b = \frac{[OH^-][HB^+]}{[B]} = \frac{[1.4\times10^{-5}][1.4\times10^{-5}]}{[0.12]} = 1.6 \times 10^{-9}$$

16.85 $HIO_4 \rightleftharpoons H^+ + IO_4^-$

$$K_a = \frac{[H^+][IO_4^-]}{[HIO_4]}$$

	[HIO₄]	**[H⁺]**	**[IO₄⁻]**
I	0.10	–	–
C	$-x$	$+x$	$+x$
E	$0.10 - x$	$+x$	$+x$

We know that at equilibrium $[H^+] = 0.038\ M = x$.
The equilibrium concentrations of the other components of the mixture are:
$[HIO_4] = 0.10 - x = 0.062\ M$ and $[IO_4^-] = x = 0.038\ M$.
Substituting the above values for equilibrium concentrations into the mass action expression gives:

$$K_a = \frac{(0.038)(0.038)}{0.062} = 2.3 \times 10^{-2}$$

$pK_a = -\log(K_a) = -\log(2.3 \times 10^{-2}) = 1.64$

16.87 $pOH = 14.00 - pH = 14.00 - 11.87 = 2.13$
$[OH^-] = 10^{-pOH} = 10^{-2.13} = 7.41 \times 10^{-3}\ M$

$$CH_3CH_2NH_2 + H_2O \rightleftharpoons CH_3CH_2NH_3^+ + OH^-$$

$$K_b = \frac{[CH_3CH_2NH_3^+][OH^-]}{[CH_3CH_2NH_2]}$$

	$[CH_3CH_2NH_2]$	$[CH_3CH_2NH_3^+]$	$[OH^-]$
I	0.10	–	–
C	$-x$	$+x$	$+x$
E	$0.10 - x$	$+x$	$+x$

In the equilibrium analysis, the value of x is, therefore, equal to $7.41 \times 10^{-3}\ M$. Therefore, our equilibrium concentrations are $[CH_3CH_2NH_3^+] = [OH^-] = 7.41 \times 10^{-3}\ M$, and $[CH_3CH_2NH_2] = 0.10\ M - 7.41 \times 10^{-3}\ M = 0.0926\ M$.
Substituting these values into the mass action expression gives:

$$K_b = \frac{(7.41 \times 10^{-3})(7.41 \times 10^{-3})}{0.0926} = 5.9 \times 10^{-4}$$

$pK_b = -\log(K_b) = -\log(5.9 \times 10^{-4}) = 3.23$

$$\text{Percentage ionization} = \frac{\text{moles ionized per liter}}{\text{moles available per liter}} \times 100\%$$

$$\text{Percentage ionization} = \frac{7.4 \times 10^{-3}}{0.10} \times 100\% = 7.4\%$$

16.89 $HC_3H_5O_2 + H_2O \rightleftharpoons H_3O^+ + C_3H_5O_2^-$

$$K_a = \frac{[H_3O^+][C_3H_5O_2^-]}{[HC_3H_5O_2]} = 1.4 \times 10^{-4}$$

	$[HC_3H_5O_2]$	$[H_3O^+]$	$[C_3H_5O_2^-]$
I	0.15	–	–
C	$-x$	$+x$	$+x$
E	$0.150 - x$	$+x$	$+x$

Assume $x << 0.15$

$$K_a = \frac{[x][x]}{[0.150]} = 1.4 \times 10^{-4}$$

$x = 4.6 \times 10^{-3} = [H_3O^+]$
pH = 2.34
$[HC_3H_5O_2] = 0.145$
$[H^+] = 4.6 \times 10^{-3}$
$[C_3H_5O_2^-] = 4.6 \times 10^{-3}$

16.91 $K_b = 10^{-pKb} = 10^{-5.80} = 1.58 \times 10^{-6}$

$$Cod + H_2O \rightleftharpoons HCod^+ + OH^-$$

$$K_b = \frac{[HCod^+][OH^-]}{[Cod]} = 1.58 \times 10^{-6}$$

	[Cod]	[HCod⁺]	[OH⁻]
I	0.020	–	–
C	$-x$	$+x$	$+x$
E	$0.020 - x$	$+x$	$+x$

Substituting these values into the mass action expression gives:

$$K_b = \frac{(x)(x)}{0.020 - x} = 1.58 \times 10^{-6}$$

If we assume that $x << 0.020$ we get; $x^2 = 3.16 \times 10^{-8}$,
$x = 1.78 \times 10^{-4}\ M = [OH^-]$
$pOH = -\log[OH^-] = -\log(1.78 \times 10^{-4}) = 3.75$
$pH = 14.00 - pOH = 14.00 - 3.75 = 10.25$

16.93 $[H^+] = 10^{-pH} = 10^{-2.54} = 2.9 \times 10^{-3}\ M$

$$HC_2H_3O_2 \rightleftharpoons H^+ + C_2H_3O_2^-$$

$$K_a = \frac{[H^+][C_2H_3O_2^-]}{[HC_2H_3O_2]} = 1.8 \times 10^{-5}$$

	$[HC_2H_3O_2]$	$[H^+]$	$[C_2H_3O_2^-]$
I	Z	–	–
C	$-x$	$+x$	$+x$
E	$Z - x$	$+x$	$+x$

Substituting these values into the mass action expression gives:

$$K_a = \frac{(x)(x)}{Z - x} = 1.8 \times 10^{-5}$$

Assuming $x << Z$ and knowing that $x = 2.9 \times 10^{-3}\ M$, we can solve for Z and find $Z = 0.47$. The initial concentration of $HC_2H_3O_2$ is 0.47 M.

16.95 NaCN dissociates to form Na^+ and CN^-. The CN^- is a conjugate base of a weak acid.

$$K_b = \frac{1.0 \times 10^{-14}}{4.9 \times 10^{-10}} = 2.0 \times 10^{-5}$$

$CN^-(aq) + H_2O \rightleftharpoons HCN(aq) + OH^-(aq)$

	$[CN^-]$	[HCN]	$[OH^-]$
I	0.0050	–	–
C	$-x$	$+x$	$+x$
E	$0.0050 - x$	$+x$	$+x$

Assume that $x << 0.0050$

$$K_b = \frac{(x)(x)}{0.0050} = 2.0 \times 10^{-5} \qquad x = 3.2 \times 10^{-4} = \left[OH^-\right]$$

Wait!!! 3.2×10^{-4} is not $<< 0.0050$

So, use the method of successive approximations.

$$K_a = \frac{(x)(x)}{0.0050 - 0.00032} = 2.0 \times 10^{-5} \qquad x = 3.1 \times 10^{-4}$$

$$K_a = \frac{(x)(x)}{0.0050 - 0.00031} = 2.0 \times 10^{-5} \qquad x = 3.1 \times 10^{-4}$$

$x = 3.1 \times 10^{-4} = [OH^-]$

pOH = 3.51

pH = 10.49

16.97 $K_a = 10^{-pKa} = 10^{-4.92} = 1.2 \times 10^{-5}$

$H\text{–}Paba \rightleftharpoons H^+ + Paba^-$

$$K_a = \frac{\left[H^+\right]\left[Paba^-\right]}{\left[H - Paba\right]} = 1.2 \times 10^{-5}$$

	[H–Paba]	$[H^+]$	$[Paba^-]$
I	0.030	–	–
C	$-x$	$+x$	$+x$
E	$0.030 - x$	$+x$	$+x$

Substituting the above values for equilibrium concentrations into the mass action expression and assuming that $x << 0.030$ gives:

$$K_a = \frac{[x][x]}{[0.030]} = 1.2 \times 10^{-5}$$

$x^2 = 3.6 \times 10^{-7}$

$x = 6.0 \times 10^{-4}\ M = [H^+]$

$pH = -\log[H^+] = -\log(6.0 \times 10^{-4}) = 3.22$

16.99 NaCN will be basic in solution since CN^- is a basic ion and Na^+ is a neutral ion.

$CN^- + H_2O \rightleftharpoons HCN + OH^-$

For HCN, $K_a = 4.9 \times 10^{-10}$, we need K_b for CN^-;

$$K_b = K_w/K_a = \frac{1.0\times10^{-14}}{4.9\times10^{-10}} = 2.0 \times 10^{-5}$$

$$K_b = \frac{[HCN][OH^-]}{[CN^-]} = 2.0 \times 10^{-5}$$

	$[CN^-]$	[HCN]	$[OH^-]$
I	0.20	–	–
C	$-x$	$+x$	$+x$
E	$0.20 - x$	$+x$	$+x$

Substituting these values into the mass action expression gives:

$$K_b = \frac{(x)(x)}{0.20 - x} = 2.0 \times 10^{-5}$$

Assuming that $x << 0.20$ we can solve for x and determine;

$x = 2.0 \times 10^{-3}\ M = [OH^-]$

$pOH = -\log[OH^-] = -\log(2.0 \times 10^{-3}) = 2.70$

$pH = 14.00 - pOH = 14.00 - 2.69 = 11.30$

Concentration of HCN is equal to that of hydroxide ion: $2.0 \times 10^{-3}\ M$

16.101 A solution of CH_3NH_3Cl will be acidic since the Cl^- ion is neutral and the $CH_3NH_3^+$ ion is acidic.

$$CH_3NH_3^+ \rightleftharpoons H^+ + CH_3NH_2$$

For CH_3NH_2, $K_b = 4.5 \times 10^{-4}$. We need K_a for $CH_3NH_3^+$;

$$K_a = K_w/K_b = \frac{1.0\times10^{-14}}{4.5\times10^{-4}} = 2.2 \times 10^{-11}$$

$$K_a = \frac{[H^+][CH_3NH_2]}{[CH_3NH_3^+]} = 2.2 \times 10^{-11}$$

	$[CH_3NH_3^+]$	$[H^+]$	$[CH_3NH_2]$
I	0.15	–	–
C	$-x$	$+x$	$+x$
E	$0.15 - x$	$+x$	$+x$

Substituting these values into the mass action expression gives:

$$K_a = \frac{(x)(x)}{0.15-x} = 2.2 \times 10^{-11}$$

Assuming that $x << 0.15$ we can solve for x and determine;

$x = 1.9 \times 10^{-6}\ M = [H_3O^+]$

$pH = -\log[H_3O^+] = -\log(1.8 \times 10^{-6}) = 5.74$

16.103 Let HNic symbolize the nicotinic acid

$Nic^- + H_2O \rightleftharpoons HNic + OH^-$

$$K_b = \frac{[HNic][OH^-]}{[Nic^-]}$$

	$[Nic^-]$	[HNic]	$[OH^-]$
I	0.18	–	–
C	$-x$	$+x$	$+x$
E	$0.18 - x$	$+x$	$+x$

Assume $x << 0.18$

$$K_b = \frac{(x)(x)}{0.18}$$

We know $x = [OH^-]$

pH = 9.05 and pOH = 4.95

and $[OH^-] = 10^{-pOH} = 10^{-4.95} = 1.1 \times 10^{-5} = x$

$$K_b = \frac{(1.1 \times 10^{-5})^2}{0.18} = 6.7 \times 10^{-10}$$

$$K_a = \frac{1 \times 10^{-14}}{6.7 \times 10^{-10}} = 1.5 \times 10^{-5}$$

16.105 $OCl^- + H_2O \rightleftharpoons HOCl + OH^-$

$$K_b = \frac{[HOCl][OH^-]}{[OCl^-]} = \frac{K_w}{K_a} = \frac{1.0 \times 10^{-14}}{3.0 \times 10^{-8}} = 3.3 \times 10^{-7}$$

$$[OCl^-] = \left(\frac{5.1 \text{ g NaOCl}}{100 \text{ g solution}}\right)\left(\frac{1 \text{ mol NaOCl}}{74.44 \text{ g NaOCl}}\right)\left(\frac{1.0 \text{ g}}{1 \text{ mL}}\right)\left(\frac{1000 \text{ mL}}{1 \text{ L}}\right)$$

$= 0.69\ M$

	$[OCl^-]$	[HOCl]	$[OH^-]$
I	0.69	–	–
C	$-x$	$+x$	$+x$
E	$0.69 - x$	$+x$	$+x$

Assume that $x << 0.69$

$$K_b = \frac{(x)(x)}{0.69} = 3.3 \times 10^{-7} \qquad x = 4.7 \times 10^{-4} = [OH^-]$$

pOH = 3.32

pH = 10.68

16.107 $HC_2H_3O_2 \rightleftharpoons H^+ + C_2H_3O_2^-$

$$K_a = \frac{[H^+][C_2H_3O_2^-]}{[HC_2H_3O_2]} = 1.8 \times 10^{-5}$$

	$[HC_2H_3O_2]$	$[H^+]$	$[C_2H_3O_2^-]$
I	0.15	–	0.25
C	$-x$	$+x$	$+x$
E	$0.15-x$	$+x$	$0.25+x$

Substituting these values into the mass action expression gives:

$$K_a = \frac{(x)(0.25+x)}{0.15-x} = 1.8 \times 10^{-5}$$

Assume that $x << 0.15\ M$ and $x << 0.25\ M$, then;

$$x\left(\frac{0.25}{0.15}\right) \approx 1.8 \times 10^{-5}$$

$$x \approx \left(\frac{0.15}{0.25}\right) \times 1.8 \times 10^{-5}$$

$x \approx 1.1 \times 10^{-5}\ M = [H^+]$

pH = –log $[H^+]$ = 4.97

16.109 $C_2H_3O_2^- + H_2O \rightleftharpoons HC_2H_3O_2 + OH^-$

$$K_b = \frac{[HC_2H_3O_2][OH^-]}{[C_2H_3O_2^-]} = \frac{K_w}{K_a} = \frac{1.0 \times 10^{-14}}{1.8 \times 10^{-5}} = 5.6 \times 10^{-10}$$

	$[C_2H_3O_2^-]$	$[HC_2H_3O_2]$	$[OH^-]$
I	0.25	0.15	–
C	$-x$	$+x$	$+x$
E	$0.25-x$	$0.15+x$	$+x$

Substituting these values into the mass action expression gives:

$$K_b = \frac{(0.15+x)(x)}{0.25-x} = 5.6 \times 10^{-10}$$

Assume that $x << 0.15\ M$ and $x << 0.25\ M$, then;

$$x\left(\frac{0.15}{0.25}\right) \approx 5.6 \times 10^{-10}$$

$$x \approx \left(\frac{0.25}{0.15}\right) \times 5.6 \times 10^{-10}$$

$x \approx 9.3 \times 10^{-10}\ M = [OH^-]$

pOH = –log $[OH^-]$ = 9.03

pH = 14.00 – pOH = 4.97

The answer is identical no matter which direction we choose to solve it.

16.111 The initial pH of the buffer is 4.97 as determined in Review Problem 16.107. The added acid, 0.025 mol, will react with the acetate ion present in the buffer solution. Assume the added acid reacts completely. For each mole of acid added, one mole of $C_2H_3O_2^-$ is converted to $HC_2H_3O_2$. Since 0.025 mol of acid is added;

$[HC_2H_3O_2]_{final} = (0.15 + 0.025)\ M = 0.175\ M$

$[C_2H_3O_2^-]_{final} = (0.25 - 0.025)\ M = 0.225\ M$

Now, substitute these values into the mass action expression to calculate the final $[H^+]$ in solution

$$\frac{[H^+](0.225)}{(0.175)} = 1.8 \times 10^{-5}$$

$[H^+] = 1.4 \times 10^{-5}$ mol L^{-1} and the pH = 4.85

The pH of the solution changes by 4.97 – 4.85 = –0.12 pH units upon addition of the acid.

16.113 Any H^+ produced will react with NH_3, reducing its concentration, to produce NH_4^+, thus increasing its concentration.

$H^+ + NH_3 \longrightarrow NH_4^+$

The amount of H^+ produced in 35 seconds = (1.8×10^{-6} mol s^{-1})(35 s) = 6.3×10^{-5} mol

The concentration of NH_3 will decrease by:

$$\frac{6.3 \times 10^{-5} \text{ mol}}{0.0025 \text{ L}} = 2.5 \times 10^{-2}\ M$$

and the NH_4^+ will increase by the same amount.

16.115 The initial pH can be calculated using the Henderson-Hasselbalch equation.

$$\text{pOH} = \text{p}K_b + \log\left(\frac{[HB^+]}{[B]}\right)$$

$$\text{pOH} = (4.74) + \log\left(\frac{0.20}{0.25}\right)$$

pOH = 4.64 pH = 14.00 – 4.64 = 9.36

For every mole of H^+ added, one mol of NH_3 will be changed to one mol of NH_4^+. Since we added 6.3×10^{-5} mol H^+:

$$[NH_4^+]_{\text{final}} = 0.20\ M + \frac{6.3 \times 10^{-5}\text{mol}}{0.0025 \text{ L}} = 0.225\ M$$

$$[NH_3]_{\text{final}} = 0.25\ M - \frac{6.3 \times 10^{-5}\text{mol}}{0.0025 \text{ L}} = 0.225\ M$$

Using these new concentrations, we can calculate a new pH:

$$K_b = \frac{[NH_4^+][OH^-]}{[NH_3]} = \frac{(0.225)[OH^-]}{0.225} = 1.8 \times 10^{-5}$$

$[OH^-] = 1.8 \times 10^{-5}$ *M*, the pOH = 4.74 and the pH = 9.26

As expected, when an acid is added, the pH decreases. In this problem, the pH decreases by 0.10 pH units from 9.36 to 9.26.

16.117 $$\text{pH} = \text{p}K_a + \log\frac{[\text{anion}]}{[\text{acid}]} = \text{p}K_a + \log\frac{[A^-]}{[HA]}$$

$$4.00 = 4.74 + \log\frac{[NaC_2H_3O_2]}{[HC_2H_3O_2]}$$

$$-0.74 = \log\frac{[NaC_2H_3O_2]}{[HC_2H_3O_2]}$$

$$\frac{[NaC_2H_3O_2]}{[HC_2H_3O_2]} = 0.18$$

$[NaC_2H_3O_2] = 0.18 \times [HC_2H_3O_2] = 0.18 \times 0.15 = 0.027\ M$
Thus to the 1 L of acetic acid solution we add:
$0.027\ \text{mol}\ NaC_2H_3O_2 \times 82.0\ \text{g/mol} = 2.2\ \text{g}\ NaC_2H_3O_2$.

16.119 The equilibrium is; $HC_2H_3O_2 \rightleftharpoons H^+ + C_2H_3O_2^-$

$$K_a = \frac{[H^+][C_2H_3O_2^-]}{[HC_2H_3O_2]} = 1.8 \times 10^{-5}$$

The initial pH is

$$\frac{[H^+](0.110)}{0.100} = 1.8 \times 10^{-5}$$

$[H^+] = 1.636 \times 10^{-5}\ M$
pH = 4.79
In this calculation we are able to use either the molar concentration or the number of moles since the volume is constant in this portion of the problem.
In order to calculate the change in pH, we need to determine the concentrations of $HC_2H_3O_2$ and $C_2H_3O_2^-$ after the complete reaction of the added acid. One mole of $C_2H_3O_2^-$ will be consumed for every mole of acid added and one mole of $HC_2H_3O_2$ will be produced. The number of moles of acid added is:

$$\text{mol H}^+ = (30.00\ \text{mL HCl})\left(\frac{0.100\ \text{mol HCl}}{1000\ \text{mL HCl}}\right)\left(\frac{1\ \text{mol H}^+}{1\ \text{mol HCl}}\right) = 3.00 \times 10^{-3}\ \text{mol H}^+$$

Since the volume of the new concentrations of $HC_2H_3O_2$ and $C_2H_3O_2^-$ are:

$$[HC_2H_3O_2]_{\text{final}} = \frac{(0.100\ \text{mol} + 0.00300\ \text{mol})}{0.130\ \text{L}} = 0.792\ M$$

$$[C_2H_3O_2^-]_{\text{final}} = \frac{(0.110\ \text{mol} - 0.00300\ \text{mol})}{0.130\ \text{L}} = 0.823\ M$$

Note: The new volume has been used in these calculations.

$$K_a = \frac{[H^+][C_2H_3O_2^-]}{[HC_2H_3O_2]} = \frac{[H^+](0.823)}{(0.792)} = 1.8 \times 10^{-5}$$

$[H^+] = 1.73 \times 10^{-5}\ M$ and pH = 4.76
Notice that the change in pH is very small in spite of adding a strong acid. If the same amount of HCl were added to water, a completely different effect would be observed.

Since HCl is a strong acid, the $[H^+]$ in a water solution will be the result of the strong acid dissociation. We do, of course, need to account for the dilution. Using the dilution equation, $M_1V_1 = M_2V_2$.

$$30\ \text{mL}\left(\frac{1\ \text{L}}{1000\ \text{mL}}\right)(0.100\ M) = M\ (125 + 30\ \text{mL})\left(\frac{1\ \text{L}}{1000\ \text{mL}}\right)$$

$M = 0.0194\ \text{mol L}^{-1}$ and the pH = 1.71. The change in pH in this case is 7.00 – 1.71 = 5.29 pH units. A significantly larger change!!

16.121 $H_2C_6H_6O_6 + H_2O \rightleftharpoons H_3O^+ + HC_6H_6O_6^-$ $K_{a1} = 8.0 \times 10^{-5}$

$HC_6H_6O_6^- + H_2O \rightleftharpoons H_3O^+ + C_6H_6O_6^{2-}$ $K_{a2} = 1.6 \times 10^{-12}$

	$[H_2C_6H_6O_6]$	$[H_3O^+]$	$[HC_6H_6O_6^-]$
I	0.15	–	–
C	$-x$	$+x$	$+x$
E	$0.15 - x$	$+x$	$+x$

Assume $x << 0.15$

$$K_a = \frac{[x][x]}{[0.15]} = 8.0 \times 10^{-5} \qquad x = 3.5 \times 10^{-3}$$

$[H_2C_6H_6O_6] \cong 0.15\ M$
$[H_3O^+] = [HC_6H_6O_6^-] = 3.5 \times 10^{-3}\ M$
The concentration of the ascorbate ion equals K_{a2}
$[C_6H_6O_6^{2-}] = 1.6 \times 10^{-12}\ M$
pH = 2.46
pOH = 11.54
$[OH^-] = 2.9 \times 10^{-12}\ M$

16.123 $H_3PO_4 + H_2O \rightleftharpoons H_2PO_4^- + H_3O^+$ $\quad K_{a1} = \frac{[H_2PO_4^-][H_3O^+]}{[H_3PO_4]} = 6.9 \times 10^{-3}$

$H_2PO_4^- + H_2O \rightleftharpoons HPO_4^{2-} + H_3O^+$ $\quad K_{a2} = \frac{[HPO_4^{2-}][H_3O^+]}{[H_2PO_4^-]} = 6.2 \times 10^{-8}$

$HPO_4^{2-} + H_2O \rightleftharpoons PO_4^{3-} + H_3O^+$ $\quad K_{a3} = \frac{[PO_4^{3-}][H_3O^+]}{[HPO_4^{2-}]} = 4.8 \times 10^{-13}$

The following assumptions are made:
$[H^+]_{total} \approx [H^+]_{first\ step}$
$[H_2PO_4^-]_{total} \approx [H_2PO_4^-]_{first\ step}$
$[HPO_4^{2-}]_{total} \approx [HPO_4^{2-}]_{second\ step}$
The first dissociation:

	$[H_3PO_4]$	$[H_2PO_4^-]$	$[H_3O^+]$
I	2.0	–	–
C	$-x$	$+x$	$+x$
E	$2.0 - x$	$+x$	$+x$

$$K_{a1} = \frac{[H_2PO_4^-][H_3O^+]}{[H_3PO_4]} = \frac{(x)(x)}{(2.0 - x)} = 6.9 \times 10^{-3}$$

Solve for x by successive approximations or by the quadratic equation:
$x = 0.11\ M = [H_3O^+] = [H_2PO_4^-]$
$[H_3PO_4] = 2.0 - 0.11 = 1.9\ M$
pH = –log(0.11) = 0.96

The second dissociation:

	$[H_2PO_4^-]$	$[HPO_4^{2-}]$	$[H_3O^+]$
I	0.11	–	0.11
C	$-x$	$+x$	$+x$
E	$0.11 - x$	$+x$	$0.11 + x$

Assume that $x << 0.11$, therefore $0.11 - x \approx 0.11$ and $0.11 + x \approx 0.11$

$$K_{a2} = \frac{(x)(0.11)}{(0.11)} = 6.2 \times 10^{-8}$$

$x = 6.2 \times 10^{-8} = [HPO_4^{2-}]$

The third dissociation:

	$[HPO_4^{2-}]$	$[PO_4^{3-}]$	$[H_3O^+]$
I	6.2×10^{-8}	–	0.11
C	$-x$	$+x$	$+x$
E	$6.2 \times 10^{-8} - x$	$+x$	$0.11 + x$

Assume that $x << 6.2 \times 10^{-8}$, therefore $6.2 \times 10^{-8} - x \approx 6.2 \times 10^{-8}$ and $0.11 + x \approx 0.12$

$$K_{a3} = \frac{(x)(0.11)}{(6.2 \times 10^{-8})} = 4.8 \times 10^{-13}$$

Solving for x we get, $x = 2.7 \times 10^{-19} = [PO_4^{3-}]$

16.125 $H_3PO_3 + H_2O \rightleftharpoons H_2PO_3^- + H_3O^+$ $\qquad K_{a_1} = \frac{[H_2PO_3^-][H_3O^+]}{[H_3PO_3]} = 5.0 \times 10^{-2}$

$H_2PO_3^- + H_2O \rightleftharpoons HPO_3^{2-} + H_3O^+$ $\qquad K_{a_2} = \frac{[HPO_3^{2-}][H_3O^+]}{[H_2PO_3^-]} = 2.0 \times 10^{-7}$

To simplify the calculation, assume that the second dissociation does not contribute a significant amount of H^+ to the final solution. Solving the equilibrium problem for the first dissociation gives:

	$[H_3PO_3]$	$[H_2PO_3^-]$	$[H_3O^+]$
I	1.0	–	–
C	$-x$	$+x$	$+x$
E	$1.0 - x$	$+x$	$+x$

$$K_{a_1} = \frac{[H_2PO_3^-][H^+]}{[H_3PO_3]} = \frac{(x)\,(x)}{(1.0 - x)} = 5.0 \times 10^{-2}$$

Because K_{a_1} is so large, a quadratic equation must be solved. On doing so we learn that

$x = 2.0 \times 10^{-1}\ M = [H_3O^+] = [H_2PO_3^-]$

$pH = -\log[H_3O^+] = -\log(0.20) = 0.70$

The $[HPO_3^{2-}]$ may be determined from the second ionization constant.

	$[H_2PO_3^-]$	$[HPO_3^{2-}]$	$[H_3O^+]$
I	0.20	–	0.20
C	$-x$	$+x$	$+x$
E	$0.20-x$	$+x$	$0.20+x$

$$K_{a_2} = \frac{[HPO_3^{2-}][H_3O^+]}{[H_2PO_3^-]} = \frac{(x)(0.20+x)}{(0.20-x)} = 2.0 \times 10^{-7}$$

If we assume that x is small then $0.20 \pm x \approx 0.20$. Then, $x = [HPO_3^{2-}] = 2.0 \times 10^{-7}\ M$.

16.127 The hydrolysis equation is:

$$SO_3^{2-} + H_2O \rightleftharpoons HSO_3^- + OH^- \qquad K_b = \frac{[HSO_3^-][OH^-]}{[SO_3^{2-}]}$$

In order to obtain K_b we will use the relationship $K_w = K_a \times K_b$

$$K_b = \frac{K_w}{K_a} = \frac{1.0 \times 10^{-14}}{1.0 \times 10^{-8}} = 1.0 \times 10^{-6}$$

	$[SO_3^{2-}]$	$[HSO_3^-]$	$[OH^-]$
I	0.24	–	–
C	$-x$	$+x$	$+x$
E	$0.24-x$	$+x$	$+x$

Since K_b is so small, assume that $x << 0.24$ and we determine $x = 4.9 \times 10^{-4}\ M = [OH^-]$
$pOH = -\log(4.9 \times 10^{-4}) = 3.31$, $pH = 14.00 - pOH = 10.69$
$[SO_3^{2-}] = 0.24\ M$ $\qquad [HSO_3^-] = [OH^-] = 4.9 \times 10^{-4}\ M$
$[Na^+] = 0.24\ M$

$$[H_3O^+] = \frac{1.0 \times 10^{-14}}{2.0 \times 10^{-4}} = 5.0 \times 10^{-11}\ M$$

The second ionization is very small compared to the first ionization so the $[HSO_3^-]$, $[OH^-]$ and $[H_3O^+]$ do not change significantly.
For the concentration of H_2SO_3

$$HSO_3^- + H_2O \rightleftharpoons H_2SO_3 + OH^- \qquad K_{b_2} = \frac{[H_2SO_3][OH^-]}{[HSO_3^-]}$$

$$K_{b_2} = \frac{K_w}{K_{a_1}} = \frac{1.0 \times 10^{-14}}{1.4 \times 10^{-2}} = 7.1 \times 10^{-13}$$

$$K_{b_2} = \frac{[H_2SO_3][4.9\times10^{-4}]}{[4.9\times10^{-4}]} = 7.1 \times 10^{-13}$$

$[H_2SO_3] = 7.1\times 10^{-13}\ M$

16.129 $C_6H_5O_7^{3-} + H_2O \rightleftharpoons HC_6H_5O_7^{2-} + OH^-$

$$K_b = \frac{[HC_6H_5O_7^{2-}][OH^-]}{[C_6H_5O_7^{3-}]} = \frac{K_w}{K_{a_3}} = \frac{1.0 \times 10^{-14}}{4.0 \times 10^{-7}} = 2.5 \times 10^{-8}$$

	$[C_6H_5O_7^{3-}]$	$[HC_6H_5O_7^{2-}]$	$[OH^-]$
I	0.10	–	–
C	$-x$	$+x$	$+x$
E	$0.10 - x$	$+x$	$+x$

Assume $x << 0.10$

$$K_b = \frac{(x)(x)}{0.10} = 2.5 \times 10^{-8} \qquad x = 5.0 \times 10^{-5} = [OH^-]$$

pOH = 4.30 pH = 9.70

16.131 $PO_4^{3-} + H_2O \rightleftharpoons HPO_4^{2-} + OH^-$ $\quad K_{b_1} = \frac{K_w}{K_{a_3}} = \frac{1.0 \times 10^{-14}}{4.8 \times 10^{-13}} = 2.1 \times 10^{-2}$

$HPO_4^{2-} + H_2O \rightleftharpoons H_2PO_4^- + OH^-$ $\quad K_{b_2} = \frac{K_w}{K_{a_2}} = \frac{1.0 \times 10^{-14}}{6.2 \times 10^{-8}} = 1.6 \times 10^{-7}$

$H_2PO_4^- + H_2O \rightleftharpoons H_3PO_4 + OH^-$ $\quad K_{b_3} = \frac{K_w}{K_{a_1}} = \frac{1.0 \times 10^{-14}}{6.9 \times 10^{-3}} = 1.4 \times 10^{-12}$

By analogy with polyprotic acids, we know that
$[H_2PO_4^-] = 1.6 \times 10^{-7}$
We need to solve the first equilibrium expression to determine $[HPO_4^{2-}]$.

$$K_{b_1} = \frac{[HPO_4^{2-}][OH^-]}{[PO_4^{3-}]} = 2.1 \times 10^{-2}$$

	$[PO_4^{3-}]$	$[HPO_4^{2-}]$	$[OH^-]$
I	0.50	–	–
C	$-x$	$+x$	$+x$
E	$0.50 - x$	$+x$	$+x$

Assume $x << 0.50$

$$K_{b_1} = \frac{(x)(x)}{0.50 - x} = 2.1 \times 10^{-2}$$

Use the quadratic equation to solve for x since K_b is so large
$x = 9.25 \times 10^{-2}\ M = [OH^-] = [HPO_4^{2-}]$
pOH = 1.03
pH = 12.97
$[PO_4^{3-}] = 0.50 - 9.3 \times 10^{-2} = 0.41\ M$
Solve the third equilibrium expression to determine $[H_3PO_4]$:

$$K_{b3} = \frac{[H_3PO_4][OH^-]}{[H_2PO_4^-]} = 1.4 \times 10^{-12}$$

Substitute the calculated values of $[H_2PO_4^-]$ and $[OH^-]$ and solve for x:

$$K_{b3} = \frac{(9.3 \times 10^{-2})x}{(1.6 \times 10^{-7})} = 1.4 \times 10^{-12}$$

$$x = [H_3PO_4] = 2.4 \times 10^{-18}$$

16.133 mol $HC_2H_3O_2$ = $(25.0 \text{ mL } HC_2H_3O_2)\left(\frac{0.180 \text{ mol } HC_2H_3O_2}{1000 \text{ mL } HC_2H_3O_2}\right) = 4.50 \times 10^{-3}$ mol $HC_2H_3O_2$

mol OH^- = $(40.0 \text{ mL } OH^-)\left(\frac{0.250 \text{ mol } OH^-}{1000 \text{ mL } OH^-}\right) = 1.00 \times 10^{-2}$ mol OH^-

excess OH^- = $(1.00 \times 10^{-2}) - (4.5 \times 10^{-3}) = 5.5 \times 10^{-3}$ mol

$$[OH^-] = \frac{5.5 \times 10^{-3} \text{ moles}}{(25.0 + 40.0 \text{ mL})\left(\frac{1 \text{ L}}{1000 \text{ mL}}\right)} = 8.46 \times 10^{-2}\ M$$

pOH = 1.07
pH = 12.93

16.135 Since HCO_2H and NaOH react in a 1:1 ratio:
$HCO_2H + NaOH \longrightarrow NaHCO_2 + H_2O$

we can use the equation $V_a \times M_a = V_b \times M_b$ to determine the volume of NaOH that is required to reach the equivalence point, i.e. the point at which the number of moles of NaOH is equal to the number of moles of HCO_2H:
V_{NaOH} = 50.0 mL × 0.050/0.050 = 50.0 mL
Thus the final volume at the equivalence point will be 50.0 + 50.0 = 100.0 mL.
The concentration of $NaHCO_2$ would then be:
$0.050 \text{ mol L}^{-1} \times 0.050 \text{ L} = 2.5 \times 10^{-3}$ mol HCO_2H = 2.5×10^{-3} mol $NaHCO_2$

$$\frac{2.5 \times 10^{-3} \text{ mol}}{0.100 \text{ L}} = 2.5 \times 10^{-2}\ M\ NaHCO_2$$

The hydrolysis of this salt at the equivalence point proceeds according to the following equilibrium:

$$HCO_2^- + H_2O \rightleftharpoons HCO_2H + OH^-$$

$$K_b = \frac{[HCO_2H][OH^-]}{[HCO_2^-]} = 5.6 \times 10^{-11}$$

	$[HCO_2^-]$	$[HCO_2H]$	$[OH^-]$
I	0.025	–	–
C	$-x$	$+x$	$+x$
E	$0.025 - x$	$+x$	$+x$

Substituting the above values for equilibrium concentrations into the mass action expression and assuming $x << 0.050$ gives:

$$K_b = \frac{(x)(x)}{0.025} = 5.6 \times 10^{-11}$$

$x^2 = 1.4 \times 10^{-12}$

$x = 1.2 \times 10^{-6}\ M = [OH^-] = [HCO_2H]$

$pOH = -\log[OH^-] = -\log(1.2 \times 10^{-6}) = 5.93$

$pH = 14.00 - pOH = 14.00 - 5.93 = 8.07$

Cresol red would be a good indicator, since it has a color change near the pH at the equivalence point.

16.137 (a) $HC_2H_3O_2 + H_2O \rightleftharpoons H_3O^+ + C_2H_3O_2^-$ $K_a = \frac{[H_3O^+][C_2H_3O_2^-]}{[HC_2H_3O_2]} = 1.8 \times 10^{-5}$

	$[HC_2H_3O_2]$	$[H_3O^+]$	$[C_2H_3O_2^-]$
I	0.1000	–	–
C	$-x$	$+x$	$+x$
E	$0.1000 - x$	$+x$	$+x$

Substituting the above values for equilibrium concentrations into the mass action expression and assuming that $x << 0.1000$ gives:

$x = [H_3O^+] = 1.342 \times 10^{-3}\ M$

$pH = -\log [H_3O^+] = -\log (1.342 \times 10^{-3}) = 2.87$

(b) When NaOH is added, it will react with the acetic acid present decreasing the amount in solution and producing additional acetate ion. Since this a one-to-one reaction, the number of moles of acetic acid will decrease by the same amount as the number of moles of NaOH added and the number of moles of acetate ion will increase by an identical amount. We must determine the number of moles of all ions present and calculate new concentrations accounting for dilution.

$$\text{mol } HC_2H_3O_2 = (0.07500 \text{ L solution})\left(\frac{0.1000 \text{ moles } HC_2H_3O_2}{1 \text{ L solution}}\right)$$

$$= 7.500 \times 10^{-3} \text{ mol } HC_2H_3O_2$$

$$\text{mol } OH^- = (0.02500 \text{ L solution})\left(\frac{0.1000 \text{ moles } OH^-}{1 \text{ L solution}}\right) = 2.500 \times 10^{-3} \text{ mol } OH^-$$

$$[HC_2H_3O_2] = \frac{7.500 \times 10^{-3} \text{ moles} - 2.500 \times 10^{-3} \text{ moles}}{0.07500 \text{ L} + 0.02500 \text{ L}} = 5.000 \times 10^{-2}\ M$$

$$[C_2H_3O_2^-] = \frac{0 \text{ moles} + 2.500 \times 10^{-3} \text{ moles}}{0.07500 \text{ L} + 0.02500 \text{ L}} = 2.500 \times 10^{-2}\ M$$

$pK_a = -\log(1.8 \times 10^{-5}) = 4.7447$ (to four sig. fig.)

$$pH = pK_a + \log\frac{[C_2H_3O_2^-]}{[HC_2H_3O_2]}$$

$$= 4.7447 + \log\frac{(2.500 \times 10^{-2})}{(5.000 \times 10^{-2})} = 4.44$$

However, since K_a values are only given to two significant figures, the pH can only be reported to two significant figures.

(c) When half the acetic acid has been neutralized, there will be equal amounts of acetic acid and acetate ion present in the solution. At this point, pH = pK_a = 4.74.

(d) At the equivalence point, all of the acetic acid will have been converted to acetate ion. The concentration of the acetate ion will be half the original concentration of acetic acid since we have doubled the volume of the solution. We then need to solve the equilibrium problem that results when we have a solution that possesses a $[C_2H_3O_2^-] = 0.05000\ M$.

$$C_2H_3O_2^- + H_2O \rightleftharpoons HC_2H_3O_2 + OH^-$$

$$K_b = \frac{[HC_2H_3O_2][OH^-]}{[C_2H_3O_2^-]} = 5.6 \times 10^{-10}$$

	$[C_2H_3O_2^-]$	$[HC_2H_3O_2]$	$[OH^-]$
I	0.05000	–	–
C	$-x$	$+x$	$+x$
E	$0.05000 - x$	$+x$	$+x$

Substituting the above values for equilibrium concentrations into the mass action expression and assuming that $x << 0.05000$ gives: $x = [OH^-] = 5.292 \times 10^{-6}\ M$.
pOH = –log [OH⁻] = –log (5.292×10^{-6}) = 5.2764.
pH = 14.0000 – pOH = 14.0000 – 5.2764 = 8.72.

Chapter Seventeen
Solubility and Simultaneous Equilibria

Practice Exercises

17.1 (a) $K_{sp} = [Pb^{2+}][Br^-]^2$

(b) $K_{sp} = [Al^{3+}][OH^-]^3$

17.2 $K_{sp} = [Hg_2{}^{2+}][Cl^-]^2$

17.3 $TlI(s) \rightleftharpoons Tl^+(aq) + I^-(aq)$

$K_{sp} = [Tl^+][I^-]$

$$\text{mol TlI} = 5.9 \times 10^{-3}\ \text{g}\left(\frac{1\ \text{mol TlI}}{331.3\ \text{g TlI}}\right) = 1.78 \times 10^{-5}\ \text{mol TlI}$$

$$[Tl^+] = [I^-] = \frac{1.78\times10^{-5}\ \text{mol}}{1\ \text{L}} = 1.78 \times 10^{-5}\ M$$

$K_{sp} = (1.8 \times 10^{-5})(1.8 \times 10^{-5}) = 3.2 \times 10^{-10}$

17.4 $AgPO_4(s) \rightleftharpoons 3Ag^+(aq) + PO_4{}^{3-}(aq)$

$K_{sp} = [Ag^+]^3[PO_4{}^{3-}]$

$K_{sp} = (3 \times 4.3 \times 10^{-5})^3(4.3 \times 10^{-5}) = 9.2 \times 10^{-17}$

17.5 (a) $AgBr(s) \rightleftharpoons Ag^+(aq) + Br^-(aq)$

$K_{sp} = [Ag^+][Br^-] = 5.4 \times 10^{-13}$

Let x = solubility of AgBr

$(x)(x) = 5.4 \times 10^{-13}$

$x^2 = 5.4 \times 10^{-13}$

$x = 7.3 \times 10^{-7}\ M$

$[Ag^+] = [Br^-] = 7.3 \times 10^{-7}\ M$

(b) $PbBr_2(s) \rightleftharpoons Pb^{2+}(aq) + 2Br^-(aq)$ $\quad K_{sp} = \left[Pb^{2+}\right]\left[Br^-\right]^2$

$K_{sp} = (s)(2s)^2 = 6.6 \times 10^{-6}$

$4s^3 = 6.6 \times 10^{-6}$

$s = 0.012\ M$

$[Pb^{2+}] = 0.012\ M$ $\qquad [Br^-] = 2s = 2 \times 0.012\ M = 0.024\ M$

17.6 $Hg_2Cl_2(s) \rightleftharpoons Hg_2{}^{2+}(aq) + 2Cl^-(aq)$ $\quad K_{sp} = \left[Hg_2{}^{2+}\right]\left[Cl^-\right]^2$

$K_{sp} = (s)(2s)^2 = 1.4 \times 10^{-18}$

$4s^3 = 1.4 \times 10^{-18}$

$s = 7.0 \times 10^{-7}\ M$

$[Hg_2^{2+}] = 7.0 \times 10^{-7}\ M$ $\qquad [Cl^-] = 2s = 2 \times 7.0 \times 10^{-7}\ M = 1.4 \times 10^{-6}\ M$

17.7 $AgI(s) \rightleftharpoons Ag^+(aq) + I^-(aq)$ $\qquad K_{sp} = [Ag^+][I^-] = 8.5 \times 10^{-17}$

The initial concentration of I^- is $2 \times 0.20\ M$ from the CaI_2.

	$[Ag^+]$	$[I^-]$
I	–	0.40
C	$+x$	$+x$
E	$+x$	$0.40 + x$

Substituting the above values for equilibrium concentrations into the expression for K_{sp} gives:
$K_{sp} = 8.5 \times 10^{-17} = [Ag^+][I^-] = (x)(0.40 + x)$
We know that the value of K_{sp} is very small, and it suggests the simplifying assumption that $(0.40 + x) \approx 0.40$:
$8.5 \times 10^{-17} = (0.40)x$, and $x = 2.125 \times 10^{-16}$
The assumption that
$(0.40 + x) \approx 0.40$ is seen to be valid indeed.
Thus, the molar concentration of AgI in a 0.20 M CaI_2 solution will be $2.1 \times 10^{-16}\ M$.
In pure water,
$K_{sp} = 8.5 \times 10^{-17} = [Ag^+][I^-] = (x)(x)$
$x = [AgI(aq)] = 9.2 \times 10^{-9}\ M$ (much more soluble)

17.8 $Ni_3(PO_4)_2(s) \rightleftharpoons 3Ni^{2+}(aq) + 2PO_4^{3-}(aq)$

$K_{sp} = [Ni^{2+}]^3[PO_4^{3-}]^2 = 4.7 \times 10^{-32}$
$[PO_4^{3-}] = 2.0 \times 10^{-6}\ M$
$[Ni^{2+}] = s$
$(s)^3(2.0 \times 10^{-6})^2 = 4.7 \times 10^{-32}$
$s = 2.3 \times 10^{-7}\ M$

$$28.3\text{ g Ni}\left(\frac{1\text{ mol Ni}}{58.69\text{ g Ni}}\right)\left(\frac{1\text{ L}}{2.3 \times 10^{-7}\text{ mol Ni}}\right)\left(\frac{1\text{ gallon seawater}}{3.79\text{ L seawater}}\right) = 5.5 \times 10^5\text{ gallons seawater}$$

17.9 $HgI_2(s) \rightleftharpoons Hg^{2+}(aq) + 2I^-(aq)$ $\qquad K_{sp} = \left[Hg^{2+}\right]\left[I^-\right]^2 = 2.9 \times 10^{-29}$

$Q = (0.0025)(0.030)^2 = 2.2 \times 10^{-6}$
$Q > K_{sp}$
A precipitate will form.

17.10 $AgBr(s) \rightleftharpoons Ag^+(aq) + Br^-(aq)$ $\qquad K_{sp} = [Ag^+][Br^-] = 5.4 \times 10^{-13}$

$Q = (3.4 \times 10^{-4})(4.8 \times 10^{-5}) = 1.6 \times 10^{-8}$
$Q > K_{sp}$
A precipitate will form.

17.11 We expect $PbBr_2(s)$ since nitrates are soluble.
Because two solutions are to be mixed together, there will be a dilution of the concentrations of the various ions, and the diluted ion concentrations must be used. In general, on dilution, the

following relationship is found for the concentrations of the initial solution (M_i) and the concentration of the final solution (M_f): $M_iV_i = M_fV_f$

Thus the final or diluted concentrations are:

$$\left[Pb^{2+}\right] = \left(1.0 \times 10^{-3}\ M\right)\left(\frac{100.0\text{ mL}}{200.0\text{ mL}}\right) = 5.0 \times 10^{-4}\ M$$

$$\left[Br^{-}\right] = \left(\frac{2.0 \times 10^{-3}\text{ mol MgBr}_2}{\text{L}}\right)\left(\frac{2\text{ mol Br}^-}{1\text{ mol MgBr}_2}\right)\left(\frac{100.0\text{ mL}}{200.0\text{ mL}}\right) = 2.0 \times 10^{-3}\ M$$

The value of the ion product for the final (diluted) solution is:

$[Pb^{2+}][Br^-] = (5.0 \times 10^{-4})(2.0 \times 10^{-3})^2 = 2.0 \times 10^{-9}$

Since this is smaller than the value of K_{sp} (6.6×10^{-6}), no precipitate is expected.

17.12 We expect a precipitate of $PbCl_2$ since nitrates are soluble.

We proceed as in Practice Exercise 17.11 $M_iV_i = M_fV_f$

$$\left[Pb^{2+}\right] = \left(0.10\ M\right)\left(\frac{50.0\text{ mL}}{70.0\text{ mL}}\right) = 0.071\ M$$

$$\left[Cl^{-}\right] = \left(0.040\ M\right)\left(\frac{20.0\text{ mL}}{70.0\text{ mL}}\right) = 0.011\ M$$

The value of the ion product for such a solution would be:

$[Pb^{2+}][Cl^-]^2 = (7.1 \times 10^{-2})(1.1 \times 10^{-2})^2 = 8.6 \times 10^{-6}$

Since the ion product is smaller than K_{sp} (1.7×10^{-5}), a precipitate of $PbCl_2$ is not expected.

17.13 (a) AgBr Ag^+ will not react with an acid.

Br^- is the conjugate base of a strong acid, so it is a very weak base and will not react with an acid. Adding acid will not increase the solubility of AgBr.

(b) $CaCO_3$ Ca^{2+} will not react with an acid.

CO_3^{2-} is the conjugate base of the weak acid, HCO_3^-, so it is a weak base and will react with an acid to from HCO_3^-. However, HCO_3^- goes on to form H_2O and CO_2. Adding acid will increase the solubility of $CaCO_3$.

(c) Ag_2CrO_4 Ag^+ will not react with an acid.

CrO_4^{2-} is the conjugate base of the weak acid, $HCrO_4^-$ so it is a weak base and will react with an acid. Adding acid will increase the solubility of Ag_2CrO_4.

(d) $Zn(CN)_2$ Zn^{2+} will not react with an acid.

CN^- is the conjugate base of a weak acid, HCN, so it is a weak base and will react with an acid to from HCN. Adding acid will increase the solubility of $Zn(CN)_2$.

17.14 $Ag_2CrO_4(s) \rightleftharpoons 2Ag^+(s) + CrO_4^{2-}(aq)$ $K_{sp} = 1.1 \times 10^{-12}$

$H_2CrO_4(aq) \rightleftharpoons H^+(aq) + HCrO_4^-(aq)$ $K_{a1} = 5.0$

$HCrO_4^-(aq) \rightleftharpoons H^+(aq) + CrO_4^{2-}(aq)$ $K_{a2} = 1.5 \times 10^{-6}$

The reaction we are looking for is

$Ag_2CrO_4(s) + H^+(aq) \rightleftharpoons 2Ag^+(aq) + HCrO_4^-(aq)$

This is equivalent to subtracting the second acid dissociation reaction from the solubility reaction

$Ag_2CrO_4(s) \rightleftharpoons 2Ag^+(s) + CrO_4^{2-}(aq)$

$H^+(aq) + CrO_4^{2-}(aq) \rightleftharpoons HCrO_4^-(aq)$

$$K = \frac{\left[Ag^+\right]^2\left[HCrO_4^-\right]}{\left[H^+\right]} = \frac{K_{sp}}{K_{a2}} = \frac{1.1 \times 10^{-12}}{1.5 \times 10^{-6}} = 7.3 \times 10^{-7}$$

	$Ag_2CrO_4(s)$ +	$H^+(aq)$	$\rightleftharpoons$	$2Ag^+(aq)$ +	$HCrO_4^-(aq)$
I		1.0		0	0
C		$-x$		$+2x$	$+x$
E		$1.0-x$		$+2x$	$+x$

$$\frac{(2x)^2(x)}{(1.0-x)} = 7.3 \times 10^{-7}$$

$$\frac{4x^3}{1.0} = 7.3 \times 10^{-7}$$

$x = 5.7 \times 10^{-3}\ M$

The molar solubility of Ag_2CrO_4 is $5.7 \times 10^{-3}\ M$.

17.15 $NiS(s) + 2H^+(aq) \rightleftharpoons Ni^{2+}(aq) + H_2S(aq)$ $\quad K_{spa} = \frac{\left[Ni^{2+}\right]\left[H_2S\right]}{\left[H^+\right]^2} = 4.0 \times 10^1$

$[Ni^{2+}] = 0.022\ M$

$[H_2S] = 0.022\ M$

$[H^+]$ = unknown

$$4.0 \times 10^1 = \frac{[0.022][0.022]}{\left[H^+\right]^2}$$

$$[H^+] = \left(\frac{[0.022][0.022]}{\left[4.0 \times 10^1\right]}\right)^{\frac{1}{2}} = 3.48 \times 10^{-3}\ M$$

pH = 2.46

The pH has to be at least 2.46 or lower.

17.16 $PbS(s) + 2H^+(aq) \rightleftharpoons Pb^{2+}(aq) + H_2S(aq)$ $\quad K_{spa} = \frac{\left[Pb^{2+}\right]\left[H_2S\right]}{\left[H^+\right]^2} = 3 \times 10^{-7}$

$$[Pb^{2+}] = \left(\frac{2.00\text{ g PbS}}{1.5 \times 10^6\text{ L}}\right)\left(\frac{1.00\text{ mol PbS}}{239.27\text{ g PbS}}\right) = 5.57 \times 10^{-9}\ M$$

$[H_2S] = [Pb^{2+}] = 5.57 \times 10^{-9}\ M$

$$K_{spa} = \frac{\left[5.57 \times 10^{-9}\right]\left[5.57 \times 10^{-9}\right]}{\left[H^+\right]^2} = 3 \times 10^{-7}$$

$$[H^+]^2 = \frac{[5.57 \times 10^{-9}][5.57 \times 10^{-9}]}{[3 \times 10^{-7}]} = 1.0 \times 10^{-10}$$

$[H^+] = 1.0 \times 10^{-5}$

$pH = -\log[H^+] = -\log(1.0 \times 10^{-5}) = 5.0$

17.17 Follow the procedure outlined in Example 17.8.

$K_{sp} = [Ca^{2+}][SO_4^{2-}]= 4.9 \times 10^{-5}$

$K_{sp} = [Ba^{2+}][SO_4^{2-}] = 1.1 \times 10^{-10}$

$CaSO_4$ is more soluble and will precipitate when:

$$[SO_4^{2-}] = \frac{K_{sp}}{[Ca^{2+}]} = \frac{4.9 \times 10^{-5}}{0.25} = 2.0 \times 10^{-4}$$

$BaSO_4$ will precipitate when:

$$[SO4^{2-}] = \frac{K_{sp}}{[Ba^{2+}]} = \frac{1.1 \times 10^{-10}}{0.05} = 2.2 \times 10^{-9}$$

$BaSO_4$ will precipitate and $CaSO_4$ will not precipitate if $[SO_4^{2-}] > 2.2 \times 10^{-9}$ and $[SO_4^{2-}] < 2.0 \times 10^{-4}$.

17.18 Follow the procedure outlined in Example 17.8.

$K_{sp} = [Ca^{2+}][OH^-]^2 = 5.0 \times 10^{-6}$

$K_{sp} = [Mg^{2+}][OH^-]^2 = 5.6 \times 10^{-12}$

$Ca(OH)_2$ is more soluble and will precipitate when:

$$[OH^-] = \left(\frac{K_{sp}}{[Ca^{2+}]}\right)^{1/2} = \left(\frac{5.0 \times 10^{-6}}{0.20}\right)^{1/2} = 5.0 \times 10^{-3}$$

$Mg(OH)_2$ will precipitate when:

$$[OH^-] = \left(\frac{K_{sp}}{[Ba^{2+}]}\right)^{1/2} = \left(\frac{5.6 \times 10^{-12}}{0.10}\right)^{1/2} = 7.5 \times 10^{-6}$$

$Mg(OH)_2$ will precipitate and $Ca(OH)_2$ will not precipitate if $7.5 \times 10^{-6}\ M < [OH^-] < 5.0 \times 10^{-3}\ M$.

Low pH:

$[OH^-] = 7.5 \times 10^{-6}\ M$

$pOH = 5.12$

$pH = 14 - pOH = 14 - 5.13 = 8.88$

High pH

$[OH^-] = 5.0 \times 10^{-3}\ M$

$pOH = 2.30$

$pH = 14 - pOH = 14 - 2.30 = 11.70$

17.19 CoS will precipitate if the H^+ concentration is too low. Solving for Q and then comparing Q to K_{spa}, we can determine whether or not CoS will precipitate.

$$K_{spa} = \frac{[Co^{2+}][H_2S]}{[H^+]^2} = 5 \times 10^{-1}$$

$$Q = \frac{[Co^{2+}][H_2S]}{[H^+]^2} = \frac{[0.005][0.10]}{[3.16 \times 10^{-4}]^2} = 5 \times 10^3$$

$Q > K_{spa}$

Since Q is greater than K_{spa}, then the reaction will move to reactants and CoS solid will form.

17.20 Consulting Table 17.2, we find that Fe^{2+} is much more soluble in acid than Hg^{2+}. We want to make the H^+ concentration large enough to prevent FeS from precipitating, but small enough that HgS *does* precipitate. First, we calculate the highest pH at which FeS will remain soluble, by using K_{spa} for FeS. (Recall that a saturated solution of $H_2S = 0.10\ M$.)

$$K_{spa} = \frac{[Fe^{2+}][H_2S]}{[H^+]^2} = \frac{[0.010][0.10]}{[H^+]^2} = 6 \times 10^2$$

$[H^+] = 0.0013\ M$

$pH = -\log [H^+] = 2.9$

Since Fe^{2+} is much more soluble in acid than Hg^{2+} we already know that this pH will precipitate HgS, but we can check it by using K_{spa} for HgS:

$$K_{spa} = \frac{[Hg^{2+}][H_2S]}{[H^+]^2} = \frac{[0.010][0.10]}{[H^+]^2} = 2 \times 10^{-32}$$

$[H^+] = 2.2 \times 10^{14}\ M$

(This concentration is impossibly high, but it tells us that this much acid would be required to dissolve HgS at these concentrations.)

17.21 $BaC_2O_4(s) \rightleftharpoons Ba^{2+}(aq) + C_2O_4^{2-}(aq)$ $\qquad K_{sp} = 1.2 \times 10^{-7} = [Ba^{2+}][C_2O_4^{2-}]$

$[Ba^{2+}] = 0.050\ M$

$1.2 \times 10^{-7} = (0.050)[C_2O_4^{2-}]$

$[C_2O_4^{2-}] = 2.4 \times 10^{-6}\ M$

$H_2C_2O_4 + H_2O \rightleftharpoons H_3O^+ + HC_2O_4^-$ $\qquad K_{a1} = 5.9 \times 10^{-2}$

$HC_2O_4^- + H_2O \rightleftharpoons H_3O^+ + C_2O_4^{2-}$ $\qquad K_{a2} = 6.4 \times 10^{-5}$

$H_2C_2O_4 + H_2O \rightleftharpoons 2H_3O^+ + C_2O_4^{2-}$ $\qquad K_a = (6.4 \times 10^{-5}) \times (5.9 \times 10^{-2}) = 3.8 \times 10^{-6}$

$[H_2C_2O_4] = 0.10$

$[C_2O_4^{2-}] = 2.4 \times 10^{-6}\ M$

$$K_a = 3.8 \times 10^{-6} = \frac{\left[H_3O^+\right]^2\left[C_2O_4^{2-}\right]}{\left[H_2C_2O_4\right]} = \frac{\left[H_3O^+\right]^2\left(2.4\times10^{-6}\right)}{(0.10)}$$

Since the amount of oxalate formed is so small, the concentration of oxalic acid is essentially unchanged.

$[H^+] = 4.0 \times 10^{-1}\ M$

This is the minimum concentration of H^+ that will prevent the formation of BaC_2O_4 precipitate.

17.22 Follow the procedure outlined in Example 17.10.

$K_{sp} = [Ca^{2+}][CO_3^{2-}] = 3.4 \times 10^{-9}$

$K_{sp} = [Ni^{2+}][CO_3^{2-}] = 1.4 \times 10^{-7}$

$NiCO_3$ is more soluble and will precipitate when:

$$[CO_3^{2-}] = \frac{K_{sp}}{[Ni^{2+}]} = \frac{1.4 \times 10^{-7}}{0.10} = 1.4 \times 10^{-6}$$

$CaCO_3$ will precipitate when:

$$[CO_3^{2-}] = \frac{K_{sp}}{[Ca^{2+}]} = \frac{3.4 \times 10^{-9}}{0.10} = 3.4 \times 10^{-8}$$

$CaCO_3$ will precipitate and $NiCO_3$ will not precipitate if
$[CO_3^{2-}] > 3.4 \times 10^{-8}$ and $[CO_3^{2-}] < 1.4 \times 10^{-6}$
Now, using the equation in example 17.10 we get:

$$[H^+]^2 = (2.4 \times 10^{-17})\left(\frac{0.030}{[CO_3^{2-}]}\right)$$

$NiCO_3$ will precipitate if:

$$[H^+]^2 = (2.4 \times 10^{-17})\left(\frac{0.030}{1.4 \times 10^{-6}}\right) = 5.14 \times 10^{-13}$$

$[H^+] = 7.17 \times 10^{-7}$ pH = 6.14
$CaCO_3$ will precipitate:

$$[H^+]^2 = (2.4 \times 10^{-17})\left(\frac{0.030}{3.4 \times 10^{-8}}\right) = 2.12 \times 10^{-11}$$

$[H^+] = 4.6 \times 10^{-6}$ pH = 5.34
So $CaCO_3$ will precipitate and $NiCO_3$ will not if the pH is maintained between pH = 5.34 and pH = 6.14

17.23 The overall equilibrium is $AgCl(s) + 2NH_3(aq) \rightleftharpoons Ag(NH_3)_2^+(aq) + Cl^-(aq)$

$$K_c = \frac{[Ag(NH_3)_2^+][Cl^-]}{[NH_3]^2}$$

In order to obtain a value for K_c for this reaction, we need to use the expressions for K_{sp} of $AgCl(s)$ and the K_{form} of $Ag(NH_3)_2^+$:

$$K_{sp} = [Ag^+][Cl^-] = 1.8 \times 10^{-10}$$

$$K_{form} = \frac{[Ag(NH_3)_2^+]}{[Ag^+][NH_3]^2} = 1.6 \times 10^7$$

$$K_c = K_{sp} \times K_{form} = \frac{[Ag(NH_3)_2^+][Cl^-]}{[NH_3]^2} = 2.9 \times 10^{-3}$$

Now we may use an equilibrium table for the reaction in question:

	$[NH_3]$	$[Ag(NH_3)_2^+]$	$[Cl^-]$
I	0.10	–	–
C	$-2x$	$+x$	$+x$
E	$0.10 - 2x$	x	x

Substituting these values into the mass action expression gives:

$$K_c = 2.9 \times 10^{-3} = \frac{(x)(x)}{(0.10 - 2x)^2}$$

Take the square root of both sides to get $0.054 = \dfrac{(x)}{(0.10 - 2x)}$

Solving for x we get, $x = 4.9 \times 10^{-3}$ M. The molar solubility of AgCl in 0.10 M NH_3 is therefore 4.9×10^{-3} M.

In order to determine the solubility in pure water, we simply look at K_{sp}

$AgCl(s) \rightleftharpoons Ag^+(aq) + Cl^-(aq)$ $\qquad K_{sp} = [Ag^+][Cl^-] = 1.8 \times 10^{-10}$

At equilibrium; $[Ag^+] = [Cl^-] = 1.3 \times 10^{-5}$ M. Hence the molar solubility of AgCl in 0.10 M NH_3 is about 380 times greater than in pure water.

17.24 We will use the information gathered for the last problem. Specifically,

$AgCl(s) + 2NH_3(aq) \rightleftharpoons Ag(NH_3)_2^+(aq) + Cl^-(aq)$

$$K_c = \frac{[Ag(NH_3)_2^+][Cl^-]}{[NH_3]^2} = 2.9 \times 10^{-3}$$

If we completely dissolve 0.20 mol of AgCl, the equilibrium $[Cl^-]$ and $[Ag(NH_3)_2^+]$ will be 0.20 M in a one liter container. The question asks, therefore, what amount of NH_3 must be initially present so that the equilibrium concentration of Cl^- is 0.20 M?

	$[NH_3]$	$[Ag(NH_3)_2^+]$	$[Cl^-]$
I	Z	–	–
C	$-2x$	$+x$	$+x$
E	$Z-2x$	x	x

$$K_c = 2.9 \times 10^{-3} = \frac{(x)(x)}{(Z-2x)^2}$$

Where x equals 0.20, take the square root of both sides to get

$$0.054 = \frac{x}{Z-2x} = \frac{0.20}{Z-0.40}$$

We have substituted the known value of x. Solving for Z we get, $Z = 4.1$ M

Consequently, we would need to add 4.1 moles of NH_3 to a one liter container of 0.20 M AgCl in order to completely dissolve the AgCl.

Review Problems

17.25 (a) $Hg_2Cl_2(s) \rightleftharpoons Hg_2^{2+}(aq) + 2Cl^-(aq)$ $\qquad K_{sp} = [Hg_2^{2+}][Cl^-]^2$

(b) $AgBr(s) \rightleftharpoons Ag^+(aq) + Br^-(aq)$ $\qquad K_{sp} = [Ag^+][Br^-]$

(c) $PbBr_2(s) \rightleftharpoons Pb^{2+}(aq) + 2Br^-(aq)$ $\qquad K_{sp} = [Pb^{2+}][Br^-]^2$

(d) $CuCl(s) \rightleftharpoons Cu^+(aq) + Cl^-(aq)$ $\qquad K_{sp} = [Cu^+][Cl^-]$

(e) $HgI_2(s) \rightleftharpoons Hg^{2+}(aq) + 2I^-(aq)$ $\qquad K_{sp} = [Hg^{2+}][I^-]^2$

17.27 (a) $CaF_2(s) \rightleftharpoons Ca^{2+}(aq) + 2F^-(aq)$ $K_{sp} = [Ca^{2+}][F^-]^2$

(b) $Ag_2CO_3(s) \rightleftharpoons 2Ag^+(aq) + CO_3^{2-}(aq)$ $K_{sp} = [Ag^+]^2[CO_3^{2-}]$

(c) $PbSO_4(s) \rightleftharpoons Pb^{2+}(aq) + SO_4^{2-}(aq)$ $K_{sp} = [Pb^{2+}][SO_4^{2-}]$

(d) $Fe(OH)_3(s) \rightleftharpoons Fe^{3+}(aq) + 3OH^-(aq)$ $K_{sp} = [Fe^{3+}][OH^-]^3$

(e) $PbF_2(s) \rightleftharpoons Pb^{2+}(aq) + 2F^-(aq)$ $K_{sp} = [Pb^{2+}][F^-]^2$

(f) $Cu(OH)_2(s) \rightleftharpoons Cu^{2+}(aq) + 2OH^-(aq)$ $K_{sp} = [Cu^{2+}][OH^-]^2$

17.29 The substances are grouped according to the number of cations and anions that are formed.
1:1 ratio cation to anion

$AgBr(s) \rightleftharpoons Ag^+(aq) + Br^-(aq)$ $K_{sp} = [Ag^+][Br^-]$

$CuCl(s) \rightleftharpoons Cu^+(aq) + Cl^-(aq)$ $K_{sp} = [Cu^+][Cl^-]$

$PbSO_4(s) \rightleftharpoons Pb^{2+}(aq) + SO_4^{2-}(aq)$ $K_{sp} = [Pb^{2+}][SO_4^{2-}]$

1:2 ratio cation to anion, or 2:1 ratio cation to anion

$Hg_2Cl_2(s) \rightleftharpoons Hg_2^{2+}(aq) + 2Cl^-(aq)$ $K_{sp} = [Hg_2^{2+}][Cl^-]^2$

$PbBr_2(s) \rightleftharpoons Pb^{2+}(aq) + 2Br^-(aq)$ $K_{sp} = [Pb^{2+}][Br^-]^2$

$CaF_2(s) \rightleftharpoons Ca^{2+}(aq) + 2F^-(aq)$ $K_{sp} = [Ca^{2+}][F^-]^2$

$PbF_2(s) \rightleftharpoons Pb^{2+}(aq) + 2F^-(aq)$ $K_{sp} = [Pb^{2+}][F^-]^2$

$Cu(OH)_2(s) \rightleftharpoons Cu^{2+}(aq) + 2OH^-(aq)$ $K_{sp} = [Cu^{2+}][OH^-]^2$

$HgI_2(s) \rightleftharpoons Hg^{2+}(aq) + 2I^-(aq)$ $K_{sp} = [Hg^{2+}][I^-]^2$

$Ag_2CO_3(s) \rightleftharpoons 2Ag^+(aq) + CO_3^{2-}(aq)$ $K_{sp} = [Ag^+]^2[CO_3^{2-}]$

1:3 ratio cation to anion

$Fe(OH)_3(s) \rightleftharpoons Fe^{3+}(aq) + 3OH^-(aq)$ $K_{sp} = [Fe^{3+}][OH^-]^3$

17.31 Compare the values of K_{sp}. The larger the value the greater the molar solubility.
$K_{sp} = 6.8 \times 10^{-8}$ for $MgCO_3$
$K_{sp} = 1.5 \times 10^{-10}$ for $ZnCO_3$
Therefore, $MgCO_3$ is more soluble in water.

17.33 $PbCl_2(s) \rightleftharpoons Pb^{2+}(aq) + 2Cl^-(aq)$ $K_{sp} = [Pb^{2+}][Cl^-]^2$
At equilibrium $[Pb^{2+}] = 0.016\ M$ and $[Cl^-] = 0.032\ M$
so $K_{sp} = (0.016)(0.032)^2 = 1.6 \times 10^{-5}$

17.35 $$\text{mol BaSO}_4 = (0.00245 \text{ g BaSO}_4)\left(\frac{1 \text{ mole BaSO}_4}{233.3906 \text{ g BaSO}_4}\right) = 1.05 \times 10^{-5} \text{ mol}$$

$[Ba^{2+}] = [SO_4^{2-}] = 1.05 \times 10^{-5}\ M$ This is the molar solubility of $BaSO_4$
$K_{sp} = [Ba^{2+}][SO_4^{2-}] = (1.05 \times 10^{-5})^2 = 1.10 \times 10^{-10}$

17.37 $BaF_2 \rightleftharpoons Ba^{2+}(aq) + 2F^-(aq)$ $K_{sp} = [Ba^{2+}][F^-]^{-2}$
First find the concentration of the Ba^{2+} and F^- that was in solution and then find the value for K_{sp}. Using the amount of BaF_2 recovered; determine the number of moles of each ion, then find the concentration of each ion.

$$\text{mole BaF}_2 = 0.132\text{ g BaF}_2\left(\frac{1\text{ mol BaF}_2}{175.32\text{ g BaF}_2}\right) = 7.53 \times 10^{-4}\text{ mol BaF}_2$$

$$[\text{Ba}^{2+}] = \left(\frac{7.53 \times 10^{-4}\text{ mol BaF}_2}{100\text{ mL solution}}\right)\left(\frac{1\text{ mol Ba}^{2+}}{1\text{ mol BaF}_2}\right)\left(\frac{1000\text{ mL}}{1\text{ L}}\right) = 7.53 \times 10^{-3}\ M\text{ Ba}^{2+}$$

$$[\text{F}^-] = \left(\frac{7.53 \times 10^{-4}\text{ mol BaF}_2}{100\text{ mL solution}}\right)\left(\frac{2\text{ mol F}^-}{1\text{ mol BaF}_2}\right)\left(\frac{1000\text{ mL}}{1\text{ L}}\right) = 1.51 \times 10^{-2}\ M\text{ F}^-$$

$K_{sp} = [Ba^{2+}][F^-] = (7.53 \times 10^{-3})(1.51 \times 10^{-2})^2 = 1.7 \times 10^{-6}$

17.39 $Ag_3PO_4(s) \rightleftharpoons 3Ag^+ + PO_4^{3-}$ $\qquad K_{sp} = [Ag^+]^3[PO_4^{3-}]$
$K_{sp} = [3(1.8 \times 10^{-5})]^3[1.8 \times 10^{-5}] = 2.8 \times 10^{-18}$

17.41 $PbBr_2(s) \rightleftharpoons Pb^{2+} + 2Br^-$ $\qquad K_{sp} = [Pb^{2+}][Br^-]^2$

	$[Pb^{2+}]$	$[Br^-]$
I	–	–
C	$+x$	$+2x$
E	$+x$	$+2x$

$K_{sp} = (x)(2x)^2 = 4x^3 = 6.6 \times 10^{-6}$, $x = \sqrt[3]{\dfrac{6.6 \times 10^{-6}}{4}} = 1.2 \times 10^{-2}\ M$

17.43 $Zn(CN)_2(s) \rightleftharpoons Zn^{2+}(aq) + 2CN^-(aq)$
For every mole of Zn^{2+} produced, 2 moles of CN^- will be produced. Let $x = [Zn^{2+}]$ at equilibrium and $[CN^-] = 2x$ at equilibrium. $K_{sp} = [Zn^{+2}][CN^-]^2 = 3.0 \times 10^{-16} = (x)(2x)^2 = 4x^3$. Solving we find $x = 4.2 \times 10^{-6}$. Thus, the molar solubility of $Zn(CN)_2$ is 4.2×10^{-6} moles/L.

17.45 To solve this problem, determine the molar solubility for each compound.
LiF: let $x = [Li^+] = [F^-]$ $\qquad K_{sp} = [Li^+][F^-] = x^2 = 1.8 \times 10^{-3}$
$x = 4.2 \times 10^{-2}$ moles/L = molar solubility of LiF.
BaF_2: let $x = [Ba^{2+}]$; $[F^-]^2 = 2x$ $\qquad K_{sp} = [Ba^{2+}][F^-]^2 = (x)(2x)^2 = 1.7 \times 10^{-6}$
$4x^3 = 1.7 \times 10^{-6}$
$x = 7.5 \times 10^{-3}\ M$ = molar solubility of BaF_2.
Because the molar solubility of LiF is greater than that of BaF_2, LiF is more soluble.

17.47 First determine the molar solubility of the *MX* salt.
Let $x = [M^+] = [X^-]$, $\qquad K_{sp} = [M^+][X^-] = (x)(x) = 3.2 \times 10^{-10}$
$x = 1.8 \times 10^{-5}\ M$. This is the equilibrium concentration of the two ions.
For the MX_3 salt, let x = equilibrium concentration of M^{3+}, $[X^-] = 3x$.
$K_{sp} = [M^+][X^-]^3 = (x)(3x)^3 = 27x^4$. The value of x in this expression is the value determined in the first part of this problem.
So, $K_{sp} = (27)(1.8 \times 10^{-5})^4 = 2.8 \times 10^{-18}$

17.49 $CaSO_4(s) \rightleftharpoons Ca^{2+}(aq) + SO_4^{2-}(aq)$ $\qquad K_{sp} = [Ca^{2+}][SO_4^{2-}] = 4.9 \times 10^{-5}$
let $x = [Ca^{2+}] = [SO_4^{2-}]$ $\qquad K_{sp} = x^2 = 4.9 \times 10^{-5}$ $\qquad$ and $x = 7.0 \times 10^{-3}\ M$
The molar solubility of $CaSO_4$ is 7.0×10^{-3} moles/L.

17.51 $BaSO_3(s) \rightleftharpoons Ba^{2+}(aq) + SO_3^{2-}(aq)$ $\quad K_{sp} = [Ba^{2+}][SO_3^{2-}]$

$K_{sp} = (0.10)(8.0 \times 10^{-6}) = 8.0 \times 10^{-7}$

In this problem, all of the Ba^{2+} comes from the $BaCl_2$.

17.53 (a) $CuCl(s) \rightleftharpoons Cu^+(aq) + Cl^-(aq)$ $\quad K_{sp} = [Cu^+][Cl^-] = 1.7 \times 10^{-7}$

	[Cu⁺]	**[Cl⁻]**
I	–	–
C	$+x$	$+x$
E	$+x$	$+x$

$K_{sp} = x^2 = 1.7 \times 10^{-7}$ $\quad \therefore x = \text{molar solubility} = 4.1 \times 10^{-4}\ M$

(b) $CuCl(s) \rightleftharpoons Cu^+(aq) + Cl^-(aq)$ $\quad K_{sp} = [Cu^+][Cl^-] = 1.7 \times 10^{-7}$

	[Cu⁺]	**[Cl⁻]**
I	–	0.0200
C	$+x$	$+x$
E	$+x$	$0.0200 + x$

$K_{sp} = (x)(0.0200 + x) = 1.7 \times 10^{-7}$ $\quad$ Assume that $x << 0.0200$

$\therefore x = \text{molar solubility} = 8.5 \times 10^{-6}\ M$

(c) $CuCl(s) \rightleftharpoons Cu^+(aq) + Cl^-(aq)$ $\quad K_{sp} = [Cu^+][Cl^-] = 1.7 \times 10^{-7}$

	[Cu⁺]	**[Cl⁻]**
I	–	0.200
C	$+x$	$+x$
E	$+x$	$0.200 + x$

$K_{sp} = (x)(0.200 + x) = 1.7 \times 10^{-7}$ Assume that $x << 0.200$

$\therefore x = \text{molar solubility} = 8.5 \times 10^{-7}\ M$

(d) $CuCl(s) \rightleftharpoons Cu^+(aq) + Cl^-(aq)$ $\quad K_{sp} = [Cu^+][Cl^-] = 1.7 \times 10^{-7}$

Note that the Cl^- concentration equals (2)(0.150 *M*) since two moles of Cl^- are produced for every mole of $CaCl_2$.

	[Cu⁺]	**[Cl⁻]**
I	–	0.300
C	$+x$	$+x$
E	$+x$	$0.300 + x$

$K_{sp} = (x)(0.300 + x) = 1.7 \times 10^{-7}$ Assume that $x << 0.300$

$\therefore x = \text{molar solubility} = 5.7 \times 10^{-7}\ M$

17.55 $Mg(OH)_2 \rightleftharpoons Mg^{2+}(aq) + 2OH^-(aq)$ $\quad K_{sp} = [Mg^{2+}][OH^-]^2 = 5.6 \times 10^{-12}$

The concentration of OH^- is determined from the pH:

pOH = 14 – 12.50 = 1.50

$[OH^-] = 0.0316\ M$

$[Mg^{2+}] = x \qquad [OH^-] = 0.0316\ M$

$K_{sp} = x(0.0316)^2 = 5.6 \times 10^{-12}$

$x = 5.6 \times 10^{-9}\ M$

The molar solubility of $Mg(OH)_2$ is $5.6 \times 10^{-9}\ M$ in a solution with a pH of 12.50.

17.57 $PbCl_2(s) \rightleftharpoons Pb^{2+}(aq) + 2Cl^-(aq)$ $K_{sp} = [Pb^{2+}][Cl^-]^2 = 1.7 \times 10^{-5}$

$[Cl^-] = 0.10\ M$

$[Pb^{2+}][Cl^-]^2 = [Pb^{2+}][0.10]^2 = 1.7 \times 10^{-5}$

$[Pb^{2+}] = 1.7 \times 10^{-3}\ M$

17.59 $Ag_2CrO_4(s) \rightleftharpoons 2Ag^+(aq) + CrO_4^{2-}(aq)$ $K_{sp} = [Ag^+]^2[CrO_4^{2-}] = 1.1 \times 10^{-12}$

(a)

	$[Ag^+]$	$[CrO_4^{2-}]$
I	0.200	–
C	+ 2x	+ x
E	0.200 + 2x	+ x

$K_{sp} = (0.200+2x)^2(x)$ Assume that $x << 0.200$

$1.1 \times 10^{-12} = (0.200)^2(x)$ $x = 2.8 \times 10^{-11}$

The molar solubility is 2.8×10^{-11} moles/L

(b)

	$[Ag^+]$	$[CrO_4^{2-}]$
I	–	0.200
C	+2x	+ x
E	+2x	0.200 + x

$K_{sp} = (2x)^2(0.200+x)$ Assume that $x << 0.200$

$1.1 \times 10^{-12} = (2x)^2(0.200)$ $x = 1.2 \times 10^{-6}$

The molar solubility is 1.2×10^{-6} moles/L.

17.61 $Fe(OH)_2(s) \rightleftharpoons Fe^{2+}(aq) + 2\ OH^-(aq)$ $K_{sp} = [Fe^{2+}][OH^-]^2 = 4.9 \times 10^{-17}$

pH = 9.50

pOH = 14.00 – pH = 4.50

$[OH^-] = 10^{-4.50} = 3.16 \times 10^{-5}\ M$

	$[Fe^{2+}]$	$[OH^-]$
I	–	3.16×10^{-5}
C	+ x	+ 2x
E	x	$(3.16 \times 10^{-5}) + 2x$

Since K_{sp} for iron(II) hydroxide is so small, we can safely assume that $2x << 3.16 \times 10^{-5}$, so that $(3.16 \times 10^{-5}) + 2x \approx 3.16 \times 10^{-5}$, then we enter the equilibrium values of the above table into the K_{sp} expression:

$K_{sp} = [Fe^{2+}][OH^-]^2$

$4.9 \times 10^{-17} = x(3.16 \times 10^{-5})^2$

x = molar solubility = $4.9 \times 10^{-8}\ M$

17.63 $Fe(OH)_2(s) \rightleftharpoons Fe^{2+}(aq) + 2\ OH^-(aq)$ $K_{sp} = [Fe^{2+}][OH^-]^2$

mol OH^- = 2.20 g NaOH(1 mol/40.01 g NaOH) = 0.0550 mol NaOH

$[OH^-]$ = mol OH^-/L solution = 0.0550 mol/0.250 L = 0.22 M

	$[Fe^{2+}]$	$[OH^-]$
I	–	0.22
C	$+x$	$+2x$
E	x	$0.22 + 2x$

We assume that $x << 0.22$, so that $0.22 + 2x \approx 0.22$, then we enter the equilibrium values of the above table into the K_{sp} expression:

$$K_{sp} = \left[Fe^{2+}\right]\left[OH^-\right]^2$$

$4.9 \times 10^{-17} = x(0.22)^2$

x = molar solubility = 1.0×10^{-15} M

Next, we must determine how many moles of $Fe(OH)_2$ are formed in the reaction.

This is a limiting reactant problem.

The number of moles of OH^- is 0.0550 (see above).

The number of moles of Fe^{2+} is (0.250 L)(0.10 mol/L) = 0.025 mol

From the balanced equation at the top, we need two OH^- for every one Fe^{2+}.

This would be 2(0.025 mol) = 0.050 mol OH^-. Looking at the molar quantities above, we have more than enough OH^- so, Fe^{2+} is our limiting reactant:

0.025 mol $Fe(OH)_2$ will form in 0.25 L solution. If dissolved, this would be a concentration of 0.025 mol/0.25 L = 0.10 M. But from above, the maximum molar solubility of is 1.0×10^{-15} M.

This means that remainder of $Fe(OH)_2$ in excess of this value precipitates:

$0.10 - 1.0 \times 10^{-15} \approx 0.10$ M.

This works out to 0.25 L(0.10 mol/L) = 0.025 mol $Fe(OH)_2$(89.8 g/mol)

= 2.2 g solid $Fe(OH)_2$ (essentially all of it).

The remaining OH^-, 0.005 mol, gives a concentration of OH^- of

$$\frac{0.005 \text{ mol } OH^-}{0.250 \text{ L}} = 0.02\ M\ OH^-$$

$4.9 \times 10^{-17} = [Fe^{2+}][0.02]^2$

$[Fe^{2+}] = 1.2 \times 10^{-13}$ M

17.65 In order for a precipitate to form, the value of the reaction quotient, Q, must be greater than the value of K_{sp}. For $PbCl_2$, $K_{sp} = 1.7 \times 10^{-5}$ (see Table 17.1).

$Q = \left[Pb^{2+}\right]\left[Cl^-\right]^2 = (0.0150)(0.0120)^2 = 2.16 \times 10^{-6}$. Since $Q < K_{sp}$, no precipitate will form.

17.67 To solve this problem, determine the value for Q and apply Le Châtelier's Principle.

(a) $\left[Pb^{2+}\right]$ = (50.0 mL)(0.0100 moles/L)/(100.0 mL) = 5.00×10^{-3}

$\left[Br^-\right]$ = (50.0 mL)(0.0100 moles/L)/(100.0 mL) = 5.00×10^{-3}

$$Q = \left[Pb^{2+}\right]\left[Br^-\right]^2 = (5.00 \times 10^{-3})(5.00 \times 10^{-3})^2 = 1.25 \times 10^{-7}$$

For $PbBr_2$, $K_{sp} = 6.6 \times 10^{-6}$

Since $Q < K_{sp}$, no precipitate will form.

(b) $\left[Pb^{2+}\right]$ = (50.0 mL)(0.0100 moles/L)/(100.0 mL) = 5.00×10^{-3}

$\left[Br^{-}\right]$ = (50.0 mL)(0.100 moles/L)/(100.0 mL) = 5.00×10^{-2}

$Q = \left[Pb^{2+}\right]\left[Br^{-}\right]^2 = (5.00 \times 10^{-3})(5.00 \times 10^{-2})^2 = 1.25 \times 10^{-5}$

For $PbBr_2$, $K_{sp} = 6.6 \times 10^{-6}$

Since $Q > K_{sp}$, a precipitate will form.

17.69 The precipitate that may form is $PbBr_2(s)$. To determine if a precipitate will form, a value for the reaction quotient, Q, must be calculated: $Q = [Pb^{2+}][Br^-]^2$. In performing this calculation, the dilution of the ions must be considered:

$[Pb^{2+}] = [Br^-] = 0.00500\ M$. $Q = [0.00500][0.00500]^2 = 1.3 \times 10^{-7}$

If $Q > K_{sp}$, a precipitate will form. K_{sp} for $PbBr_2(s)$ is 6.6×10^{-6}. Therefore, a precipitate will not form. Hence, the concentrations calculated are the diluted concentrations. Since no precipitate forms, the concentrations are not equilibrium values.

17.71 In order to answer this question, we need the $[OH^-]$ at equilibrium. $Ca(OH)_2$ is a sparingly soluble compound. According to Table 17.1, $K_{sp} = 5.0 \times 10^{-6}$.

$Ca(OH)_2(s) \rightleftharpoons Ca^{2+}(aq) + 2OH^-(aq)$ $K_{sp} = \left[Ca^{2+}\right]\left[OH^-\right]^2$

	$[Mg^{2+}]$	$[OH^-]$
I	–	–
C	$+x$	$+2x$
E	$+x$	$+2x$

$K_{sp} = (x)(2x)^2 = 4x^3 = 5.0 \times 10^{-6}$

$x = 1.1 \times 10^{-2}\ M$, $[OH^-] = 2x = 2.2 \times 10^{-2}\ M$

pOH = –log$[OH^-]$ = 1.66

The pH = 14.00 – pOH = 12.34.

17.73 (a) $Mg(OH)_2(s) \rightleftharpoons Mg^{2+}(aq) + 2OH^-(aq)$

$NH_4^+(aq) + OH^-(aq) \rightleftharpoons NH_3(aq) + H_2O$

$Mg(OH)_2(s) + 2NH_4^+(aq) \rightleftharpoons Mg^{2+}(aq) + 2H_2O + 2NH_3(aq)$

(b) We want 0.1 mole of the $Mg(OH)_2$ to go into solution. The NH_4^+ reacts with any OH^- produced in the dissociation of $Mg(OH)_2$ thereby shifting the equilibrium to the right. Using the K_{sp} value for $Mg(OH)_2$, we may find the hydroxide ion concentration under these conditions:

$Mg(OH)_2(s) \rightleftharpoons Mg^{2+}(aq) + 2OH^-(aq)$

$K_{sp} = \left[Mg^{2+}\right]\left[OH^-\right]^2$

$5.6 \times 10^{-12} = \left[0.10\right]\left[OH^-\right]^2$

$\left[OH^-\right] = 7.5 \times 10^{-6}$

Now we can use this value in the following, simultaneous equilibrium:

$NH_4^+(aq) + OH^-(aq) \rightleftharpoons H_2O + NH_3(aq)$

$K_c = 1/K_{bNH_3} = 1/1.8 \times 10^{-5} = 5.6 \times 10^4$

$$K_c = \frac{[NH_3]}{[OH^-][NH_4^+]} = \frac{[0.20]}{[7.5 \times 10^{-6}][NH_4^+]} = 5.6 \times 10^4$$

We know that $[NH_3] = 0.20\ M$ because in the equation below 2 moles of ammonia are formed for every one mole of magnesium ion:

$Mg(OH)_2(s) + 2NH_4^+(aq) \rightleftharpoons Mg^{2+}(aq) + 2H_2O + 2NH_3(aq)$
Solving for $[NH_4^+]$, we get 0.48 M.
So the total $[NH_3] + [NH_4^+] = 0.20 + 0.48 = 0.68\ M$
One must therefore add 0.68 mol NH_4Cl to a liter of solution.

(c) The resulting solution will contain 0.20 mol of NH_3. Solve the weak base equilibrium problem for NH_3. The pH = 11.28.

17.75 Initially, both Ag^+ and $HC_2H_3O_2$ are at 1.0 M concentrations. These values will be used to determine if the $AgC_2H_3O_2$ will precipitate.
First, determine the concentration of the acetate ion from the equilibrium:

$HC_2H_3O_2(aq) \rightleftharpoons H^+(aq) + C_2H_2O_2^-(aq)$

$$K_a = \frac{[H^+][C_2H_3O_2^-]}{[HC_2H_3O_2]} = 1.8 \times 10^{-5}$$

	$[HC_2H_3O_2]$	$[H_3O^+]$	$[C_2H_3O_2^-]$
I	1.0	–	–
C	$-x$	$+x$	$+x$
E	$1.0 - x$	x	x

Assume $x << 1.0$

$$K_a = \frac{[x][x]}{[1.0]} = 1.8 \times 10^{-5} \quad x = 4.2 \times 10^{-3} = [C_2H_3O_2^-]$$

Next, using the concentration of the acetate ion, determine whether or not a precipitate will form.

$AgC_2H_3O_2(s) \rightleftharpoons Ag^+(aq) + C_2H_3O_2^-(aq)$
$K_{sp} = [Ag^+][C_2H_3O_2^-]$
$Q = [Ag^+][C_2H_3O_2^-]$ before equilibrium is established
$Q = (1.0)(4.2 \times 10^{-3}) = 4.2 \times 10^{-3}$
$Q > K_{sp}$ therefore a precipitate will form.

17.77 We must first calculate the solubility in terms of number of moles/L, i.e.,

$$\text{mol}/\text{L} = \left(7.05 \times 10^{-3}\ \text{g}/\text{L}\right)\left(\frac{1\ \text{mol Mg(OH)}_2}{58.32\ \text{g Mg(OH)}_2}\right) = 1.21 \times 10^{-4}\ M$$

Next, use this to establish the individual ion concentrations based on the equilibrium:

$Mg(OH)_2(s) \rightleftharpoons Mg^{2+}(aq) + 2OH^-(aq)$

$$[Mg^{2+}] = 1.21 \times 10^{-4}\ M$$

$$[OH^-] = 2.42 \times 10^{-4}\ M$$

Finally, calculate K_{sp} using the standard expression:

$$K_{sp} = [Mg^{2+}][OH^-]^2 = (1.21 \times 10^{-4})(2.42 \times 10^{-4})^2 = 7.09 \times 10^{-12}$$

17.79 $BaC_2O_4(s) \rightleftharpoons Ba^{2+}(aq) + C_2O_4^{2-}(aq)$

$K_{sp} = [Ba^{2+}][C_2O_4^{2-}] = 1.2 \times 10^{-7}$

$[Ba^{2+}] = [C_2O_4^{2-}] = x$

$x^2 = 1.2 \times 10^{-7}$

$x = 3.5 \times 10^{-4}$

The molar solubility of BaC_2O_4 in water is 3.5×10^{-4} *M*.

In 1.0 *M* HCl, the reactions are

$BaC_2O_4(s) \rightleftharpoons Ba^{2+}(aq) + C_2O_4^{2-}(aq)$

$C_2O_4^{2-}(aq) + H^+(aq) \rightleftharpoons HC_2O_4^-(aq)$

The overall reaction is

$BaC_2O_4(s) + H^+(aq) \rightleftharpoons Ba^{2+}(aq) + HC_2O_4^-(aq)$

The overall equilibrium constant is

$$K = \frac{K_{sp}}{K_{a_2}} = \frac{1.2 \times 10^{-7}}{6.4 \times 10^{-5}} = 1.9 \times 10^{-3}$$

$$K = \frac{[Ba^{2+}][HC_2O_4^-]}{[H^+]}$$

The concentration table is

	$BaC_2O_4(s)$ +	$H^+(aq)$	$\rightleftharpoons$	$Ba^{2+}(aq)$ +	$HC_2O_4^-(aq)$
I	-	1.0		0	0
C	-	$-x$		$+x$	$+x$
E	-	$1.0 - x$		x	x

$$K = \frac{[Ba^{2+}][HC_2O_4^-]}{[H^+]} = \frac{(x)(x)}{(1.0 - x)} = 3.16 \times 10^{-2}$$

$[Ba^{2+}] = 0.13$ *M*

17.81 $[Ag^+] = 0.200$ *M*

$[H^+] = 0.10$ *M*

First, the concentration of acetate ion needs to be determined at the point that the silver acetate precipitates:

$AgC_2H_3O_2(s) \rightleftharpoons Ag^+(aq) + C_2H_3O_2^-(aq)$

$K_{sp} = [Ag^+][C_2H_3O_2^-] = 2.3 \times 10^{-3}$

Let $x = [C_2H_3O_2^-]$

$2.3 \times 10^{-3} = (0.200)(x)$

$x = 1.2 \times 10^{-2}\ M = [C_2H_3O_2^-]$

When $NaC_2H_3O_2$ is added to the solution, the $C_2H_3O_2^-$ will react with the H^+ from the nitric acid to form $HC_2H_3O_2$. This will give a concentration of 0.10 *M* $HC_2H_3O_2$.

First, we need to determine the number of moles of $NaC_2H_3O_2$ to add to react with the nitric acid

$$0.200 \text{ L solution}\left(\frac{0.100 \text{ mol } C_2H_3O_2^-}{1 \text{ L solution}}\right)\left(\frac{1 \text{ mol } NaC_2H_3O_2}{1 \text{ mol } C_2H_3O_2^-}\right)\left(\frac{82.03 \text{ g } NaC_2H_3O_2}{1 \text{ mol } NaC_2H_3O_2}\right) = 1.64 \text{ g } NaC_2H_3O_2$$

Now, we need to add enough acetate to start precipitating the silver, which will occur when the acetate ion concentration is 1.2×10^{-2} *M*.

$$0.200 \text{ L solution}\left(\frac{0.012 \text{ mol } C_2H_3O_2^-}{1 \text{ L solution}}\right)\left(\frac{1 \text{ mol } NaC_2H_3O_2}{1 \text{ mol } C_2H_3O_2^-}\right)\left(\frac{82.03 \text{ g } NaC_2H_3O_2}{1 \text{ mol } NaC_2H_3O_2}\right)$$

$= 0.20$ g $NaC_2H_3O_2$

Now we can add these two together

1.64 g $NaC_2H_3O_2$ + 0.20 g $NaC_2H_3O_2$ = 1.84 g $NaC_2H_3O_2$

17.83 Step 1: Determine the $[OH^-]$ from the NH_3 reaction with water:

$NH_4^+(aq) + OH^-(aq) \rightleftharpoons NH_3(aq) + H_2O$

$$K_b = \frac{[NH_4^+][OH^-]}{[NH_3]}$$

$$1.8 \times 10^{-5} = \frac{[NH_4^+][OH^-]}{[NH_3]}$$

	$[NH_3]$	$[OH^-]$	$[NH_4^+]$
I	0.10	–	–
C	$-x$	$+x$	$+x$
E	$0.10 - x$	x	x

Assume $x << 0.10$

$$1.8 \times 10^{-5} = \frac{[x][x]}{[0.10]}$$

$x = 1.3 \times 10^{-3} = [OH^-]$

Step 2: Find the concentration of Mg^{2+} at the given concentration of OH^-.

The concentration of OH^- from the NH_3 is 1.3×10^{-3} *M*, there is an additional amount of OH^- from the equilibrium of the $Mg(OH)_2$, which makes the calculation:

$K_{sp} = 5.6 \times 10^{-12} = [Mg^{2+}][OH^-]^2$

$5.6 \times 10^{-12} = (x)(1.3 \times 10^{-3} + 2x)^2$

The additional amount of OH^- can be ignored since it will be less than 1.3×10^{-3}.

Using the molar solubility of $Mg(OH)_2$ in distilled water:

$5.6 \times 10^{-12} = [Mg^{2+}][OH^-]^2$

$s = [Mg^{2+}]$ and $2s = [OH^-]$

$4s^3 = 5.6 \times 10^{-12}$

$s = 1.1 \times 10^{-4}$

The solubility of $Mg(OH)_2$ in distilled water is less than the amount of OH^- supplied by the ammonia so we are justified in ignoring its contribution. We may now solve for x

$x = 3.3 \times 10^{-6}$ $M = [Mg^{2+}]$

The molar solubility of $Mg(OH)_2 = 3.3 \times 10^{-6}$ *M*

17.85 In water: $CoS(s) + H_2O \rightleftharpoons Co^{2+}(aq) + HS^-(aq) + OH^-(aq)$ $K_{sp} = [Co^{2+}][S^{2-}][OH^-] = 5 \times 10^{-22}$

Define the solubility of CoS as s.

$[Co^{2+}] = [S^{2-}] = [OH^-]$ s

$s \times s \times s = 5 \times 10^{-22}$

$s = 8 \times 10^{-8}$

Molar solubility of CoS is 2×10^{-11} *M* in water.

In 6.0 *M* HNO_3

$$CoS(s) + 2H^+(aq) \rightleftharpoons Co^{2+}(aq) + H_2S(aq) \qquad K_{spa} = \frac{[Co^{2+}][H_2S]}{[H^+]^2} = 5 \times 10^{-1}$$

Define the solubility of CoS in 6 *M* H^+ as *x*

$[Co^{2+}] = x$

$[H_2S] = x$

	$CoS(s)$ +	$2H^+(aq)$	$\rightleftharpoons$	$Co^{2+}(aq)$ +	$H_2S(aq)$
I	-	6.0		0	0
C	-	$-2x$		$+x$	$+x$
E	-	$6.0 - 2x$		x	x

$$K_{spa} = \frac{[Co^{2+}][H_2S]}{[H^+]^2} = \frac{(x)(x)}{(6.0-2x)^2} = 5 \times 10^{-1}$$

$x = 1.8$ *M*

The molar solubility of CoS in 6.0 *M* HNO_3 is 1.8 *M*.

17.87 $NiS(s) + 2H^+(aq) \rightleftharpoons Ni^{2+}(aq) + H_2S(aq) \qquad K_{spa} = \dfrac{[Ni^{2+}][H_2S]}{[H^+]^2}$

	$[H^+]$	**$[Ni^{2+}]$**	**$[H_2S]$**
I	4	–	–
C	$-2x$	$+x$	$+x$
E	$4-2x$	$+x$	$+x$

$$K_{spa} = \frac{[Ni^{2+}][H_2S]}{[H^+]^2} = \frac{(x)(x)}{(4-2x)^2} = 40 \qquad \text{(from Table 17.2)}$$

Take the square root of both sides to get

$$\frac{x}{(4-2x)} = 6.3$$

Solving gives $x = 1.85$ *M*. NiS is very soluble in 4 *M* acid.

17.89 $AgCl(s) \rightleftharpoons Ag^+(aq) + Cl^-(aq) \qquad K_{sp} = [Ag^+][Cl^-] = 1.8 \times 10^{-10}$

$AgI(s) \rightleftharpoons Ag^+(aq) + I^-(aq) \qquad K_{sp} = [Ag^+][I^-] = 8.5 \times 10^{-17}$

When $AgNO_3$ is added to the solution, AgI will precipitate before any AgCl does due to the lower solubility of AgI. In order to answer the question, i.e., what is the $[I^-]$ when AgCl first precipitates, we need to find the minimum concentration of Ag^+ that must be added to precipitate AgCl.

Let $x = [Ag^+]$; $K_{sp} = (x)(0.050) = 1.8 \times 10^{-10}$; $x = 3.6 \times 10^{-9}$ *M*

When the AgCl starts to precipitate, the solution will have a $[Ag^+]$ of 3.6×10^{-9} *M*. Now we ask, what is the $[I^-]$ if $[Ag^+] = 3.6 \times 10^{-9}$ *M*?

So, $K_{sp} = [Ag^+][I^-] = (3.6 \times 10^{-9})(x) = 8.5 \times 10^{-17}$; $x = 2.3 \times 10^{-8}\ M = [I^-]$

17.91 The less soluble substance is PbS. We need to determine the minimum $[H^+]$ at which CoS will precipitate.

$$K_{spa} = \frac{[Co^{2+}][H_2S]}{[H^+]^2} = \frac{(0.010)(0.1)}{[H^+]^2} = 0.5 \qquad \text{(from Table 17.2)}$$

$$[H^+] = \sqrt{\frac{(0.010)(0.1)}{0.5}} = 0.045$$

pH = $-\log[H^+]$ = 1.35. At a pH lower than 1.35, PbS will precipitate and CoS will not. At larger values of pH, both PbS and CoS will precipitate.

17.93 $Cu(OH)_2(s) \rightleftharpoons Cu^{2+}(aq) + 2\ OH^-(aq)$

$$K_{sp} = [Cu^{2+}][OH^-]^2$$

$$4.8 \times 10^{-20} = [0.10][OH^-]^2$$

$[OH^-] = 6.9 \times 10^{-10}$ *M*

pOH = $-\log[OH^-] = -\log[6.9 \times 10^{-10}] = 9.2$

pH = 14.00 –pOH = 4.8

4.8 is the pH *below* which all the $Cu(OH)_2$ will be soluble.

$Mn(OH)_2(s) \rightleftharpoons Mn^{2+}(aq) + 2\ OH^-(aq)$

$$K_{sp} = [Mn^{2+}][OH^-]^2$$

$$1.6 \times 10^{-13} = [0.10][OH^-]^2$$

$[OH^-] = 1.3 \times 10^{-6}$ *M*

pOH = $-\log[OH^-] = -\log[1.3 \times 10^{-6}] = 5.9$

pH = 14.00 – pOH = 8.1

8.1 is the pH *below* which all the $Mn(OH)_2$ will be soluble.

Therefore, from pH = 4.8–8.1 $Mn(OH)_2$ will be soluble, but some $Cu(OH)_2$ will precipitate out of solution.

17.95 In this problem, we have two simultaneous equilibria occurring:

$Mn(OH)_2(s) \rightleftharpoons Mn^{2+}(aq) + 2OH^-(aq) \qquad K_{sp} = [Mn^{2+}][OH^-]^2 = 1.6 \times 10^{-13}$

$Fe^{2+}(aq) + 2OH^-(aq) \rightleftharpoons Fe(OH)_2(s) \qquad K_c = \dfrac{1}{[Fe^{2+}][OH^-]^2} = 1/(4.9 \times 10^{-17}) = 2.04 \times 10^{16}$

The second equilibrium represents the opposite equation from that of K_{sp}. Therefore, its value is $1/K_{sp}$ for $Fe(OH)_2$.

Combined, and omitting spectator ions, this is:

$Mn(OH)_2(s) + Fe^{2+}(aq) \rightleftharpoons Mn^{2+}(aq) + Fe(OH)_2(s)$

$$K_c = \frac{[Mn^{2+}]}{[Fe^{2+}]} = K_{sp\,(Mn)} \cdot K_{c\,(Fe)} = (1.6\times10^{-13})(2.04\times10^{16}) = 3265$$

	$[Fe^{2+}]$	$[Mn^{2+}]$
I	0.100	–
C	$-x$	$+x$
E	$0.100-x$	x

$$K_c = \frac{[Mn^{2+}]}{[Fe^{2+}]}$$

$$3265 = \frac{[x]}{[0.100-x]}$$

$326.5 - 3265x = x$

$326.5 = 3266x$

$x = 0.099969 = 0.100\ M$

Therefore, $[Fe^{2+}] = 0.100 - 0.099969 = 3 \times 10^{-5}\ M$ and $[Mn^{2+}] = 0.100\ M$

To calculate the pH of the solution go back to either K_{sp} expression and solve for $[OH^-]$

$Mn(OH)_2(s) \rightleftharpoons Mn^{2+}(aq) + 2OH^-(aq)$ $\quad K_{sp} = [Mn^{2+}][OH^-]^2 = 1.6 \times 10^{-13}$

$$K_{sp} = (0.1)[OH^-]^2 = 1.6 \times 10^{-13}$$

$[OH^-] = 1.3 \times 10^{-6}\ M$

$pOH = 5.89$

$pH = 14 - pOH = 14 - 5.90 = 8.11$

17.97 $K_{sp} = [Ag^+]^2[CO_3^{2-}] = 8.5 \times 10^{-12}$

$K_{sp} = [Ni^{2+}][CO_3^{2-}] = 1.4 \times 10^{-7}$

Assume the solutions are equal molar with concentration of 0.10 M.

$NiCO_3$ is more soluble and will precipitate when:

$$[CO_3^{2-}] = \frac{K_{sp}}{[Ni^{2+}]} = \frac{1.4 \times 10^{-7}}{0.10} = 1.4 \times 10^{-6}$$

Ag_2CO_3 will precipitate when:

$$[CO_3^{2-}] = \frac{K_{sp}}{[Ag^+]^2} = \frac{8.5 \times 10^{-12}}{(0.10)^2} = 8.5 \times 10^{-10}$$

Ag_2CO_3 will precipitate and $NiCO_3$ will not precipitate if $[CO_3^{2-}] > 8.5 \times 10^{-10}$ and $[CO_3^{2-}] < 1.4 \times 10^{-6}$. Now, using the equation in example 17.10 we get:

$$[H^+]^2 = (2.4 \times 10^{-17})\left(\frac{0.030}{[CO_3^{2-}]}\right)$$

$NiCO_3$ will precipitate if:

$$[H^+]^2 = (2.4 \times 10^{-17})\left(\frac{0.030}{1.4 \times 10^{-6}}\right) = 5.1 \times 10^{-13}$$

$[H^+] = 7.2 \times 10^{-7}$ $\qquad$ $pH = 6.14$

Ag_2CO_3 will precipitate:

$$[H^+]^2 = (2.4 \times 10^{-17})\left(\frac{0.030}{8.5 \times 10^{-10}}\right) = 8.5 \times 10^{-10}$$

$[H^+] = 2.9 \times 10^{-5}$ $\quad$ pH = 4.54

So Ag_2CO_3 will precipitate and $NiCO_3$ will not if the pH is maintained between pH = 4.54 and pH = 6.14.

17.99 (a) $Cu^{2+}(aq) + 4Cl^-(aq) \rightleftharpoons CuCl_4^{2-}(aq)$ $\quad K_{form} = \frac{[CuCl_4^{2-}]}{[Cu^{2+}][Cl^-]^4}$

(b) $Ag^+(aq) + 2I^-(aq) \rightleftharpoons AgI_2^-(aq)$ $\quad K_{form} = \frac{[AgI_2^-]}{[Ag^+][I^-]^2}$

(c) $Cr^{3+}(aq) + 6NH_3(aq) \rightleftharpoons Cr(NH_3)_6^{3+}(aq)$ $\quad K_{form} = \frac{[Cr(NH_3)_6^{3+}]}{[Cr^{3+}][NH_3]^6}$

17.101 (a) $Co^{3+}(aq) + 6NH_3(aq) \rightleftharpoons Co(NH_3)_6^{3+}(aq)$ $\quad K_{form} = \frac{[Co(NH_3)_6^{3+}]}{[Co^{3+}][NH_3]^6}$

(b) $Hg^{2+}(aq) + 4I^-(aq) \rightleftharpoons HgI_4^{2-}(aq)$ $\quad K_{form} = \frac{[HgI_4^{2-}]}{[Hg^{2+}][I^-]^4}$

(c) $Fe^{2+}(aq) + 6CN^-(aq) \rightleftharpoons Fe(CN)_6^{4-}(aq)$ $\quad K_{form} = \frac{[Fe(CN)_6^{4-}]}{[Fe^{2+}][CN^-]^6}$

17.103 (a) $Co(NH_3)_6^{3+}(aq) \rightleftharpoons Co^{3+}(aq) + 6NH_3(aq)$ $\quad K_{inst} = \frac{[Co^{3+}][NH_3]^6}{[Co(NH_3)_6^{3+}]}$

(b) $HgI_4^{2-}(aq) \rightleftharpoons Hg^{2+}(aq) + 4I^-(aq)$ $\quad K_{inst} = \frac{[Hg^{2+}][I^-]^4}{[HgI_4^{2-}]}$

(c) $Fe(CN)_6^{4-}(aq) \rightleftharpoons Fe^{2+}(aq) + 6CN^-(aq)$ $\quad K_{inst} = \frac{[Fe^{2+}][CN^-]^6}{[Fe(CN)_6^{4-}]}$

17.105 $K_c = K_{sp} \times K_{form} = (1.7 \times 10^{-5})(2.5 \times 10^1) = 4.3 \times 10^{-4}$

17.107 There are two events in this net process: one is the formation of a complex ion (an equilibrium which has an appropriate value for K_{form}), and the other is the dissolving of $Fe(OH)_3$, which is governed by K_{sp} for the solid.

$Fe(OH)_3(s) \rightleftharpoons Fe^{3+}(aq) + 3OH^-(aq)$ $\qquad K_{sp} = [Fe^{3+}][OH^-]^3 = 2.8\times 10^{-39}$

$Fe^{3+}(aq) + 6CN^-(aq) \rightleftharpoons Fe(CN)_6^{3-}(aq)$ $\qquad K_{form} = \dfrac{[Fe(CN)_6^{3-}]}{[Fe^{3+}][CN^-]^6} = 1.0 \times 10^{31}$

The net process is:

$Fe(OH)_3(s) + 6CN^-(aq) \rightleftharpoons Fe(CN)_6^{3-}(aq) + 3OH^-(aq)$

The equilibrium constant for this process should be:

$$K_c = \frac{[Fe(CN)_6^{3-}][OH^-]^3}{[CN^-]^6}$$

The numerical value for the above K_c is equal to the product of K_{sp} for $Fe(OH)_3(s)$ and K_{form} for $Fe(CN)_6^{3-}$, as can be seen by multiplying the mass action expressions for these two equilibria:

$K_c = K_{form} \times K_{sp} = 2.8 \times 10^{-8}$

Because K_{form} is so very large, we can assume that all of the dissolved iron ion is present in solution as the complex, thus:

$[Fe(CN)_6^{3-}] = 0.11$ mol/1.2 L = 0.092 M

Also the reaction stoichiometry shows that each iron ion that dissolves gives 3 OH^- ions in solution, and we have: $[OH^-] = 0.092 \times 3 = 0.28\ M$. We substitute these values into the K_c expression and rearrange to get:

$$[CN-] = \sqrt[6]{\frac{[Fe(CN)_6^{3-}][OH^-]^3}{K_c}}$$

$$= \sqrt[6]{\frac{(0.092)(0.28)^3}{2.8 \times 10^{-8}}}$$

Thus we arrive at the concentration of cyanide ion that is required in order to satisfy the mass action requirements of the equilibrium: $[CN^-] = 6.45$ mol L^{-1}. Since this concentration of CN^- must be present in 1.2 L, the number of moles of cyanide that are required is:

6.45 mol L^{-1} × 1.2 L = 7.74 mol CN^-

Additionally, a certain amount of cyanide is needed to form the complex ion. The stoichiometry requires six times as much cyanide ion as iron ion. This is 0.11 moles × 6 = 0.66 mol. This brings the total required cyanide to (7.74 + 0.66) = 8.4 mol.

8.4 mol × 49.0 g/mol = 412 g NaCN are required.

17.109 The applicable equilibria are as follows:

$AgI(s) \rightleftharpoons Ag^+(aq) + I^-(aq)$ $\qquad K_{sp} = [Ag^+][I^-] = 8.5 \times 10^{-17}$

$Ag^+(aq) + 2I^-(aq) \rightleftharpoons AgI_2^-(aq)$ $\qquad K_{form} = \dfrac{[AgI_2^-]}{[Ag^+][I^-]^2} = 1 \times 10^{11}$

When a solution of AgI_2^- is diluted, all of the concentrations of the species in K_{form} above decrease. However, the decrease of $[I^-]$ has more effect on equilibrium because its expression is *squared.* Hence, the denominator is decreased more than the numerator in the reaction quotient,

Q. The system reacts according to Le Châtelier's Principle, by moving to the left (toward reactants) to increase the value of $[I^-]$.
As the system moves to the left, more Ag^+ is created, which has an effect on the first equilibrium above. Again, Le Châtelier's Principle causes the reaction to move to the left to re-establish equilibrium, which produces $AgI(s)$ precipitate.
The two equations above may be combined and K_c found as follows:

$$AgI(s) + I^-(aq) \rightleftharpoons AgI_2^-(aq) \quad K_c = \frac{[AgI_2^-]}{[I^-]} = K_{sp} \times K_{form} = 8.5 \times 10^{-6}$$

To answer the second question, we make a table and fill in what we know. We begin with 1.0 *M* I^-. This is reduced by some amount (x) as it reacts with the silver ions, and $[AgI_2^-]$ is increased by the same amount:

	$[I^-]$	$[AgI_2^-]$
I	1.0	–
C	$-x$	$+x$
E	$1.0 - x$	$+x$

Now we insert the equilibrium values into the above equation:

$$K_c = \frac{[AgI_2^-]}{[I^-]} = 8.5 \times 10^{-6}$$

$$K_c = \frac{[x]}{[1.0 - x]} = 8.5 \times 10^{-6}$$

$x = 8.5 \times 10^{-6}$
This value represents the change in concentration of I^- which, from the balanced equation, equals the change in concentration of $AgI(s)$. The given volume is 0.100 L, which allows us to find the amount of AgI reacting:

$$0.100\ \text{L}\left(\frac{8.5 \times 10^{-6}\ \text{mol AgI}}{\text{L}}\right)\left(\frac{234.8\ \text{g AgI}}{\text{mol AgI}}\right) = 2.0 \times 10^{-4}\ \text{g AgI}$$

17.111 The applicable equilibria are as follows:

$$AgI(s) \rightleftharpoons Ag^+(aq) + I^-(aq) \qquad K_{sp} = [Ag^+][I^-] = 8.5 \times 10^{-17}$$

$$Ag^+(aq) + 2CN^-(aq) \rightleftharpoons Ag(CN)_2^-(aq) \qquad K_{form} = \frac{[Ag(CN)_2^-]}{[Ag^+][CN^-]^2} = 5.3 \times 10^{18}$$

The two equations above may be combined and K_c found as follows:

$$AgI(s) + 2CN^-(aq) \rightleftharpoons Ag(CN)_2^-(aq) + I^-(aq) \quad K_c = \frac{[Ag(CN)_2^-][I^-]}{[CN^-]^2} = K_{sp} \times K_{form} = 4.5 \times 10^2$$

We begin with 0.010 *M* CN^-. This is reduced by some amount ($2x$) as it reacts with the silver ions, and $[Ag(CN)_2^-]$ is increased by x:

	$[CN^-]$	$[Ag(CN)_2^-]$	I^-
I	0.010	–	–
C	$-2x$	$+x$	$+x$
E	$0.010-2x$	x	x

Now we insert the equilibrium values into the above equation:

$$K_c = \frac{\left[Ag(CN)_2^-\right]\left[I^-\right]}{\left[CN^-\right]^2} = 4.5 \times 10^2$$

$$K_c = \frac{[x][x]}{[0.010-2x]^2} = 4.5 \times 10^2$$

Take the square root of both sides and solve for x:

$x = 4.9 \times 10^{-3}$

This value represents the change in concentration of I^- which, from the balanced equation, equals the change in concentration of AgI(s).

The molar solubility of AgI in 0.010 M KCN is 4.9×10^{-3} M.

17.113 $Cu(OH)_2(s) \rightleftharpoons Cu^{2+}(aq) + 2OH^-(aq)$ $\qquad K_{sp} = \left[Cu^{2+}\right]\left[OH^-\right]^2 = 4.8 \times 10^{-20}$

$Cu^{2+}(aq) + 4NH_3(aq) \rightleftharpoons Cu(NH_3)_4^{2+}(aq)$ $\qquad K_{form} = 1.1 \times 10^{13}$

Combined, this is:

$Cu(OH)_2(s) + 4NH_3(aq) \rightleftharpoons Cu(NH_3)_4^{2+}(aq) + 2OH^-(aq)$

$$K_c = \frac{\left[Cu(NH_3)_4^{2+}\right]\left[OH^-\right]^2}{\left[NH_3\right]^4} = 5.3 \times 10^{-7}$$

	$[NH_3]$	$[Cu(NH_3)_4^{2+}]$	$[OH^-]$
I	2.0	–	–
C	$-4x$	$+x$	$+2x$
E	$2.0-4x$	x	$2x$

$$K_c = \frac{[x][2x]^2}{[2.0-4x]^4} = 5.3 \times 10^{-7}$$

Solve for x using successive approximations.

$x = 1.2 \times 10^{-2}$

$[Cu(NH_3)_4^{2+}] = 1.2 \times 10^{-2}$

Since all of the Cu^{2+} comes from the $Cu(OH)_2$, the molar solubility of $Cu(OH)_2$ is 1.2×10^{-2} M

17.115 Assume the more soluble compound is 1.00 M. Then the less soluble compound is 0.0001 M.

$\left[M_1^{q+}\right] = 1.00\ M \qquad \left[M_2^{q+}\right] = 0.0001\ M$

The counter ion has molar concentration $= \left[X^{q-}\right]$

$$\frac{K_{sp(1)}}{K_{sp(2)}} = \frac{1.00\left[X^{q-}\right]}{0.0001\left[X^{q-}\right]} = 1 \times 10^4$$

17.117 (a) The number of moles of the two reactants are:
$0.12\ M\ Ag^+ \times 0.050\ L = 6.0 \times 10^{-3}$ moles Ag^+
$0.048\ M\ Cl^- \times 0.050\ L = 2.4 \times 10^{-3}$ moles Cl^-
The precipitation of AgCl proceeds according to the following stoichiometry
$Ag^+(aq) + Cl^-(aq) \longrightarrow AgCl(s)$.

If we assume that the product is completely insoluble, then 2.4×10^{-3} moles of AgCl will be formed because Cl^- is the limiting reagent (see above.)

$$\text{g AgCl} = \left(2.4 \times 10^{-3} \text{ mol AgCl}\right)\left(\frac{143.3 \text{ g AgCl}}{1 \text{ mol AgCl}}\right) = 0.34 \text{ g AgCl}$$

(b) The silver ion concentration may be determined by calculating the amount of excess silver added to the solution:
$[Ag^+] = (6.0 \times 10^{-3} \text{ moles} - 2.4 \times 10^{-3} \text{ moles})/0.100\ L = 3.6 \times 10^{-2}\ M$
The concentrations of nitrate and sodium ions are easily calculated since they are spectators in this reaction:
$[NO_3^-] = (0.12\ M)(50.0\ mL)/(100.0\ mL) = 6.0 \times 10^{-2}\ M$
$[Na^+] = (0.048\ M)(50.0\ mL)/(100.0\ mL) = 2.4 \times 10^{-2}\ M$
In order to determine the chloride ion concentration, we need to solve the equilibrium expression. Specifically, we need to ask what is the chloride ion concentration in a saturated solution of AgCl that has a $[Ag^+] = 3.6 \times 10^{-2}\ M$.

$AgCl(s) \rightleftharpoons Ag^+ + Cl^-$ $\qquad K_{sp} = 1.8 \times 10^{-10}$

	$[Ag^+]$	$[Cl^-]$
I	0.036	–
C	$+x$	$+x$
E	$0.036 + x$	$+x$

$$K_{sp} = \left[Ag^+\right]\left[Cl^-\right] = (0.036 + x)(x) = 1.8 \times 10^{-10}$$

$x = 5.0 \times 10^{-9}\ M$ if we assume that $x << 0.036$
Therefore, $[Cl^-] = 5.0 \times 10^{-9}\ M$.

(c) The percentage of the silver that has precipitated is:
$(2.4 \times 10^{-3} \text{ moles})/(6.0 \times 10^{-3} \text{ moles}) \times 100\% = 40\%$

17.119 Let x = mol of PbI_2 that dissolve per liter;
Let y = mol of $PbBr_2$ that dissolve per liter.
Then, at equilibrium, we have
$[Pb^{2+}] = x + y$, $[I^-] = 2x$ and $[Br^-] = 2y$
We know

$PbBr_2(s) \rightleftharpoons Pb^{2+} + 2Br^-$ $\qquad K_{sp} = 6.6 \times 10^{-6} = [Pb^{2+}][Br^-]^2$

$PbI_2(s) \rightleftharpoons Pb^{2+} + 2I^-$ $\qquad K_{sp} = 9.8 \times 10^{-9} = [Pb^{2+}][I^-]^2$

Substituting we get: $6.6 \times 10^{-6} = (x + y)(2y)^2$ and $9.8 \times 10^{-9} = (x + y)(2x)^2$

Solving for x and y we find
$x = 4.85 \times 10^{-4}$
$y = 7.91 \times 10^{-3}$
Thus, $[Pb^{2+}] = 8.40 \times 10^{-3}\ M$, $[I^-] = 9.70 \times 10^{-4}\ M$ and $[Br^-] = 1.58 \times 10^{-2}\ M$
Note: $[Br^-] > [I^-]$ because $PbBr_2$ is more soluble than PbI_2.

17.121 $[PbCl_4]^{2-}$ is formed.

$PbCl_2(s) \rightleftharpoons Pb^{2+}(aq) + 2Cl^-(aq)$ $\qquad K_{sp} = 1.7 \times 10^{-5}$

$Pb^{2+}(aq) + 4Cl^-(aq) \rightleftharpoons [PbCl_4]^{2-}(aq)$ $\qquad K_{form} = 2.5 \times 10^{15}$

Adding these equation gives

$$PbCl_2(s) + 2Cl^-(aq) \rightleftharpoons [PbCl_4]^{2-}(aq) \qquad K = K_{sp} \times K_{form} = 4.3 \times 10^{10} = \frac{\left[PbCl_4^{\ 2-}\right]}{\left[Cl^-\right]^2}$$

	$PbCl_2(s)$ +	$2Cl^-(aq)$	$\rightleftharpoons$	$[PbCl_4]^{2-}(aq)$
I	-	6.0		0
C	-	$-2x$		$+x$
E	-	$6.0 - 2x$		x

$[Cl^-] = 6.0\ M$
$[PbCl_4^{2-}] = x$

$$\frac{\left[PbCl_4^{\ 2-}\right]}{\left[Cl^-\right]^2} = \frac{x}{(6.0 - 2x)^2} = 4.3 \times 10^{10}$$

$x = 1.2 \times 10^5\ M$. This is incredibly high. $[PbCl_4]^{2-}$ is essentially soluble.
$K_{sp} = [Pb^{2+}][Cl^-]^2 = (0.016)(0.032)^2 = 1.6 \times 10^{-5}$

Chapter Eighteen
Thermodynamics

Practice Exercises

18.1 $w = -P\Delta V = -(14.0 \text{ atm})(12.0 \text{ L} - 1.0 \text{ L}) = -154 \text{ L atm}$
$\Delta E = w + q = 0$
$0 = -154 + q$
Therefore, $q = +154$ L atm
(The energy is converted into heat; since the heat does not leave the system the temperature increases.)

18.2 $\Delta E = q - P\Delta V$ since $q = 0$
$\Delta E = -P\Delta V$
but ΔV is negative for a compression so ΔE increases and T increases.
Energy is added to the system in the form of work.

18.3 $\Delta H = \Delta H_f^0 [NO_2(g)] - \{ \Delta H_f^0 N_2O(g)] + \Delta H_f^0 [O_2(g)]\}$
$\Delta H = \{4 \text{ mol } NO_2(g) \times (34 \text{ kJ/mol})\} - \{2 \text{ mol } N_2O \times (81.5 \text{ kJ/mol}) + 3 \text{ mol } O_2 \times (0 \text{ kJ/mol})\}$
$\Delta H = -27$ kJ
$\Delta E - \Delta H = \Delta nRT = (-1 \text{ mol})(8.314 \text{ J mol}^{-1} \text{ K}^{-1})(318 \text{ K}) = -2.64 \times 10^3 \text{ J}$
$\Delta E - \Delta H = -2.64$ kJ
ΔE is more exothermic.

18.4 $\Delta E° = \Delta H° - \Delta nRT = -217.1 \text{ kJ} - (-1 \text{ mol})(8.314 \text{ J mol}^{-1} \text{ K}^{-1})(298 \text{ K})$
$= -217.1 \text{ kJ} + 2.48 \text{ kJ}$
$= -214.6$ kJ

$$\% \text{ Difference} = \frac{2.48 \text{ kJ}}{217 \text{ kJ}} \times 100\% = 1\%$$

18.5 (a) Yes, spontaneous
(b) No, nonspontaneous
(c) No, nonspontaneous

18.6 (a) Push the stone uphill, or let it roll downhill.
(b) The ice in the iced tea can melt, or the glass of iced tea can be put in the freezer.
(c) Mix sodium hydroxide and hydrochloric acid in water to form sodium chloride.

18.7 ΔS should be negative since the reaction starts from ions in solution free to move in any direction to a solid in which the movement of the ions is constrained.

18.8 (a) ΔS is negative since the products have lower entropy, i.e. a lower freedom of movement.
(b) ΔS is positive since the products have higher entropy, i.e. a higher freedom of movement.

18.9 (a) ΔS is negative since there are less gas molecules. (The product is also more complex, indicating an increase in order.)
(b) ΔS is negative since the gases become a liquid and there are fewer molecules in the product.

(c) ΔS is negative since there are less gas molecules. (The product is also more complex, indicating an increase in order.)

(d) ΔS is positive since the particles go from an ordered, crystalline state to a more disordered, aqueous state.

18.10 (a) ΔH is positive and ΔS is negative since there are fewer molecules in the gas products as compared to the reactants. This reaction is expected to be nonspontaneous.

(b) ΔH is negative and ΔS is positive since there are more molecules in the gas products as compared to the reactants. This reaction is expected to be spontaneous.

(c) ΔH is negative and ΔS is positive since there are more molecules in the gas products as compared to the reactants. This reaction is expected to be spontaneous.

18.11 (a) ΔS is positive since there are more molecules in the gas phase as products as compared to the reactants, but ΔH is positive since heat has to be absorb to change a liquid into a gas. High temperatures will make this process spontaneous.

(b) ΔS is negative since there are fewer molecules in solution in the products as compared to the reactants, but ΔH is negative since heat has to be released as the ions form a solid. Low temperatures will make this process spontaneous.

18.12 $\frac{1}{2}N_2(g) + \frac{3}{2}H_2(g) \longrightarrow NH_3(g)$

$$\Delta S^\circ_f = \left[S^\circ_{NH_3(g)}\right] - [\left[\frac{1}{2}S^\circ_{N_2(g)} - \frac{3}{2}S^\circ_{H_2(g)}\right]$$

$$\Delta S^\circ_f = \left[(1 \text{ mol NH}_3) \times \left(\frac{192.5 \text{ J}}{\text{mol K}}\right)\right] - \left[\left(\frac{1}{2} \text{ mol N}_2\right) \times \left(\frac{191.5 \text{ J}}{\text{mol K}}\right) + \left(\frac{3}{2} \text{ mol H}_2\right) \times \left(\frac{130.6 \text{ J}}{\text{mol K}}\right)\right]$$

$$\Delta S^\circ_f = -99.1 \text{ J K}^{-1}$$

18.13 $\Delta S^\circ = (\text{sum } S^\circ[\text{products}]) - (\text{sum } S^\circ[\text{reactants}])$

(a) $\Delta S^\circ = \{S^\circ[H_2O(l)] + S^\circ[CaCl_2(s)]\} - \{S^\circ[CaO(s)] + 2S^\circ[HCl(g)]\}$

$\Delta S^\circ = \{1 \text{ mol} \times (69.96 \text{ J mol}^{-1} \text{ K}^{-1}) + 1 \text{ mol} \times (114 \text{ J mol}^{-1} \text{ K}^{-1})\}$
$- \{1 \text{ mol} \times (40 \text{ J mol}^{-1} \text{ K}^{-1}) + 2 \text{ mol} \times (186.7 \text{ J mol}^{-1} \text{ K}^{-1})\}$

$\Delta S^\circ = -229 \text{ J/K}$

(b) $\Delta S^\circ = \{S^\circ[C_2H_6(g)]\} - \{S^\circ[H_2(g)] + S^\circ[C_2H_4(g)]\}$

$\Delta S^\circ = \{1 \text{ mol} \times (229.5 \text{ J mol}^{-1} \text{ K}^{-1})\}$
$- \{1 \text{ mol} \times (130.6 \text{ J mol}^{-1} \text{ K}^{-1}) + 1 \text{ mol} \times (219.8 \text{ J mol}^{-1} \text{ K}^{-1})\}$

$\Delta S^\circ = -120.9 \text{ J/K}$

18.14 $N_2(g) + 2O_2(g) \longrightarrow N_2O_4(g)$

$$\Delta H^\circ_{N_2O_4(g)} = 9.67 \text{ kJ mol}^{-1}$$

$$\Delta S^\circ_{N_2O_4(g)} = (1 \text{ mol } N_2O_4)(304 \text{ J mol}^{-1} \text{ K}^{-1}) - (1 \text{ mol } N_2)(191.5 \text{ J mol}^{-1} \text{ K}^{-1}) - (2 \text{ mol } O_2)(205.0 \text{ J mol}^{-1} \text{ K}^{-1}) = -297.5 \text{ J K}^{-1}$$

$$\Delta G^\circ_f = \Delta H^\circ_f - T\Delta S^\circ_f$$

$$\Delta G^\circ_f = 9.67 \text{ kJ mol}^{-1} - (298 \text{ K})(-0.2975 \text{ kJ mol}^{-1} \text{ K}^{-1})$$

$$\Delta G^\circ_f = 98.3 \text{ kJ mol}^{-1}$$

18.15 First, we calculate $\Delta S°$, using the data:

$\Delta S° = \{2S°[Fe_2O_3(s)]\} - \{3S°[O_2(g)] + 4S°[Fe(s)]\}$

$\Delta S° = \{2 \text{ mol} \times (90.0 \text{ J mol}^{-1} \text{ K}^{-1})\} - \{3 \text{ mol} \times (205.0 \text{ J mol}^{-1} \text{ K}^{-1}) + 4 \text{ mol} \times (27 \text{ J mol}^{-1} \text{ K}^{-1})\}$

$\Delta S° = -543 \text{ J/K} = -0.543 \text{ kJ/mol}$

Next, we calculate $\Delta H°$ using the data

$\Delta H° = (\text{sum } \Delta H_f°[\text{products}]) - (\text{sum } \Delta H_f°[\text{reactants}])$

$\Delta H° = \{2\Delta H_f°[Fe_2O_3(s)]\} - \{3\Delta H_f°[O_2(g)] + 4\Delta H_f°[Fe(s)]\}$

$\Delta H° = \{2 \text{ mol} \times (-822.2 \text{ kJ/mol})\} - \{3 \text{ mol} \times (0.0 \text{ kJ/mol}) + 4 \times (0.0 \text{ kJ/mol})\}$

$\Delta H° = -1644 \text{ kJ}$

The temperature is 25.0 + 273.15 = 298.15 K, and the calculation of $\Delta G°$ is as follows:

$\Delta G° = \Delta H° - T\Delta S° = -1644 \text{ kJ} - (298.15 \text{ K})(-0.543 \text{ kJ/K}) = -1482 \text{ kJ}$

18.16 $Fe_2O_3(s) + 3CO(g) \longrightarrow 2Fe(s) + 3CO_2(g)$

$\Delta G° = \{2 \text{ mol Fe}(s) \times \Delta G_f°[Fe(s)] + 3 \text{ mol } CO_2(g) \times \Delta G_f°[CO_2(g)]\}$
$- \{1 \text{ mol } Fe_2O_3(s) \times \Delta G_f°[Fe_2O_3(s)] + 3 \text{ mol } CO(g) \times \Delta G_f°[CO(g)]\}$

$\Delta G° = \{2 \text{ mol Fe}(s) \times 0 \text{ kJ mol}^{-1} + 3 \text{ mol } CO_2(g) \times -394.4 \text{ kJ mol}^{-1}\}$
$- \{1 \text{ mol } Fe_2O_3(s) \times -741.0 \text{ kJ mol}^{-1} + 3 \text{ mol } CO(g) \times -137.3 \text{ kJ mol}^{-1}\}$

$\Delta G° = -30.3 \text{ kJ}$

18.17 We calculate $\Delta G°_{rxn}$, using the data from Table 18.2:

(a) $\Delta G°_{rxn} = \{2\Delta G_f°[NO_2(g)]\} - \{2\Delta G_f°[NO(g)] + \Delta G_f°[O_2(g)]\}$

$\Delta G°_{rxn} = \{2 \text{ mol} \times (+51.84 \text{ kJ mol}^{-1})\}$
$- \{2 \text{ mol} \times (+86.69 \text{ kJ mol}^{-1}) + 1 \text{ mol} \times (0 \text{ kJ mol}^{-1})\}$

$\Delta G°_{rxn} = -69.7 \text{ kJ/mol}$

(b) $\Delta G°_{rxn} = \{\Delta G_f°[CaCl_2(s)] + 2\Delta G_f°[H_2O(g)]\} - \{\Delta G_f°[Ca(OH)_2(s)] + 2\Delta G_f°[HCl(g)]\}$

$\Delta G°_{rxn} = \{1 \text{ mol} \times (-750.2 \text{ kJ mol}^{-1}) + 2 \text{ mol} \times (-228.6 \text{ kJ mol}^{-1})\}$
$- \{1 \text{ mol} \times (-896.76 \text{ kJ mol}^{-1}) + 2 \text{ mol} \times (-95.27 \text{ kJ mol}^{-1})\}$

$\Delta G°_{rxn} = -120.1 \text{ kJ/mol}$

18.18 $C_2H_5OH(l) + 3O_2(g) \longrightarrow 2CO_2(g) + 3H_2O(l)$

$\Delta G°_{rxn} = \{2\Delta G_f°[CO_2(g)] + 3\Delta G_f°[H_2O(l)]\} - \{\Delta G_f°[C_2H_5OH(l)] + 3\Delta G_f°[O_2(g)]\}$

$\Delta G°_{rxn} = \{2 \text{ mol} \times (-394.4 \text{ kJ mol}^{-1}) + 3 \text{ mol} \times (-237.2 \text{ kJ mol}^{-1})\}$
$- \{1 \text{ mol} \times (-174.8 \text{ kJ mol}^{-1}) + 3 \text{ mol} \times (0 \text{ kJ mol}^{-1})\}$

$G°_{rxn} = -1325.6 \text{ kJ}$

For 125 g C_2H_5OH:

$$\text{mol } C_2H_5OH = 125 \text{ g } C_2H_5OH\left(\frac{1 \text{ mol } C_2H_5OH}{46.08 \text{ g } C_2H_5OH}\right) = 2.71 \text{ mol } C_2H_5OH$$

Maximum work:

$(1325.6 \text{ kJ mol}^{-1})(2.71 \text{ mol } C_2H_5OH) = 3590 \text{ kJ}$

$C_8H_{18}(l) + \frac{25}{2}O_2(g) \longrightarrow 8CO_2(g) + 9H_2O(l)$

$\Delta G°_{rxn} = \{8\Delta G_f°[CO_2(g)] + 9\Delta G_f°[H_2O(l)]\} - \{\Delta G_f°[C_8H_{18}(l)] + \frac{25}{2}\Delta G_f°[O_2(g)]\}$

$\Delta G°_{rxn} = \{8 \text{ mol} \times (-394.4 \text{ kJ mol}^{-1}) + 9 \text{ mol} \times (-237.2 \text{ kJ mol}^{-1})\}$
$- \{1 \text{ mol} \times (+17.3 \text{ kJ mol}^{-1}) + \frac{25}{2} \text{ mol} \times (0 \text{ kJ mol}^{-1})\}$

$G°_{rxn} = -5304.1 \text{ kJ}$

For 125 g C_8H_{18}:

$$\text{mol } C_8H_{18} = 125 \text{ g } C_8H_{18}\left(\frac{1 \text{ mol } C_8H_{18}}{114.26 \text{ g } C_8H_{18}}\right) = 1.09 \text{ mol } C_8H_{18}$$

Maximum work:

$(5304.1 \text{ kJ mol}^{-1})(1.09 \text{ mol } C_8H_{18}) = 5810 \text{ kJ}$

The octane is a better fuel on both a per gram and per mole basis.

18.19 The maximum amount of work that is available is the free energy change for the process, in this case, the standard free energy change, $\Delta G°$, since the process occurs at 25 °C.

$4Al(s) + 3O_2(g) \longrightarrow 2Al_2O_3(s)$

$\Delta G° = (\text{sum } \Delta G_f°[\text{products}]) - (\text{sum } \Delta G_f°[\text{reactants}])$

$\Delta G° = 2\Delta G_f°[Al_2O_3(s)] - \{3\Delta G_f°[O_2(g)] + 4\Delta G_f°[Al(s)]\}$

$\Delta G° = 2 \text{ mol} \times (-1576.4 \text{ kJ/mol}) - \{3 \text{ mol} \times (0.0 \text{ kJ/mol}) + 4 \text{ mol} \times (0.0 \text{ kJ/mol})\}$

$\Delta G° = -3152.8$ kJ, for the reaction as written.

This calculation conforms to the reaction *as written*. This means that the above value of $\Delta G°$ applies to the equation involving *4 mol* of Al. The conversion to give energy *per mole* of aluminum is then: −3152.8 kJ/4 mol Al = −788 kJ/mol

The maximum amount of energy that may be obtained is thus 788 kJ.

18.20 $T \approx \dfrac{\Delta H°}{\Delta S°}$

$$\Delta H° = 21.7 \text{ kJ mol}^{-1}\left(\frac{1000 \text{ J}}{1 \text{ kJ}}\right) = 21.7 \times 10^3 \text{ J}$$

$$239.9 \text{ K} \approx \frac{21.7 \times 10^3 \text{ J}}{\Delta S°}$$

$\Delta S° = 90.5 \text{ J K}^{-1}$

18.21 For the vaporization process in particular, and for any process in general, we have:

$\Delta G = \Delta H - T\Delta S$

If the temperature is taken to be that at which equilibrium is obtained, that is the temperature of the boiling point (where liquid and vapor are in equilibrium with one another), then we also have the result that ΔG is equal to zero:

$$\Delta G = 0 = \Delta H - T\Delta S, \text{ or } T_{eq} = \frac{\Delta H}{\Delta S}$$

We know ΔH to be 60.7 kJ/mol; we need the value for ΔS in units kJ $\text{mol}^{-1}\text{ K}^{-1}$:

$\Delta S° = (\text{sum } S°[\text{products}]) - (\text{sum } S°[\text{reactants}])$

$\Delta S° = S°[Hg(g)] - S°[Hg(l)]$

$\Delta S° = (175 \times 10^{-3} \text{ kJ mol}^{-1}\text{ K}^{-1}) - (76.1 \times 10^{-3} \text{ kJ mol}^{-1}\text{ K}^{-1})$

$\Delta S° = 98.9 \times 10^{-3} \text{ kJ mol}^{-1}\text{ K}^{-1}$

$$T_{eq} = \frac{60.7 \text{ kJ mol}^{-1}}{98.9 \times 10^{-3} \text{ kJ mol}^{-1}\text{ K}^{-1}} = 614 \text{ K} \qquad \text{or } 341 \text{ °C}$$

18.22 $\Delta G° = (\text{sum } \Delta G_f°[\text{products}]) - (\text{sum } \Delta G_f°[\text{reactants}])$

$$\Delta G° = \Delta G_f°[SO_3(g)] - \{\Delta G_f°[SO_2(g)] + \Delta G_f°\left[\frac{1}{2}O_2(g)\right]\}$$

$$\Delta G° = 1 \text{ mol} \times (-370.4 \text{ kJ/mol}) - \{1 \text{ mol} \times (-300.4 \text{ kJ/mol}) + \frac{1}{2}(0.0 \text{ kJ/mol})\}$$

$\Delta G° = -70.0$ kJ/mol

Since the sign of $\Delta G°$ is negative, the reaction should be spontaneous.

18.23 $\Delta G° = \Delta H° - T\Delta S°$

$\Delta G° = \{2\Delta G_f°[HCl(g)] + \Delta G_f°[CaCO_3(s)]\} - \{\Delta G_f°[CaCl_2(s)] + \Delta G_f°[H_2O(g)] + \Delta G_f°[CO_2(g)]\}$

$\Delta G° = \{2 \text{ mol} \times (-95.27 \text{ kJ/mol}) + 1 \text{ mol} \times (-1128.8 \text{ kJ/mol})\}$
$- \{1 \text{ mol} \times (-750.2 \text{ kJ/mol}) + 1 \text{ mol} \times (-228.6 \text{ kJ/mol}) + 1 \text{ mol} \times (-394.4 \text{ kJ/mol})\}$

$\Delta G° = +53.9$ kJ

$\Delta G°$ is positive, the reaction is not spontaneous, and we do not expect to see products formed from reactants.

18.24 $T = 75\ °C + 273.15\ K = 348\ K$

$\Delta G° = 119.2 \text{ kJ} - (348 \text{ K})(0.3548 \text{ kJ/K}) = -4.3 \text{ kJ}$

18.25 $\Delta G° = \{\Delta G_f°[Na_2CO_3(s)] + \Delta G_f°[CO_2(g)] + \Delta G_f°[H_2O(g)]\} - \{2\Delta G_f°[NaHCO_3(s)]\}$

$\Delta G° = \{1 \text{ mol} \times (-1048 \text{ kJ/mol}) + 1 \text{ mol} \times (-394.4 \text{ kJ/mol}) + 1 \text{ mol} \times (-228.6 \text{ kJ/mol})\}$
$- \{2 \text{ mol} \times (-851.9 \text{ kJ/mol})\}$

$\Delta G° = +32.8$ kJ

$\Delta H° = \{\Delta H_f°[Na_2CO_3(s)] + \Delta H_f°[CO_2(g)] + \Delta H_f°[H_2O(g)]\} - \{2\Delta H_f°[NaHCO_3(s)]\}$

$\Delta H° = \{1 \text{ mol} \times (-1131 \text{ kJ/mol}) + 1 \text{ mol} \times (-393.5 \text{ kJ/mol}) + 1 \text{ mol} \times (-241.8 \text{ kJ/mol})\}$
$- \{2 \text{ mol} \times (-947.7 \text{ kJ/mol})\}$

$\Delta H° = +129.1$ kJ

$\Delta S° = \{S_f°[Na_2CO_3(s)] + S_f°[CO_2(g)] + S_f°[H_2O(g)]\} - \{2S_f°[NaHCO_3(s)]\}$

$\Delta S° = \{1 \text{ mol} \times (136 \text{ J/mol K}) + 1 \text{ mol} \times (213.6 \text{ J/mol K}) + 1 \text{ mol} \times (188.7 \text{ J/mol K})\}$
$- \{2 \text{ mol} \times (102 \text{ J/mol K})\}$

$\Delta S° = +334.3 \text{ J K}^{-1} = 0.3343 \text{ kJ K}^{-1}$

$\Delta G^{\circ}_{490} = \Delta H° - T\Delta S°$

$\Delta G^{\circ}_{490} = 129.1 \text{ kJ} - (490 \text{ K})(0.3343 \text{ kJ K}^{-1}) = -34.7 \text{ kJ}$

The equilibrium shifts to products.

18.26 $$\Delta G = \Delta G^{\circ}_{298} + RT \ln\left(\frac{P_{N_2O_4}}{P_{NO_2}{}^2}\right)$$

$$\Delta G = -5.40 \times 10^3 \text{ J mol}^{-1} + \left(8.314 \text{ J mol}^{-1} \text{ K}^{-1}\right)\left(298 \text{ K}\right)\ln\left(\frac{0.598 \text{ atm}}{\left(0.260 \text{ atm}\right)^2}\right)$$

$\Delta G = 0 \text{ J mol}^{-1} = 0 \text{ kJ mol}^{-1}$

The system is at equilibrium, so that the reaction will not move.

18.27 Using the data provided we may write:

$$\Delta G = \Delta G^{\circ} + RT \ln\left(\frac{P_{N_2O_4}}{\left(P_{NO_2}\right)^2}\right)$$

$$= -5.40 \times 10^3 \text{ J mol}^{-1} + \left(8.314 \text{ J mol}^{-1} \text{ K}^{-1}\right)\left(298 \text{ K}\right) \ln\left(\frac{0.25 \text{ atm}}{\left(0.60 \text{ atm}\right)^2}\right)$$

$$= -5.40 \times 10^3 \text{ J mol}^{-1} + \left(-9.03 \times 10^2 \text{ J mol}^{-1}\right)$$

$$= -6.30 \times 10^3 \text{ J mol}^{-1}$$

Since ΔG is negative, the forward reaction is spontaneous and the reaction will proceed in the forward direction.

18.28 $\Delta G^{\circ} = -RT \ln K_p$
$\Delta G^{\circ} = -(8.314 \text{ J K}^{-1} \text{ mol}^{-1})(25 + 273 \text{ K}) \times \ln(6.9 \times 10^5) = -33 \times 10^3 \text{ J}$
$\Delta G^{\circ} = -33 \text{ kJ}$

18.29 $\Delta G^{\circ} = -RT \ln K_p$
$3.3 \times 10^3 \text{ J mol}^{-1} = -(8.314 \text{ J K}^{-1} \text{ mol}^{-1})(25.0 + 273.15 \text{ K}) \times \ln(K_p)$
$K_p = 0.26$

18.30 $3C(g) + 8H(g) + O(g)$

ΔH_f^0 (atoms) ΔH (Bond Formation)

$3C(\text{graphite}) + 4H_2(g) + ½O_2(g) \longrightarrow CH_3CH_2CH_2OH(g)$ ΔH_f^0

The heat of formation of the gaseous reactants is endothermic, and therefore positive.
3 mol C × 716.7 kJ mol^{-1} = 2150.1 kJ
8 mol H × 217.9 kJ mol^{-1} = 1743.2 kJ
1 mol O × 249.2 kJ mol^{-1} = 249.2 kJ
Total energy input is 4142.5 kJ
Now determine the energy released during bond formation.

2 C—C bonds formed	2 mol × 348 kJ mol^{-1} = 696 kJ
7 C—H bonds formed	7 mol × 412 kJ mol^{-1} = 2884 kJ
1 C—O bond formed	1 mol × 360 kJ mol^{-1} = 360 kJ
1 O—H bond formed	1 mol × 463 kJ mol^{-1} = 463 kJ

Total energy released is 4403 kJ
The heat of formation is: 4142.5 kJ + (–4403 kJ) = –260.5 kJ
$4C(s) + 9/2H_2(g) + Br_2(l) \longrightarrow CH_3CHBrCH_2CH_3(g)$ ΔH_f^0

The heat of formation of the gaseous reactants is endothermic, and therefore positive.
4 mol C × 716.7 kJ mol^{-1} = 2866.8 kJ
9 mol H × 217.9 kJ mmol^{-1} = 1961.1 kJ
1 mol Br × 112.4 kJ mol^{-1} = 112.4 kJ
Total energy input is 4940.3 kJ

Now determine the energy released during bond formation.

3 C–C bonds formed 3 mol × 348 kJ mol^{-1} = 1044 kJ

9 C–H bonds formed 9 mol × 412 kJ mol^{-1} = 3708 kJ

1 C–Br bond formed 1 mol × 276 kJ mol^{-1} = 276 kJ

Total energy released is 5028 kJ

The heat of formation is: 4940.3 kJ + (– 5028 kJ) = –88 kJ

18.31 $6C(s) + 6H_2(g) \longrightarrow C_6H_{12}(g) \qquad \Delta H_f^0$

The heat of formation of the gaseous reactants is endothermic, and therefore positive.

6 mol C × 716.7 kJ mol^{-1} = 4300.2 kJ

12 mol H × 217.9 kJ mol^{-1} = 2614.8 kJ

Total energy input is 6915 kJ

Now determine the energy released during bond formation.

6 C—C bonds formed 6 mol × 348 kJ mol^{-1} = 2088 kJ

12 C—H bonds formed 12 mol × 412 kJ mol^{-1} = 4944 kJ

Total energy released is 7032 kJ

The heat of formation is: 6915 kJ + (–7032 kJ) = –117 kJ

$6C(g) + 6H(g) \longrightarrow C_6H_6(g)$

The heat of formation of the gaseous reactants is endothermic, and therefore positive.

6 mol C × 716.7 kJ mol^{-1} = 4300.2 kJ

6 mol H × 217.9 kJ mmol^{-1} = 1307.4 kJ

Total energy input is 5607.6 kJ

Now determine the energy released during bond formation.

Benzene bonding is neither single nor double bonding between carbon atoms but we will approximate the bonding as 3 single C–C bonds and 3 double C–C bonds.

3 C—C bonds formed 3 mol × 348 kJ mol^{-1} = 1044 kJ

3 C=C bonds formed 3 mol × 612 kJ mol^{-1} = 1836

6 C—H bonds formed 6 mol × 412 kJ mol^{-1} = 2472 kJ

Total energy released is 5352 kJ

The heat of formation is: 5607.6 kJ + (– 5352 kJ) = +255.6 kJ

Review Problems

18.55 $\Delta E = q + w$ = 0.300 kJ + 0.700 kJ = +1.000 kJ

The overall process is endothermic, meaning that the internal energy of the system increases. Notice that both terms, q and w, contribute to the increase in internal energy of the system; the system gains heat (+q) and has work done on it (+w).

18.57 work = $P \times \Delta V$

The total pressure caused by the hand pump:

P = 30.0 lb/in^2

Converting to atmospheres we get:

P = 30.0 lb/in^2 × 1 atm/14.7 lb/in^2 = 2.04 atm

Next we convert the volume change in units in^3 to units L:

24.0 in^3 × (2.54 cm/in)^3 × 1 L/1000 cm^3 = 0.393 L

Hence $P \times \Delta V = (2.04 \text{ atm})(0.393 \text{ L}) = 0.802 \text{ L·atm}$
$0.802 \text{ L·atm} \times 101.3 \text{ J/L·atm} = 81 \text{ J}$

18.59 We use the data supplied in Appendix C.

(a) $3PbO(s) + 2NH_3(g) \longrightarrow 3Pb(s) + N_2(g) + 3H_2O(g)$

$$\Delta H° = \{ \Delta H_f^o[Pb(s)] + \Delta H_f^o[N_2(g)] + 3\,\Delta H_f^o[H_2O(g)]\} - \{3\,\Delta H_f^o[PbO(s)] + 2\,\Delta H_f^o NH_3(g)]\}$$

$$\Delta H° = \{3 \text{ mol} \times (0 \text{ kJ/mol}) + 1 \text{ mol} \times (0 \text{ kJ/mol}) + 3 \text{ mol} \times (-241.8 \text{ kJ/mol})\} - \{3 \text{ mol} \times (-219.2 \text{ kJ/mol}) + 2 \text{ mol} \times (-46.19 \text{ kJ/mol})\}$$

$\Delta H° = +\,24.58 \text{ kJ}$

$\Delta E = \Delta H° - \Delta nRT$

$\Delta E = 24.58 \text{ kJ} - (+2 \text{ mol})(8.314 \text{ J/mol K})(10^{-3} \text{ kJ/J})(298 \text{ K}) = 19.6 \text{ kJ}$

(b) $NaOH(s) + HCl(g) \longrightarrow NaCl(s) + H_2O(l)$

$$\Delta H° = \{ \Delta H_f^o[NaCl(s)] + \Delta H_f^o[H_2O(l)]\} - \{ \Delta H_f^o[NaOH(s)] + \Delta H_f^o[HCl(g)]\}$$

$$\Delta H° = \{1 \text{ mol} \times (-411.0 \text{ kJ/mol}) + 1 \text{ mol} \times (-285.9 \text{ kJ/mol})\} - \{1 \text{ mol} \times (-426.8 \text{ kJ/mol}) + 1 \text{ mol} \times (-92.3)\}$$

$\Delta H° = -178 \text{ kJ}$

$\Delta E = \Delta H° - \Delta nRT$

$\Delta E = -178 \text{ kJ} - (-1)(8.314 \text{ J/mol K})(10^{-3} \text{ kJ/J})(298 \text{ K}) = -175 \text{ kJ}$

(c) $Al_2O_3(s) + 2Fe(s) \longrightarrow Fe_2O_3(s) + 2Al(s)$

$$\Delta H° = \{ \Delta H_f^o[Fe_2O_3(s)] + 2\,\Delta H_f^o[Al(s)]\} - \{ \Delta H_f^o[Al_2O_3(s)] + 2\,\Delta H_f^o[Fe(s)]\}$$

$$\Delta H° = \{1 \text{ mol} \times (-822.2 \text{ kJ/mol}) + 2 \text{ mol} \times (0 \text{ kJ/mol})\} - \{1 \text{ mol} \times (-1669.8 \text{ kJ/mol}) + 2 \text{ mol} \times (0 \text{ kJ/mol})\}$$

$\Delta H° = 847.6 \text{ kJ}$

$\Delta E = \Delta H°$, since the value of Δn for this reaction is zero.

(d) $2CH_4(g) \longrightarrow C_2H_6(g) + H_2(g)$

$$\Delta H° = \{ \Delta H_f^o[C_2H_6(g)] + \Delta H_f^o[H_2(g)]\} - \{2\,\Delta H_f^o[CH_4(g)]\}$$

$$\Delta H° = \{1 \text{ mol} \times (-84.667 \text{ kJ/mol}) + 1 \text{ mol} \times (0.0 \text{ kJ/mol})\} - \{2 \text{ mol} \times (-74.848 \text{ kJ/mol})\}$$

$\Delta H° = 65.029 \text{ kJ}$

$\Delta E = \Delta H°$, since the value of Δn for this reaction is zero.

18.61 $\Delta H = \Delta E + \Delta n_{gas}RT$

$\Delta E = \Delta H - \Delta n_{gas}RT$

$\Delta H = -163.14 \text{ kJ}$

$\Delta n_{gas} = 3 \text{ mol} - 2 \text{ mol} = 1 \text{ mol}$

$R = 8.314 \text{ J mol}^{-1} \text{ K}^{-1}$

$T = 25\ °C + 273 \text{ K} = 298 \text{ K}$

For 186 g N_2O, first calculate the number of moles of N_2O and then how many kJ of energy will be released for that many moles of N_2O

$\Delta E = -163.14 \text{ kJ} - (1 \text{ mol})(8.314 \times 10^{-3} \text{ kJ mol}^{-1} \text{ K}^{-1})(298 \text{ K})$

$\Delta E = -165.62 \text{ kJ}$

$$\text{mol } N_2O = 186 \text{ g}\left(\frac{1 \text{ mol } N_2O}{44.02 \text{ g } N_2O}\right) = 4.23 \text{ mol } N_2O$$

$$\text{Amount of Energy} = 4.23 \text{ mol } N_2O\left(\frac{-165.62 \text{ kJ}}{2 \text{ mol } N_2O}\right) = -350 \text{ kJ}$$

For $T = 217$ °C
$T = 217$ °C + 273 K = 490 K
$\Delta E = -163.14$ kJ – (1 mol)(8.314×10^{-3} kJ mol^{-1} K^{-1})(490 K)
$\Delta E = -163.14$ kJ –4.07 kJ
$\Delta E_{217\,°C} = -167.21$ kJ for 2 moles of N_2O
For 186 g N_2O, 4.23 mol N_2O:

$$\text{Amount of Energy at 217 °C} = 4.23 \text{ mol } N_2O\left(\frac{-167.21 \text{ kJ}}{2 \text{ mol } N_2O}\right) = -354 \text{ kJ}$$

18.63 (a) ΔS will increase; a solid is becoming a gas.
(b) ΔS will decrease; three particles of a gas are becoming one particle.
(c) ΔS will decrease; two solid particles and one gas particle are becoming a crystalline solid.

18.65 In general, we have the equation: $\Delta H° = (\text{sum } \Delta H_f^\circ[\text{products}]) - (\text{sum } \Delta H_f^\circ[\text{reactants}])$

(a) $\Delta H° = \{\Delta H_f^\circ[CaCO_3(s)]\} - \{\Delta H_f^\circ[CO_2(g)] + \Delta H_f^\circ[CaO(s)]\}$
$\Delta H° = \{1 \text{ mol} \times (-1207 \text{ kJ/mol})\}$
$- \{1 \text{ mol} \times (-393.5 \text{ kJ/mol}) + 1 \text{ mol} \times (-635.5 \text{ kJ/mol})\}$
$\Delta H° = -178$ kJ ∴ favored.

(b) $\Delta H° = \{\Delta H_f^\circ[C_2H_6(g)]\} - \{\Delta H_f^\circ[C_2H_2(g)] + 2\,\Delta H_f^\circ[H_2(g)]\}$
$\Delta H° = \{1 \text{ mol} \times (-84.5 \text{ kJ/mol})\} - \{1 \text{ mol} \times (226.75 \text{ kJ/mol}) + 2 \text{ mol} \times (0.0 \text{ kJ/mol})\}$
$\Delta H° = -311$ kJ ∴ favored.

(c) $\Delta H° = \{\Delta H_f^\circ[Fe_2O_3(s)] + 3\,\Delta H_f^\circ[Ca(s)]\} - \{2\,\Delta H_f^\circ[Fe(s)] + 3\,\Delta H_f^\circ[CaO(s)]\}$
$\Delta H° = \{1 \text{ mol} \times (-822.2 \text{ kJ/mol}) + 3 \text{ mol} \times (0.0 \text{ kJ/mol})\}$
$- \{2 \text{ mol} \times (0.0 \text{ kJ/mol}) + 3 \text{ mol} \times (-635.5 \text{ kJ/mol})\}$
$\Delta H° = +1084.3$ kJ ∴ not favorable from the standpoint of enthalpy alone.

18.67 The system with fewer ways to arrange the particles will have the least entropy. That would be the second system.

18.69 System (*b*) has the highest entropy since it has the greatest number of particles.

18.71 $2N_2O(g) \longrightarrow 2N_2(g) + O_2(g)$

The factors needed in order to determine the sign of ΔS are
(1) the number of moles of products versus reactants
in this case the number of moles increases
(2) the state of the products versus reactants
both products and reactants are gases
(3) the complexity of the molecules
N_2O is more complex than either N_2 or O_2
ΔS is expected to be positive.

18.73 (a) Negative – since the number of moles of gaseous material decreases.
(b) Negative – since the number of moles of gaseous material decreases.
(c) Negative – since the number of moles of gas decreases.
(d) Positive – since a gas appears where there formerly was none.

18.75 (a) At low temperatures the reaction will be spontaneous.
(b) At high temperatures the reaction will be spontaneous.

18.77 $\Delta S° = (\text{sum } S°[\text{products}]) - (\text{sum } S°[\text{reactants}])$

(a) $\Delta S° = \{2S°[NH_3(g)]\} - \{3S°[H_2(g)] + S°[N_2(g)]\}$

$\Delta S° = \{2 \text{ mol} \times (192.5 \text{ J mol}^{-1} \text{ K}^{-1})\} - \{3 \text{ mol} \times (130.6 \text{ J mol}^{-1} \text{ K}^{-1}) + 1 \text{ mol} \times (191.5 \text{ J mol}^{-1} \text{ K}^{-1})\}$

$\Delta S° = -198.3$ J/K ∴ not spontaneous from the standpoint of entropy.

(b) $\Delta S° = \{S°[CH_3OH(l)]\} - \{2S°[H_2(g)] + S°[CO(g)]\}$

$\Delta S° = \{1 \text{ mol} \times (126.8 \text{ J mol}^{-1} \text{ K}^{-1})\} - \{2 \text{ mol} \times (130.6 \text{ J mol}^{-1} \text{ K}^{-1}) + 1 \text{ mol} \times (197.9 \text{ J mol}^{-1} \text{ K}^{-1})\}$

$\Delta S° = -332.3$ J/K ∴ not favored from the standpoint of entropy alone.

(c) $\Delta S° = \{6S°[H_2O(g)] + 4S°[CO_2(g)]\} - \{7S°[O_2(g)] + 2S°[C_2H_6(g)]\}$

$\Delta S° = \{6 \text{ mol} \times (188.7 \text{ J mol}^{-1} \text{ K}^{-1}) + 4 \text{ mol} \times (213.6 \text{ J mol}^{-1} \text{ K}^{-1})\} - \{7 \text{ mol} \times (205.0 \text{ J mol}^{-1} \text{ K}^{-1}) + 2 \text{ mol} \times (229.5 \text{ J mol}^{-1} \text{ K}^{-1})\}$

$\Delta S° = +92.6$ J/K ∴ favorable from the standpoint of entropy alone.

(d) $\Delta S° = \{2S°[H_2O(l)] + S°[CaSO_4(s)]\} - \{S°[H_2SO_4(l)] + S°[Ca(OH)_2(s)]\}$

$\Delta S° = \{2 \text{ mol} \times (69.96 \text{ J mol}^{-1} \text{ K}^{-1}) + 1 \text{ mol} \times (107 \text{ J mol}^{-1} \text{ K}^{-1})\} - \{1 \text{ mol} \times (157 \text{ J mol}^{-1} \text{ K}^{-1}) + 1 \text{ mol} \times (76.1 \text{ J mol}^{-1} \text{ K}^{-1})\}$

$\Delta S° = +14$ J/K ∴ favorable from the standpoint of entropy alone.

(e) $\Delta S° = \{2S°[N_2(g)] + S°[SO_2(g)]\} - \{2S°[N_2O(g)] + S°[S(s)]\}$

$\Delta S° = \{2 \text{ mol} \times (191.5 \text{ J mol}^{-1} \text{ K}^{-1}) + 1 \text{ mol} \times (248 \text{ J mol}^{-1} \text{ K}^{-1})\} - \{2 \text{ mol} \times (220.0 \text{ J mol}^{-1} \text{ K}^{-1}) + 1 \text{ mol} \times (31.9 \text{ J mol}^{-1} \text{ K}^{-1})\}$

$\Delta S° = +159$ J/K ∴ favorable from the standpoint of entropy alone.

18.79 The entropy change that is designated ΔS_f^o is that which corresponds to the reaction in which one mole of a substance is formed from elements in their standard states. Since the value is understood to correspond to the reaction forming one mole of a single pure substance, the units may be written either J K^{-1} or J mol^{-1} K^{-1}.

(a) $2C(s) + 2H_2(g) \longrightarrow C_2H_4(g)$

$\Delta S° = \{S°[C_2H_4(g)]\} - \{2S°[C(s)] + 2S°[H_2(g)]\}$

$\Delta S° = \{1 \text{ mol} \times (219.8 \text{ J mol}^{-1} \text{ K}^{-1})\} - \{2 \text{ mol} \times (5.69 \text{ J mol}^{-1} \text{ K}^{-1}) + 2 \text{ mol} \times (130.6 \text{ J mol}^{-1} \text{ K}^{-1})\}$

$\Delta S° = -52.8$ J/K or -52.8 J mol^{-1} K^{-1}

(b) $2H_2(g) + 2C(s) + O_2(g) \longrightarrow HC_2H_3O_2(l)$

$\Delta S° = \{S°[HC_2H_3O_2(l)]\} - \{2S°[H_2(g)] + 2S°[C(s)] + S°[O_2(g)]\}$

$\Delta S° = \{1 \text{ mol} \times (160 \text{ J mol}^{-1} \text{ K}^{-1})\} - \{2 \text{ mol} \times (130.6 \text{ J mol}^{-1} \text{ K}^{-1}) + 2 \text{ mol} \times (5.69 \text{ J mol}^{-1} \text{ K}^{-1}) + 1 \text{ mol} \times (205.0 \text{ J mol}^{-1} \text{ K}^{-1})\}$

$\Delta S° = -318$ J/K or -318 J mol^{-1} K^{-1}

(c) $Ca(s) + S(s) + 3O_2(g) + 2H_2(g) \longrightarrow CaSO_4{\cdot}2H_2O(s)$

$\Delta S° = \{S°[CaSO_4{\cdot}2H_2O(s)]\} - \{2S°[H_2(g)] + 3S°[O_2(g)] + S°[S(s)] + S°[Ca(s)]\}$

$\Delta S° = \{1 \text{ mol} \times (194.0 \text{ J mol}^{-1} \text{ K}^{-1})\} - \{2 \text{ mol} \times (130.6 \text{ J mol}^{-1} \text{ K}^{-1}) + 3 \text{ mol} \times (205.0 \text{ J mol}^{-1} \text{ K}^{-1}) + 1 \text{ mol} \times (31.9 \text{ J mol}^{-1} \text{ K}^{-1}) + 1 \text{ mol} \times (154.8 \text{ J mol}^{-1} \text{ K}^{-1})\}$

$\Delta S° = -868.9$ J mol^{-1} K^{-1} J/K or -868.9 J mol^{-1} K^{-1}

18.81 $\Delta S° = (\text{sum } S°[\text{products}]) - (\text{sum } S°[\text{reactants}])$

$\Delta S° = \{2S°[HNO_3(l)] + S°[NO(g)]\} - \{3S°[NO_2(g)] + S°[H_2O(l)]\}$

$\Delta S° = \{2 \text{ mol} \times (155.6 \text{ J mol}^{-1} \text{ K}^{-1}) + 1 \text{ mol} \times (210.6 \text{ J mol}^{-1} \text{ K}^{-1})\}$
$- \{3 \text{ mol} \times (240.5 \text{ J mol}^{-1} \text{ K}^{-1}) + 1 \text{ mol} \times (69.96 \text{ J mol}^{-1} \text{ K}^{-1})\}$

$\Delta S° = -269.7 \text{ J/K}$

18.83 The quantity ΔG_f^o applies to the equation in which one mole of pure phosgene is produced from the naturally occurring forms of the elements:

$C(s) + 1/2O_2(g) + Cl_2(g) \longrightarrow COCl_2(g),\ \Delta G_f^o = ?$

We can determine ΔG_f^o if we can find values for ΔH_f^o and ΔS_f^o, because:

$\Delta G° = \Delta H° - T\Delta S°$

The value of ΔS_f^o is determined using $S°$ for phosgene in the following way:

$\Delta S_f^o = \{S°[COCl_2(g)]\} - \{S°[C(s)] + 1/2S°[O_2(g)] + S°[Cl_2(g)]\}$

$\Delta S_f^o = \{1 \text{ mol} \times (284 \text{ J mol}^{-1} \text{ K}^{-1})\} - \{1 \text{ mol} \times (5.69 \text{ J mol}^{-1} \text{ K}^{-1})$
$+ 1/2 \text{ mol} \times (205.0 \text{ J mol}^{-1} \text{ K}^{-1}) + 1 \text{ mol} \times (223.0 \text{ J mol}^{-1} \text{ K}^{-1})\}$

$\Delta S_f^o = -47 \text{ J mol}^{-1} \text{ K}^{-1}$ or -47 J/K

$\Delta G_f^o = \Delta H_f^o - T\Delta S_f^o = -223 \text{ kJ/mol} - (298 \text{ K})(-0.047 \text{ kJ/mol K}) = -209 \text{ kJ/mol}$

18.85 $\Delta G° = (\text{sum } \Delta G_f^o[\text{products}]) - (\text{sum } \Delta G_f^o[\text{reactants}])$

(a) $\Delta G° = \{\Delta G_f^o[H_2SO_4(l)]\} - \{\Delta G_f^o[H_2O(l)] + \Delta G_f^o[SO_3(g)]\}$

$\Delta G° = \{1 \text{ mol} \times (-689.9 \text{ kJ/mol})\} - \{1 \text{ mol} \times (-237.2 \text{ kJ/mol}) + 1 \text{ mol} \times (-370 \text{ kJ/mol})\}$

$\Delta G° = -82.3 \text{ kJ}$

(b) $\Delta G° = \{2\Delta G_f^o[NH_3(g)] + \Delta G_f^o[H_2O(l)] + \Delta G_f^o[CaCl_2(s)]\}$
$- \{\Delta G_f^o[CaO(s)] + 2\Delta G_f^o[NH_4Cl(s)]\}$

$\Delta G° = \{2 \text{ mol} \times (-16.7 \text{ kJ/mol}) + 1 \text{ mol} \times (-237.2 \text{ kJ/mol})$
$+ 1 \text{ mol} \times (-750.2 \text{ kJ/mol})\} - \{1 \text{ mol} \times (-604.2 \text{ kJ/mol}) + 2 \text{ mol} \times (-203.9 \text{ kJ/mol})\}$

$\Delta G° = -8.8 \text{ kJ}$

(c) $\Delta G° = \{\Delta G_f^o[H_2SO_4(l)] + \Delta G_f^o[CaCl_2(s)]\} - \{\Delta G_f^o[CaSO_4(s)] + \Delta G_f^o[HCl(g)]\}$

$\Delta G° = \{1 \text{ mol} \times (-689.9 \text{ kJ/mol}) + 1 \text{ mol} \times (-750.2 \text{ kJ/mol})\}$
$- \{1 \text{ mol} \times (-1320.3 \text{ kJ/mol}) + 2 \text{ mol} \times (-95.27 \text{ kJ/mol})\}$

$\Delta G° = +70.7 \text{ kJ}$

18.87 Multiply second equation by 2 (remembering to multiply the associated free energy change by 2), and add the result to the reverse of the first equation

$4NO(g) \longrightarrow 2N_2O(g) + O_2(g),\quad \Delta G° = -139.56 \text{ kJ}\quad (1)$

$2NO(g) + O_2(g) \longrightarrow 2NO_2(g),\quad \Delta G° = -69.70 \text{ kJ}\quad (2)$

Reverse reaction (1) double reaction (2)

$\Delta G° = 139.56 \text{ kJ} + 2 \times (-69.70) \text{ kJ}$

$N_2O(g) + 3O_2(g) \longrightarrow 4NO_2(g),\quad \Delta G° = +0.16 \text{ kJ}$

18.89 The maximum work obtainable from a reaction is equal in magnitude to the value of ΔG for the reaction. Thus, we need only determine $\Delta G°$ for the process

$\Delta G° = (\text{sum } \Delta G_f^o \text{ [products]}) - (\text{sum } \Delta G_f^o \text{ [reactants]})$

$\Delta G° = \{3\,\Delta G_f^o[H_2O(g)] + 2\,\Delta G_f^o[CO_2(g)]\} - \{3\,\Delta G_f^o[O_2(g)] + \Delta G_f^o[C_2H_5OH(l)]\}$

$\Delta G° = \{3 \text{ mol} \times (-228.6 \text{ kJ/mol}) + 2 \text{ mol} \times (-394.4 \text{ kJ/mol})\}$
$- \{3 \text{ mol} \times (0.0 \text{ kJ/mol}) + 1 \text{ mol} \times (-174.8 \text{ kJ/mol})\}$

$\Delta G° = -1299.8 \text{ kJ}$

18.91 At equilibrium, $\Delta G = 0 = \Delta H - T\Delta S$
$T_{eq} = \Delta H/\Delta S$, and assuming that ΔS is independent of temperature, we have
$T_{eq} = (31.4 \times 10^3 \text{ J mol}^{-1}) \div (94.2 \text{ J mol}^{-1} \text{ K}^{-1}) = 333 \text{ K}$

18.93 At equilibrium, $\Delta G = 0 = \Delta H - T\Delta S$
Thus $\Delta H = T\Delta S$, and if we assume that both ΔH and ΔS are independent of temperature, we have
$\Delta S = \Delta H/T_{eq} = (37.7 \times 10^3 \text{ J/mol}) \div (99.3 + 273.15 \text{ K})$
$\Delta S = 101 \text{ J mol}^{-1} \text{ K}^{-1}$

18.95 Balance the equation

$$C_2H_4(g) + 2HNO_3(l) \longrightarrow HC_2H_3O_2(l) + H_2O(l) + NO(g) + NO_2(g)$$

The reaction is spontaneous if its associated value for $\Delta G°$ is negative.

$\Delta G° = (\text{sum } \Delta G_f^o \text{ [products]}) - (\text{sum } \Delta G_f^o \text{ [reactants]})$

$\Delta G° = \{\Delta G_f^o[HC_2H_3O_2(l)] + \Delta G_f^o[H_2O(l)] + \Delta G_f^o[NO(g)] + \Delta G_f^o[NO_2(g)]\}$
$- \{\Delta G_f^o[C_2H_4(g)] + 2\,\Delta G_f^o[HNO_3(l)]\}$

$\Delta G° = \{1 \text{ mol} \times (-392.5 \text{ kJ/mol}) + 1 \text{ mol} \times (-237.2 \text{ kJ/mol})$
$+ 1 \text{ mol} \times (86.69 \text{ kJ/mol}) + 1 \text{ mol} \times (51.84 \text{ kJ/mol})\}$
$- \{1 \text{ mol} \times (68.12 \text{ kJ/mol}) + 2 \text{ mol} \times (-79.91 \text{ kJ/mol})\}$

$\Delta G° = -399.5 \text{ kJ}$
Yes, the reaction is spontaneous.

18.97 (a) $PCl_3(g) + \frac{1}{2}O_2(g) \longrightarrow POCl_3(g)$

$\Delta G° = \{1 \text{ mol} \times \Delta G_f^o[POCl_3(g)]\} - \{1 \text{ mol} \times \Delta G_f^o[PCl_3(g)] + 1/2\text{mol} \times \Delta G_f^o[O_2(g)]\}$

$\Delta G° = \{1 \text{ mol} \times (-512.9\text{kJ/mol})\} - \{1 \text{ mol} \times (-267.8 \text{ kJ/mol}) + 1/2 \text{ mol} \times (0 \text{ kJ/mol})\}$
$\Delta G° = -245.1 \text{ kJ} = -2.45 \times 10^5 \text{ J}$
$\Delta G° = -2.45 \times 10^5 \text{ J} = -RT\ln K_p = -(8.314 \text{ J/K mol})(298\ K) \times \ln K_p$
$\ln K_p = 98.9$ and $K_p = 9.20 \times 10^{42}$

(b) $2SO_3(g) \rightleftharpoons 2SO_2(g) + O_2(g)$

$SO_3(g) \rightleftharpoons SO_2(g) + ½\, O_2(g)$

$\Delta G° = \{1 \text{ mol} \times \Delta G_f^o[SO_2(g)] + ½ \text{ mol} \times \Delta G_f^o[O_2(g)]\} - \{1 \text{ mol} \times \Delta G_f^o[SO_3(g)]\}$

$\Delta G° = \{1 \text{ mol} \times (-300.4 \text{ kJ/mol}) + ½ \text{ mol} \times (0 \text{ kJ/mol})\} - \{1 \text{ mol} \times (-370.4 \text{ kJ/mol})\}$
$\Delta G° = 70 \text{ kJ} = 7.\,0 \times 10^4 \text{ J}$
$7.0 \times 10^4 \text{ J} = -RT\ln K_p = -(8.314 \text{ J/K mol})(298 \text{ K}) \times \ln K_p$
$\ln K_p = 28.25$ and $K_p = 5.37 \times 10^{-13}$

18.99 $\Delta G° = -RT \ln K_p$

-9.67×10^3 J = –(8.314 J/K mol)(1273 K) × ln K_p

$\ln K_p = 0.914 \therefore K_p = 2.49$

$$Q = \frac{[N_2O][O_2]}{[NO_2][NO]} = \frac{(0.015)(0.0350)}{(0.0200)(0.040)} = 0.66$$

Since the value of Q is less than the value of K, the system is not at equilibrium and must shift to the right to reach equilibrium.

18.101 $\Delta G° = -RT \ln K_p$

-50.79×10^3 J = –(8.314 J K^{-1} mol^{-1})(298 K) × ln K_p

$\ln K_p = 20.50$

Taking the exponential of both sides of this equation gives: $K_p = 8.000 \times 10^8$

This is a favorable reaction, since the equilibrium lies far to the side favoring products and is worth studying as a method for methane production.

18.103 If $\Delta G° = 0$, $K_c = 1$. If we start with pure products, the value of Q will be infinite (there are zero reactants) and, since $Q > K_c$, the equilibrium will shift towards the reactants, i.e., the pure products will decompose to their reactants.

18.105 This requires the breaking of three N—H single bonds:

$NH_3 \longrightarrow N + 3H$

The enthalpy of atomization of NH_3 is thus three times the average N—H single bond energy:

3×388 kJ/mol = 1.16×10^3 kJ/mol

18.107 The heat of formation for ethanol vapor describes the following change:

$2C(s) + 3H_2(g) + 1/2O_2(g) \longrightarrow C_2H_5OH(g)$

This can be arrived at by adding the following thermochemical equations, using data from Table 18.3:

$3H_2(g) \longrightarrow 6H(g)$	$\Delta H_1° = (6)217.89$ kJ = 1307.34 kJ
$2C(s) \longrightarrow 2C(g)$	$\Delta H_2° = (2)716.67$ kJ = 1,433.34 kJ
$1/2O_2(g) \longrightarrow O(g)$	$\Delta H_3° = (1)249.17$ kJ = 249.17 kJ
$6H(g) + 2C(g) + O(g) \longrightarrow C_2H_5OH(g)$	$\Delta H_4° = x$
$3H_2(g) + 2C(s) + 1/2O_2(g) \longrightarrow C_2H_5OH(g)$	$\Delta H_f° = (2989.85 + x)$ kJ

Since $\Delta H_f°$ is given as –235.3 kJ

$-235.3 = 2989.85 + x$

$x = -3225.2$ kJ

$\Delta H°_{atom}$ is the reverse reaction, so the sign will change:

$\Delta H°_{atom} = 3225.2$ kJ

The sum of all the bond energies in the molecule should be equal to the atomization energy:

$\Delta H°_{atom}$ = 1(C—C bond) + 5(C—H bonds) + 1(O—H bond) + 1(C—O bond)

We use bond energy values from Table 18.4:

3225.2 kJ = 1(348 kJ) + 5(412 kJ) + 1(463 kJ) + 1(C–O bond)

C—O bond energy = 354 kJ/mol

18.109 There are two C=S double bonds to be considered:

ΔH_f° = sum(ΔH_f° [gaseous atoms]) – sum(average bond energies in the molecule)

ΔH_f° [$CS_2(g)$] = 115.3 kJ/mol = [716.67 + 2 × 276.98] – [2 × C=S]

The C=S double bond energy is therefore given by the equation:

C=S = –(115.3 – 716.67 – 2 × 276.98) ÷ 2 = 577.7 kJ/mol

18.111 There are six S—F bonds in the molecule:

ΔH_f° = sum(ΔH_f° [gaseous atoms]) – sum(average bond energies in the molecule)

ΔH_f° [$SF_6(g)$] = –1096 kJ/mol = [277.0 + 6 × 79.14] – [6 × S—F]

S—F = (1096 + 277.0 + 6 × 79.14) ÷ 6 = 308.0 kJ/mol

18.113 ΔH_f° = sum(ΔH_f° [gaseous atoms]) – sum(average bond energies in the molecule)

ΔH_f° [$C_2H_2(g)$] = [2 × 716.7 + 2 × 218.0] – [2 × 412 + 960]

= 85 kJ/mol

18.115 The heat of formation of CF_4 should be more exothermic than that of CCl_4 because more energy is released on formation of a C—F bond than on formation of a C—Cl bond. Also, less energy is needed to form gaseous F atoms than to form gaseous Cl atoms.

Chapter Nineteen
Electrochemistry

Practice Exercises

19.1 anode: $Mg(s) \longrightarrow Mg^{2+}(aq) + 2e^-$

cathode: $Fe^{2+}(aq) + 2e^- \longrightarrow Fe(s)$

cell notation: $Mg(s)|Mg^{2+}(aq)||Fe^{2+}(aq)|Fe(s)$

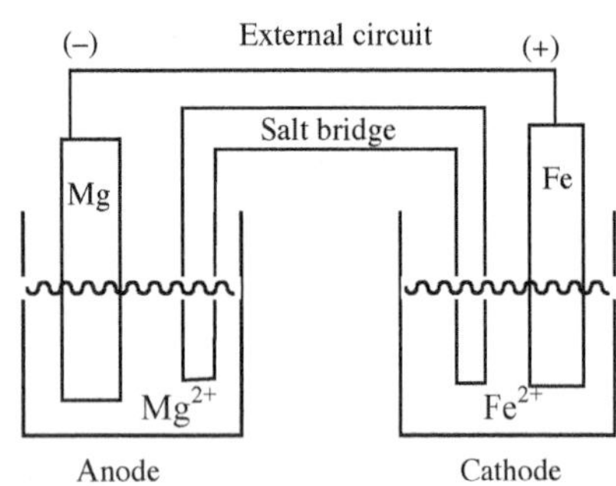

19.2 anode: $Al(s) \longrightarrow Al^{3+}(aq) + 3e^-$

cathode: $Ni^{2+}(aq) + 2e^- \longrightarrow Ni(s)$

overall: $3Ni^{2+}(aq) + 2Al(s) \longrightarrow 2Al^{3+}(aq) + 3Ni(s)$

19.3 $E°_{cell} = E°_{substance\ reduced} - E°_{substance\ oxidized}$

$2Ag^+(aq) + Cu(s) \longrightarrow 2Ag(s) + Cu^{2+}(aq)$

$$E^o_{cell} = E^o_{Ag^+} - E^o_{Cu^{2+}}$$

$E°_{cell} = 0.80\ V - 0.34\ V = 0.46\ V$

$2Ag^+(aq) + Zn(s) \longrightarrow 2Ag(s) + Zn^{2+}(aq)$

$$E^o_{cell} = E^o_{Ag^+} - E^o_{Zn^{2+}}$$

$E°_{cell} = 0.80\ V - (-0.76\ V) = 1.56\ V$

Zinc will have the larger value for $E°_{cell}$.

19.4 $E°_{cell} = E°_{substance\ reduced} - E°_{substance\ oxidized}$

$Mg^{2+}(aq) + 2e^- \longrightarrow Mg(s) \quad E° = -2.37\ V$

Magnesium will be oxidized.

$1.93\ V = E° - (-2.37\ V)$

$E° = -0.44\ V$

Fe^{2+} has a reduction of –0.44 V. The half reaction is

$Fe^{2+}(aq) + 2e^- \longrightarrow Fe(s)$

19.5 Either nickel(II) or iron(III) will be reduced, depending on which way the reaction proceeds. Iron(III) is listed higher than nickel(II) in Table 19.1 (it has a greater reduction potential), so we would expect that the reaction would not be spontaneous in the direction shown.

The spontaneous reaction is:

$Ni(s) + 2Fe^{3+}(aq) \longrightarrow Ni^{2+}(aq) + 2Fe^{2+}(aq)$

19.6 The half-reaction with the more positive value of $E°$ (listed higher in Table 19.1) will occur as a reduction. The half-reaction having the less positive (more negative) value of $E°$ (listed lower in Table 19.1) will be reversed and occur as an oxidation.

$Br_2(aq) + 2e^- \longrightarrow 2Br^-(aq)$ reduction

$SO_4^{2-}(aq) + 4H^+(aq) + 2e^- \longrightarrow H_2SO_3(aq) + H_2O$ oxidation

$Br_2(aq) + H_2SO_3(aq) + H_2O \longrightarrow 2Br^-(aq) + SO_4^{2-}(aq) + 4H^+(aq)$

19.7 The half–reaction having the more positive value for $E°$ will occur as a reduction. The other half–reaction should be reversed, so as to appear as an oxidation.

$NiO_2(s) + 2H_2O + 2e^- \longrightarrow Ni(OH)_2(s) + 2OH^-(aq)$ reduction

$Fe(s) + 2OH^-(aq) \longrightarrow 2e^- + Fe(OH)_2(s)$ oxidation

$NiO_2(s) + Fe(s) + 2H_2O \longrightarrow Ni(OH)_2(s) + Fe(OH)_2(s)$ net reaction

$E°_{cell} = E°_{substance\ reduced} - E°_{substance\ oxidized}$

$E°_{cell} = E°_{NiO_2} - E°_{Fe}$

$E°_{cell} = 0.49 - (-0.89) = 1.38\ V$

19.8 $5\{Cr(s) \rightleftharpoons Cr^{3+}(aq) + 3e^-\}$ oxidation

$3\{MnO_4^-(aq) + 8H^+(aq) + 5e^- \rightleftharpoons Mn^{2+}(aq) + 4H_2O(l)\}$ reduction

Net reaction

$5Cr(s) + 3MnO_4^-(aq) + 24H^+(aq) \rightleftharpoons 3Mn^{2+}(aq) + 12H_2O(l) + 5Cr^{3+}(aq)$

$E°_{cell} = E°_{substance\ reduced} - E°_{substance\ oxidized}$

$E^o_{cell} = E^o_{MnO_4^-} - E^o_{Cr^{3+}}$

$E°_{cell} = 1.51 - (-0.74) = 2.25\ V$

19.9 The half–reaction having the more positive value for $E°$ will occur as a reduction. The other half–reaction should be reversed, so as to appear as an oxidation.

$3Cu^{2+}(aq) + 2Cr(s) \longrightarrow 3Cu(s) + 2Cr^{3+}(aq)$

$E°_{cell} = E°_{substance\ reduced} - E°_{substance\ oxidized}$

$E^o_{cell} = E^o_{Cu^+} - E^o_{Cr^{3+}}$

$E°_{cell} = 0.34 - (-0.74) = 1.08\ V$

19.10 The possible reactions and the reduction potentials are

$Fe^{3+}(aq) + e^- \longrightarrow Fe^{2+}(aq)$ $E°_{Fe3+/Fe2+} = +0.77\ V$

$Sn^{4+}(aq) + 2e^- \longrightarrow Sn^{2+}(aq)$ $E°_{Sn4+/Sn2+} = +0.15\ V$ (See Appendix ♦♦♦)

Since Fe^{3+} has a higher reduction potential than Sn^{4+}, it will be reduced and Sn^{2+} will be oxidized. The overall reaction will be:

$2Fe^{3+}(aq) + Sn^{2+}(aq) \longrightarrow 2Fe^{2+}(aq) + Sn^{4+}(aq)$

$E°_{cell} = E°_{substance\ reduced} - E°_{substance\ oxidized}$

$E°_{cell} = 0.77\ V - 0.15\ V = 0.62\ V$

19.11 A reaction will occur spontaneously in the forward direction if the value of $E°$ is positive. We therefore evaluate $E°$ for each reaction using:

$E°_{cell} = E°_{substance\ reduced} - E°_{substance\ oxidized}$

(a) $Br_2(aq) + 2e^- \longrightarrow 2Br^-(aq)$ oxidation, when reverse

$I_2(s) + 2e^- \longrightarrow 2I^-(aq)$ reduction

$E°_{cell} = E°_{I_2} - E°_{Br_2}$
$E°_{cell} = 0.54 \text{ V} - 1.07 \text{ V} = -0.53 \text{ V}$, ∴ not spontaneous
$Br_2(aq) + 2I^-(aq) \longrightarrow 2Br^-(aq) + I_2(s)$

(b) $MnO_4^-(aq) + 8H^+ + 5e^- \longrightarrow Mn^{2+}(aq) + 4H_2O$ oxidation, when reversed
$5Ag^+(aq) + 5e^- \longrightarrow 5Ag(s)$ reduction

$E°_{cell} = E°_{Ag^+} - E°_{MnO_4^-}$
$E°_{cell} = 0.80 \text{ V} - (1.51 \text{ V}) = -0.71 \text{ V}$, ∴ non-spontaneous
$MnO_4^-(aq) + 8H^+ + 5Ag^+(aq) \longrightarrow Mn^{2+}(aq) + 5Ag(s) + 4H_2O$

19.12 A reaction will occur spontaneously in the forward direction if the value of $E°$ is positive. We therefore evaluate $E°$ for each reaction using:
$E°_{cell} = E°_{\text{substance reduced}} - E°_{\text{substance oxidized}}$

(a) $2Br^-(aq) \longrightarrow Br_2(aq) + 2e^-$ oxidation
$2HOCl(aq) + 2H^+(aq) + 2e^- \longrightarrow Cl_2(aq) + 2H_2O$ reduction

$E°_{cell} = E°_{HOCl} - E°_{Br_2}$
$E°_{cell} = 1.63 \text{ V} - (1.07 \text{ V}) = 0.56 \text{ V}$, ∴ spontaneous

(b) $2Cr(s) \longrightarrow 2Cr^{3+}(aq) + 6e^-$ oxidation
$3Zn^{2+}(aq) + 6e^- \longrightarrow 3Zn(s)$ reduction
$E°_{cell} = E°_{Zn^{2+}} - E°_{Cr^{3+}}$
$E°_{cell} = -0.76 \text{ V} - (-0.74 \text{ V}) = -0.02 \text{ V}$, ∴ non-spontaneous

19.13 From the equation $\Delta G° = -nFE°_{cell}$
$\Delta G° = -30.9 \text{ kJ}$ or $-30{,}900 \text{ J}$

$$F = \frac{96{,}500 \text{ C}}{\text{mol } e^-}$$

$E°_{cell} = 0.107 \text{ V}$

$$-30{,}900 \text{ J} = -n\left(\frac{96{,}500 \text{ C}}{\text{mol } e^-}\right)(0.107 \text{ V})$$

$n = 2.99$ which rounds to 3
Therefore, 3 moles of electrons are transferred in the reaction.

19.14 $\Delta G° = -nFE°_{cell}$
From Practice Exercise 19.11 (a): $n = 2\ e^-$, $E°_{cell} = -0.53 \text{ V}$

$$\Delta G° = -nFE°_{cell} = -(2 \text{ mol } e^-)\left(\frac{96{,}500 \text{ C}}{\text{mol } e^-}\right)\left(\frac{-0.53 \text{ J}}{\text{C}}\right) = 102{,}000 \text{ J} = 102 \text{ kJ}$$

From Practice Exercise 19.11 (b): $n = 5\ e^-$, $E°_{cell} = -0.71 \text{ V}$

$$\Delta G° = -nFE°_{cell} = -(5 \text{ mol } e^-)\left(\frac{96{,}500 \text{ C}}{\text{mol } e^-}\right)\left(\frac{-0.71 \text{ J}}{\text{C}}\right) = 342{,}600 \text{ J} = 343 \text{ kJ}$$

From Practice Exercise 19.12 (a): $n = 2\ e^-$, $E°_{cell} = +0.56 \text{ V}$

$$\Delta G° = -nFE°_{cell} = -(2 \text{ mol } e^-)\left(\frac{96{,}500 \text{ C}}{\text{mol } e^-}\right)\left(\frac{0.56 \text{ J}}{\text{C}}\right) = -108{,}080 \text{ J} = -108 \text{ kJ}$$

From Practice Exercise 19.12 (b): $n = 6\ e^-$, $E°_{cell} = -0.02 \text{ V}$

$$\Delta G° = -nFE°_{cell} = -(6\ e^-)\left(\frac{96{,}500\ C}{mol\ e^-}\right)\left(\frac{-0.02\ J}{C}\right) = 11{,}600\ J = 11.6\ kJ$$

19.15 Using Equation 19.7,

$$E°_{cell} = \frac{RT}{nF}\ln K_c$$

$$-0.46\ V = \frac{(8.314\ J\ mol^{-1}K^{-1})(298\ K)}{2(96{,}500\ C\ mol^{-1})}\ \ln K_c$$

$$\ln K_c = -35.83$$

Taking the antilog (e^x) of both sides of the above equation gives

$K_c = 2.7 \times 10^{-16}$

This very small value for the equilibrium constant means that the products of the reaction are not formed spontaneously. The equilibrium lies far to the left, favoring reactants, and we do not expect much product to form.

The reverse reaction will be spontaneous, therefore, the value for K_c for the spontaneous reaction will be

$Cu(s) + 2Ag^+(aq) \longrightarrow Cu^{2+}(aq) + 2Ag(s)$

$$K_c' = \frac{1}{K_c} = \frac{1}{2.7\times10^{-16}} = 3.7 \times 10^{15}$$

19.16 $Ag^+(aq) + e^- \longrightarrow Ag(s)$ $\quad E° = 0.80\ V$

$AgBr(s) + e^- \longrightarrow Ag(s) + Br^-(aq)$ $\quad E° = 0.07\ V$

Equation for the spontaneous reaction:

$Ag^+(aq) + Br^-(aq) \longrightarrow AgBr(s)$ $\quad E°_{cell} = 0.73\ V$

$$K = \frac{1}{[Ag^+][Br^-]}$$

$$\ln K_c = \frac{E°_{cell}nF}{RT} = \frac{(0.73\ V)(1\ e^-)(96{,}500\ C\ mol^{-1})}{(8.314\ J\ mol^{-1}\ K^{-1})(298\ K)} = 28.43$$

$K_c = 2.23 \times 10^{12}$

K_{sp} for AgBr is 5.4×10^{-13}

$$\frac{1}{K_c} = \frac{1}{2.23\times10^{12}} = 4.5 \times 10^{-13}$$

The K_c is the inverse of the K_{sp}. So the cell potential value agrees with K_{sp} table.

19.17 $Cu^{2+}(aq) + Mg(s) \longrightarrow Cu(s) + Mg^{2+}(aq)$ $\quad E°_{cell} = 0.34 - (-2.37\ V) = 2.71\ V$

$$E_{cell} = E°_{cell} - \frac{RT}{nF}\ln\frac{[Mg^{2+}]}{[Cu^{2+}]}$$

$$E_{cell} = 2.71\ V - \frac{(8.314\ J\,mol^{-1}\ K^{-1})(298\ K)}{(2)(96{,}500\ C\ mol^{-1})}\ \ln\frac{[2.2\times10^{-6}]}{[0.015]} = 2.82\ V$$

19.18 $E_{cell} = E^{o}_{cell} - \dfrac{RT}{nF} \ln \dfrac{(0.01)(0.965)}{[H^+]^2}$

$$0.00 = 0.14\text{ V} - \frac{(8.314\text{ J mol}^{-1}\text{ K}^{-1})(298\text{ K})}{(2)(96{,}500\text{ C mol}^{-1})} \ln \frac{(0.01)(0.965)}{[H^+]^2}$$

$$-0.14\text{ V} = -1.28 \times 10^{-2} \ln \frac{0.00965}{[H^+]^2}$$

$$10.91 = \ln \frac{0.00965}{[H^+]^2}$$

$$e^{10.91} = 5.45 \times 10^4 = \frac{0.00965}{[H^+]^2}$$

$[H^+]^2 = 1.77 \times 10^{-7}$
$[H^+] = 4.21 \times 10^{-4}$
$pH = -\log[H^+] = -\log[4.21 \times 10^{-4}] = -(-3.38) = 3.38$

19.19 $Cu(s) \longrightarrow Cu^{2+}(aq) + 2e^-$ oxidation
$Ag^+(aq) + e^- \longrightarrow Ag(s)$ reduction
$E^{o}_{cell} = E^{o}_{Ag^+} - E^{o}_{Cu^{2+}}$
$E^{o}_{cell} = +0.80\text{ V} - (+0.34\text{ V}) = +0.46\text{ V}$

$$E_{cell} = E^{o}_{cell} - \frac{RT}{nF} \ln \frac{[Cu^{2+}]}{[Ag^+]^2}$$

$$\ln \frac{[Cu^{2+}]}{[Ag^+]^2} = \frac{E^{o}_{cell} - E_{cell}}{\frac{RT}{nF}}$$

$$\ln \frac{[Cu^{2+}]}{[Ag^+]^2} = \frac{(0.46\text{ V} - 0.57\text{ V})}{0.01284}$$

$$\ln \frac{[Cu^{2+}]}{[Ag^+]^2} = -8.5670$$

$$\frac{[Cu^{2+}]}{[Ag^+]^2} = e^{-8.5670} = 1.9 \times 10^{-4}$$

Since the $[Ag^+] = 0.225\ M$, $[Cu^{2+}] = 9.6 \times 10^{-6}\ M$
Substituting the second value of 0.82 V into the same expression gives $[Cu^{2+}] = 3.4 \times 10^{-14}\ M$

19.20 $Cu^{2+}(aq) + Mg(s) \longrightarrow Cu(s) + Mg^{2+}(aq)$ $\qquad E°_{cell} = 0.34 - (-2.37\text{ V}) = 2.71\text{ V}$

$$E_{cell} = E^{o}_{cell} - \frac{RT}{nF}\ln\frac{[Mg^{2+}]}{[Cu^{2+}]}$$

$$2.79 = 2.71\text{ V} - \frac{(8.314\text{ J mol}^{-1}\text{ K}^{-1})(298\text{ K})}{(2)(96{,}500\text{ C mol}^{-1})}\ln\frac{[Mg^{2+}]}{[0.015]}$$

$$0.08\text{ V} = -0.0128\ln\frac{[Mg^{2+}]}{[0.015]}$$

$$-6.25 = \ln\frac{[Mg^{2+}]}{[0.015]}$$

$$e^{-6.25} = \frac{[Mg^{2+}]}{[0.015]}$$

$[Mg^{2+}] = 2.9 \times 10^{-5}\ M$

19.21 $Fe^{3+}(aq) + e^{-} \longrightarrow Fe^{2+}(aq)$ $\qquad E° = 0.77\text{ V}$

$2I_2(s) + e^{-} \longrightarrow 2I^{-}(aq)$ $\qquad E° = +0.54\text{ V}$

$O_2(g) + 4H^{+}(aq) + 4e^{-} \longrightarrow 2H_2O$ $\qquad E° = +1.23\text{ V}$

The reaction with the least positive reduction potential will be the easiest to oxidize, and its product will be the product at the anode. I_2 will be produced.

19.22 The cathode is always where reduction occurs. We must consider which species could be candidates for reduction, then choose the species with the highest reduction potential from Table 19.1.

$Cd^{2+}(aq) + 2e^{-} \longrightarrow 2Cd(s)$ $\qquad E° = -0.40\text{ V}$

$Sn^{2+}(aq) + 2e^{-} \longrightarrow 2Sn(s)$ $\qquad E° = -0.14\text{ V}$

$2H_2O + 2e^{-} \longrightarrow H_2(g) + 2OH^{-}(aq)$ $\qquad E° = -0.83\text{ V}$

Tin(II) has the highest reduction potential, so we would expect it to be reduced in this environment. We expect Sn(s) at the cathode.

19.23 The number of Coulombs is: $4.00\text{ A} \times \left(13\text{ min}\left(\frac{60\text{ sec}}{1\text{ min}}\right)\right) = 3120\text{ C}$

The number of moles is:

$$\text{mol OH}^{-} = 3120\text{ C} \times \frac{1\ F}{96{,}500\text{ C}} \times \frac{1\text{ mol OH}^{-}}{1\ F} = 3.23 \times 10^{-2}\text{ mol OH}^{-}$$

19.24 The number of moles of Au to be deposited is: 3.00 g Au ÷ 197 g/mol = 0.0152 mol Au. The number of Coulombs (A × s) is:

$$\text{Coulombs} = 0.0152\text{ mol Au} \times \frac{3\ F}{1\text{ mol Au}} \times \frac{96{,}500\text{ C}}{1\ F} = 4.40 \times 10^{3}\text{ C}$$

The number of minutes is:

$$\text{min} = \frac{4.40 \times 10^{3}\text{ A s}}{14.0\text{ A}} \times \frac{1\text{ min}}{60\text{ s}} = 5.24\text{ min}$$

19.25 $0.0500 \text{ g Au}\left(\frac{1 \text{ mol Au}}{196.97 \text{ g Au}}\right)\left(\frac{3 \text{ mol } e^-}{1 \text{ mol Au}}\right)\left(\frac{9.65\times 10^4 \text{ C}}{1 \text{ mol } e^-}\right)\left(\frac{1 \text{ A s}}{1 \text{ C}}\right)\left(\frac{1}{1200 \text{ s}}\right) = 0.0612 \text{ A}$

19.26 The number of Coulombs is:

$0.550 \text{ A} \times 3.46 \text{ hr.}\left(\frac{3600 \text{ s}}{1 \text{ hr}}\right) = 6851 \text{ C}$

The number of moles of copper ions produced is:

$\text{mol Cu}^{2+} = 6851 \text{ C} \times \frac{1 \text{ mol } e^-}{96{,}500 \text{ C}} \times \frac{1 \text{ mol Cu}^{2+}}{2 \text{ mol } e^-} = 0.0355 \text{ mol Cu}^{2+}$

Therefore, the increase in concentration is:

$M = \frac{\text{mol Cu}^{2+}}{\text{L solution}} = \frac{0.0355 \text{ mol Cu}^{2+}}{0.125 \text{ L solution}} = +0.284\ M$

Review Problems

19.55 (a) anode: $Zn(s) \longrightarrow Zn^{2+}(aq) + 2e^-$

cathode: $Cr^{3+}(aq) + 3e^- \longrightarrow Cr(s)$

cell: $3Zn(s) + 2Cr^{3+}(aq) \longrightarrow 3Zn^{2+}(aq) + 2Cr(s)$

(b) anode: $Pb(s) + HSO_4^-(aq) \longrightarrow PbSO_4(s) + H^+(aq) + 2e^-$

cathode: $PbO_2(s) + HSO_4^-(aq) + 3H^+(aq) + 2e^- \longrightarrow PbSO_4(s) + 2H_2O$

cell: $Pb(s) + PbO_2(s) + 2HSO_4^{2-}(aq) + 2H^+(aq) \longrightarrow 2PbSO(s) + 2H_2O$

(c) anode: $Mg(s) \longrightarrow Mg^{2+}(aq) + 2e^-$

cathode: $Sn^{2+}(aq) + 2e^- \longrightarrow Sn(s)$

cell: $Mg(s) + Sn^{2+}(aq) \longrightarrow Mg^{2+}(aq) + Sn(s)$

19.57 (a) $Pt(s)|Fe^{2+}(aq),Fe^{3+}(aq)||NO_3^-(aq), H^+(aq)|NO(g)|Pt(s)$

(b) $Pt(s)|Br_2(aq),Br^-(aq)||Cl^-(aq)|Cl_2(g)|Pt(s)$

(c) $Ag(s)|Ag^+(aq)||Au^{3+}(aq)|Au(s)$

19.59 (a) $Sn(s)$ (b) $Br^-(aq)$ (c) $Zn(s)$ (d) $I^-(aq)$

19.61 (a) $E^o_{cell} = 0.96 \text{ V} - (0.77) \text{ V} = 0.19 \text{ V}$

(b) $E^o_{cell} = 1.07 \text{ V} - (1.36 \text{ V}) = -0.29 \text{ V}$

(c) $E^o_{cell} = 1.42 \text{ V} - (0.80 \text{ V}) = 0.62 \text{ V}$

19.63 The reactions are spontaneous if the overall cell potential is positive.

$E^o_{cell} = E_{\text{substance reduced}} - E_{\text{substance oxidized}}$

(a) $E^o_{cell} = 1.42 \text{ V} - (0.54 \text{ V}) = 0.88 \text{ V}$ spontaneous

(b) $E^o_{cell} = 1.07 \text{ V} - (0.17 \text{ V}) = 0.90 \text{ V}$ spontaneous

(c) $E^o_{cell} = -0.74 \text{ V} - (-2.76 \text{ V}) = 2.02 \text{ V}$ spontaneous

19.65 The half–cell with the more positive E^o_{cell} will appear as a reduction, and the other half–reaction is reversed, to appear as an oxidation:

$BrO_3^-(aq) + 6H^+(aq) + 6e^- \longrightarrow Br^-(aq) + 3H_2O$ reduction

$3 \times (2I^-(aq) \longrightarrow I_2(s) + 2e^-)$ oxidation

$BrO_3^-(aq) + 6I^-(aq) + 6H^+(aq) \longrightarrow 3I_2(s) + Br^-(aq) + 3H_2O$ net reaction

$E^\circ_{cell} = E_{substance\ reduced} - E_{substance\ oxidized}$ or

$E^\circ_{cell} = E^\circ_{reduction} - E^\circ_{oxidation} = 1.44\ V - (0.54\ V) = 0.90\ V$

19.67 The half–reaction having the more positive standard reduction potential is the one that occurs as a reduction, and the other one is written as an oxidation:

$2 \times (2HOCl(aq) + 2H^+(aq) + 2e^- \longrightarrow Cl_2(g) + 2H_2O)$ reduction

$3H_2O + S_2O_3^{2-}(aq) \longrightarrow 2H_2SO_3(aq) + 2H^+(aq) + 4e^-$ oxidation

$4HOCl(aq) + 4H^+(aq) + 3H_2O + S_2O_3^{2-}(aq) \longrightarrow 2Cl_2(g) + 4H_2O + 2H_2SO_3(aq) + 2H^+(aq)$

which simplifies to give the following net reaction:

$4HOCl(aq) + 2H^+(aq) + S_2O_3^{2-}(aq) \longrightarrow 2Cl_2(g) + H_2O + 2H_2SO_3(aq)$

19.69 The two half–reactions are:

$SO_4^{2-}(aq) + 2e^- + 4H^+(aq) \longrightarrow H_2SO_3(aq) + H_2O(\ell)$ reduction

$2I^-(aq) \longrightarrow I_2(s) + 2e^-$ oxidation

$E^\circ_{cell} = E^\circ_{reduction} - E^\circ_{oxidation} = 0.17\ V - (0.54\ V) = -0.37\ V$

Since the overall cell potential is negative, we conclude that the reaction is not spontaneous in the direction written.

19.71 First, separate the overall reaction into its two half–reactions:

$2Br^-(aq) \longrightarrow Br_2(aq) + 2e^-$ oxidation

$I_2(s) + 2e^- \longrightarrow 2I^-(aq)$ reduction

$E^\circ_{cell} = E^\circ_{reduction} - E^\circ_{oxidation} = 0.54\ V - (1.07\ V) = -0.53\ V$

The value of n is 2:

$\Delta G^\circ = -nFE^\circ_{cell} = -(2)(96{,}500\ C)(-0.53\ J/C)$

$\Delta G^\circ = 1.0 \times 10^5\ J = 1.0 \times 10^2\ kJ$

19.73 (a) $E^\circ_{cell} = E^\circ_{reduction} - E^\circ_{oxidation} = 2.01\ V - (1.47\ V) = 0.54\ V$

(b) Since $n = 10$, $\Delta G^\circ = -nFE^\circ_{cell} = -(10)(96{,}500\ C)(0.54\ J/C) = -5.2 \times 10^5\ J$

$\Delta G^\circ = -5.2 \times 10^2\ kJ$

(c) $E^\circ_{cell} = \dfrac{RT}{nF} \ln K_c$

$$0.54\ V = \frac{(8.314\ J\ mol^{-1}K^{-1})(298\ K)}{10(96{,}500\ C\ mol^{-1})} \ln K_c$$

$\ln K_c = 210.3$

Taking the exponential of both sides of this equation:

$K_c = 2.1 \times 10^{91}$

19.75 Sn is oxidized by two electrons and Ag is reduced by two electrons:

$$E^\circ_{cell} = \frac{RT}{nF} \ln K_c$$

$$-0.015\ V = \frac{(8.314\ J\ mol^{-1}\ K^{-1})(298\ K)}{(2)(9.65 \times 10^4\ C/mol\ e^-)} \ln K_c$$

$\ln K_c = -1.17$

$K_c = \text{antiln}(-1.17) = 0.31$

19.77 This reaction involves the oxidation of Ag by two electrons and the reduction of Ni by two electrons. The concentration of the hydrogen ion is derived from the pH of the solution: $[H^+]$ = antilog (–pH) = antilog (–2) = 1×10^{-2} M

$$E_{cell} = E°_{cell} - \frac{RT}{nF} \ln Q$$

$$+0.13 \text{ V} = \frac{(8.314 \text{ J mol}^{-1} \text{ K}^{-1})(298 \text{ K})}{(2)(9.65 \times 10^4 \text{ C/mol } e^-)} \ln K_c$$

$$E_{cell} = 2.48 \text{ V} - \frac{(8.314 \text{ J mol}^{-1} \text{ K}^{-1})(298 \text{ K})}{(2)(9.65 \times 10^4 \text{ C/mol } e^-)} \ln \frac{[Ag^+]^2[Ni^{2+}]}{[H^+]^4}$$

$$= 2.48 \text{ V} - \frac{(8.314 \text{ J mol}^{-1} \text{ K}^{-1})(298 \text{ K})}{(2)(9.65 \times 10^4 \text{ C/mol } e^-)} \ln \frac{[3.0 \times 10^{-2}]^2[3.0 \times 10^{-2}]}{[1.0 \times 10^{-2}]^4}$$

$E_{cell} = 2.48 \text{ V} - 0.101 \text{ V} = 2.38 \text{ V}$

19.79 $$E_{cell} = E°_{cell} - \frac{RT}{nF} \ln \frac{[Mg^{2+}]}{[Cd^{2+}]}$$

$$E_{cell} = 1.97 - \frac{(8.314 \text{ J mol}^{-1} \text{ K}^{-1})(298 \text{ K})}{2(96{,}500 \text{ C mol}^{-1})} \ln \frac{[1.00]}{[Cd^{2+}]}$$

$$1.54 \text{ V} = 1.97 \text{ V} - 0.01284 \ln \frac{1}{[Cd^{2+}]}$$

$$\ln \frac{1}{[Cd^{2+}]} = 33.489$$

Taking (e^x) of both sides:

$$\frac{1}{[Cd^{2+}]} = 3.50 \times 10^{14}$$

$[Cd^{2+}] = 2.86 \times 10^{-15}$ M

19.81 In the iron half–cell, we are initially given

0.0500 L × 0.100 mol/L = 5.00×10^{-3} mol $Fe^{2+}(aq)$

The precipitation of $Fe(OH)_2(s)$ consumes some of the added OH^-, as well as some of the iron ion

$Fe^{2+}(aq) + 2OH^-(aq) \longrightarrow Fe(OH)_2(s)$

The number of moles of OH^- that have been added to the iron half–cell is

0.500 mol/L × 0.0500 L = 2.50×10^{-2} mol OH^-

The stoichiometry of the precipitation reaction requires that the following number of moles of OH^- be consumed on precipitation of 5.00×10^{-3} mol of $Fe(OH)_2(s)$

5.00×10^{-3} mol $Fe(OH)_2 \times$ (2 mol OH^-/mol $Fe(OH)_2$) = 1.00×10^{-2} mol OH^-
The number of moles of OH^- that are unprecipitated in the iron half-cell is
2.50×10^{-2} mol – 1.00×10^{-2} mol = 1.50×10^{-2} mol OH^-
Since the resulting volume is 50.0 mL + 50.0 mL, the concentration of hydroxide ion in the iron half–cell becomes, upon precipitation of the $Fe(OH)_2$
$[OH^-] = 1.50 \times 10^{-2}$ mol/0.100 L = 0.150 M OH^-
The standard cell potential is
$E^o_{cell} = E^o_{reduction} - E^o_{oxidation} = 0.3419\text{ V} - (-0.447\text{ V}) = 0.7889\text{ V}$
The Nernst equation is

$$E_{cell} = E^o_{cell} - \frac{RT}{nF}\ln\frac{\left[Fe^{2+}\right]}{\left[Cu^{2+}\right]}$$

$$1.175 = 0.7889 - \frac{(8.314\text{ J mol}^{-1}\text{K}^{-1})(298\text{ K})}{2(96{,}500\text{ C mol}^{-1})}\ln\frac{\left[Fe^{2+}\right]}{[1.00]}$$

$$1.175 = 0.7889 - 0.01284\ln\left[Fe^{2+}\right]$$

$\ln[Fe^{2+}] = -30.08$
$[Fe^{2+}] = 8.66 \times 10^{-14}$ M

19.83 $$E^o = E^o_{cell} - \frac{RT}{nF}\ln\frac{\left[Ag^+\right]_{dilute}}{\left[Ag^+\right]_{conc}}$$

$$E^o = 0 - \frac{\left(8.314\text{ J mol}^{-1}\text{ K}^{-1}\right)(298\text{ K})}{(1\text{ mol})\left(9.65\times10^4\text{ C mol}^{-1}\right)}\ln\frac{[0.015]}{[0.50]}$$

$E^o = 0.090$ V

$$E^o = E^o_{cell} - \frac{RT}{nF}\ln\frac{\left[Ag^+\right]_{dilute}}{\left[Ag^+\right]_{conc}}$$

$$E^o = 0 - \frac{\left(8.314\text{ J mol}^{-1}\text{ K}^{-1}\right)(348\text{ K})}{(1\text{ mol})\left(9.65\times10^4\text{ C mol}^{-1}\right)}\ln\frac{[0.015]}{[0.50]}$$

$E^o = 0.105$ V

19.85 Possible cathode reactions:

$Al^{3+} + 3e^- \rightleftharpoons Al(s)$ $E° = -1.66$ V

$2H_2O + 2e^- \rightleftharpoons H_2(g) + 2OH^-(aq)$ $E° = -0.83$ V

Possible anode reactions:

$S_2O_8^{2-} + 2e^- \rightleftharpoons 2SO_4^{2-}$ $E° = +2.05$ V

$O_2 + 4H^+ + 4e^- \rightleftharpoons 2H_2O$ $E° = +1.23$ V

Cathode reaction: $2H_2O + 2e^- \rightleftharpoons H_2(g) + 2OH^-(aq)$ $E° = -0.83$ V

Anode reactions: $2H_2O \rightleftharpoons O_2 + 4H^+ + 4e^-$ $E° = -1.23$ V

Net cell reaction: $2H_2O \rightleftharpoons 2H_2(g) + O_2(g)$ $E° = -2.06$ V

19.87 The answers to the previous Review Problems guide us here:

Possible cathode reactions:

$K^+ + e^- \rightleftharpoons K(s)$ $E° = -2.92$ V

$Cu^{2+} + 2e^- \rightleftharpoons Cu(s)$ $E° = +0.34$ V

$2H_2O + 2e^- \rightleftharpoons H_2(g) + 2OH^-(aq)$ $E° = -0.83$ V

Cathode reaction: $Cu^{2+} + 2e^- \rightleftharpoons Cu(s)$

Possible anode reactions:

$2SO_4^{2-} \rightleftharpoons S_2O_8^{2-} + 2e^-$ $E° = -2.01$ V

$2Br^- \rightleftharpoons Br_2 + 2e^-$ $E° = -1.07$ V

$2H_2O \rightleftharpoons O_2(g) + 4H^+(aq) + 4e^-$ $E° = -1.23$ V

Anode reaction: $2Br^- \rightleftharpoons Br_2 + 2e^-$

Overall reaction: $Cu^{2+} + 2Br^- \rightleftharpoons Br_2 + Cu(s)$

19.89 In aqueous solution the following reduction of water can occur:

$2H_2O(l) + 2e^- \longrightarrow H_2(g) + 2OH^-(aq)$ $E^o = -0.83$ V

Reactions that are less positive than this cannot occur at the cathode.

Therefore, Al^{3+}, Mg^{2+}, Na^+, Ca^{2+}, K^+, and Li^+ would not be reduced at the cathode.

19.91 (a) $Fe^{2+}(aq) + 2e^- \longrightarrow Fe(s)$

$$0.20 \text{ mol } Fe^{2+} \times \left(\frac{2 \text{ mol } e^-}{1 \text{ mol } Fe^{2+}}\right) = 0.40 \text{ mol } e^-$$

(b) $Cl^-(aq) \longrightarrow 1/2Cl_2(g) + e^-$

$$0.70 \text{ mol } Cl^- \times \left(\frac{1 \text{ mol } e^-}{1 \text{ mol } Cl^-}\right) = 0.70 \text{ mol } e^-$$

(c) $Cr^{3+}(aq) + 3e^- \longrightarrow Cr(s)$

$$1.50 \text{ mol } Cr^{3+} \times \left(\frac{3 \text{ mol } e^-}{1 \text{ mol } Cr^{3+}}\right) = 4.50 \text{ mol } e^-$$

(d) $Mn^{2+}(aq) + 4H_2O(l) \longrightarrow MnO_4^-(aq) + 8H^+(aq) + 5e^-$

$$1.0 \times 10^{-2} \text{ mol } Mn^{2+} \times \left(\frac{5 \text{ mol } e^-}{1 \text{ mol } Mn^{2+}}\right) = 5.0 \times 10^{-2} \text{ mol } e^-$$

19.93 $Fe(s) + 2OH^-(aq) \longrightarrow Fe(OH)_2(s) + 2e^-$

The number of Coulombs is: 12.0 min × 60 s/min × 8.00 C/s = 5.76×10^3 C. The number of grams of $Fe(OH)_2$ is:

$$\text{g } Fe(OH)_2 = (5.76 \times 10^3 \text{ C})\left(\frac{1 \text{ mol } e^-}{96500 \text{ C}}\right)\left(\frac{1 \text{ mol } Fe(OH)_2}{2 \text{ mol } e^-}\right)\left(\frac{89.86 \text{ g } Fe(OH)_2}{1 \text{ mol } Fe(OH)_2}\right)$$

$$= 2.68 \text{ g } Fe(OH)_2$$

19.95 $Cr^{3+}(aq) + 3e^- \longrightarrow Cr(s)$

The number of Coulombs that will be required is:

$$\text{Coulombs} = (75.0\text{ g Cr})\left(\frac{1\text{ mol Cr}}{52.00\text{ g Cr}}\right)\left(\frac{3\text{ mol }e^-}{1\text{ mol Cr}}\right)\left(\frac{96{,}500\text{ C}}{1\text{ mol }e^-}\right) = 4.18\times10^5\text{ C}$$

The time that will be required is:

$$\text{hr} = (4.18\times10^5\text{ C})\left(\frac{1\text{ s}}{2.25\text{ C}}\right)\left(\frac{1\text{ hr}}{3600\text{ s}}\right) = 51.5\text{ hr}$$

19.97 $Mg^{2+} + 2e^- \longrightarrow Mg(l)$

The number of Coulombs that will be required is:

$$\text{Coulombs} = (60.0\text{ g Mg})\left(\frac{1\text{ mol Mg}}{24.31\text{ g Mg}}\right)\left(\frac{2\text{ mol }e^-}{1\text{ mol Mg}}\right)\left(\frac{96{,}500\text{ C}}{1\text{ mol }e^-}\right) = 4.76\times10^5\text{ C}$$

The number of amperes is: 4.76×10^5 C ÷ 7200 s = 66.2 amp

19.99 The electrolysis of NaCl solution results in the reduction of water, together with the formation of hydroxide ion

$2H_2O + 2e^- \longrightarrow H_2(g) + 2OH^-(aq)$

The number of Coulombs is

2.00 A × 20.0 min × 60 s/min = 2.40×10^3 C

The number of moles of OH^- is

$$\text{mol OH}^- = (2.40\times10^3\text{ C})\left(\frac{1\text{ mol }e^-}{96{,}500\text{ C}}\right)\left(\frac{2\text{ mol OH}^-}{2\text{ mol }e^-}\right) = 0.0249\text{ mol OH}^-$$

$[OH^-]$ = 0.0249 mol/0.250 L = 0.0996 *M*

Chapter Twenty
Nuclear Reactions and Their Role in Chemistry

Practice Exercises

20.1 $$m = \frac{m_0}{\sqrt{1-\left(\frac{v}{c}\right)^2}}$$

$$m = \frac{57.7\ \text{g}}{\sqrt{1 - \left(\frac{53.6\ \text{m s}^{-1}}{3.00 \times 10^8\text{m s}^{-1}}\right)^2}} = 57.7\ \text{g}$$

The tennis ball is too large and travelling too slow to have any relativistic effect on its mass.

20.2

$$m = \frac{57.7\ \text{g}}{\sqrt{1 - \left(\frac{5.36 \times 10^7\ \text{m s}^{-1}}{3.00 \times 10^8\text{m s}^{-1}}\right)^2}} = 58.6\ \text{g}$$

A slight relativistic effect on mass occurs at the higher velocity.

20.3 $^{56}_{26}Fe$ has 26 protons and 30 neutrons.

26 protons have a mass of	26×1.0072764668 u = 26.1891881368 u
30 neutrons have a mass of	30×1.0086649160 u = 30.2599474748 u
The total rest mass of the isotope is	56.4491356168 u
The mass defect is	56.4491356168 u – 55.934939 u = 0.514197 u

20.4 $^{12}_{6}C$ has 6 protons and 6 neutrons.

6 protons have a mass of 6×1.0072764668 u = 6.0436588008 u

6 neutrons have a mass of 6×1.0086649160 u = 6.0519894960 u

The total rest mass of the isotope is 12.0956482968 u

The mass defect is 12.0956482968 u – 11.996708 u = 0.098940 u

The binding energy is given by $E = mc^2$

$$E = 0.098940\ \text{u} \times \left(\frac{1.66065389 \times 10^{-27}\text{kg}}{\text{u}}\right)\left(2.9979 \times 10^8\ \text{m s}^{-1}\right)^2$$

$E = 1.47668 \times 10^{-11}$ J

There are twelve nucleons in this isotope. Therefore, the bonding per nucleon is given as 1.47668×10^{-11} J/12 nucleons = 1.23057×10^{-12} J/nucleon

20.5 $^{226}_{88}Ra \longrightarrow {}^{222}_{86}Rn + {}^{4}_{2}He + {}^{0}_{0}\gamma$ An alpha particle is emitted.

20.6 $^{131}_{53}I \longrightarrow {}^{131}_{54}Xe + {}^{0}_{-1}e + \bar{\nu}$

20.7 $^{11}_{6}C \longrightarrow {}^{11}_{5}B + {}^{0}_{1}e + \nu$

20.8 $^{13}_{4}Be \longrightarrow {}^{12}_{4}Be + {}^{1}_{0}n$

20.9 $^{72}_{34}Se + ^{0}_{-1}e \longrightarrow ^{72}_{33}As + X\ ray + v$

20.10 $^{242}_{96}Cm \longrightarrow ^{238}_{94}Pu + ^{4}_{2}He$

20.11 $^{262}_{106}Sg \longrightarrow ^{258}_{104}Rf + ^{4}_{2}He$

20.12 Using the value of k for Pu from Example 20.2

Activity = kN

Activity = 6.22×10^{11} Bq = 6.22×10^{11} disintegrations/second

$k = 2.50 \times 10^{-10}$ seconds^{-1}

6.22×10^{11} disintegrations/second = $(2.50 \times 10^{-10}$ seconds$^{-1})N$

$N = 2.49 \times 10^{21}$ atoms Pu

Now find the mass of Pu

$$\text{mass Pu} = (2.49 \times 10^{21} \text{ atoms Pu})\left(\frac{1 \text{ mole Pu}}{6.022 \times 10^{23} \text{ atoms Pu}}\right)\left(\frac{238 \text{ g Pu}}{1 \text{ mole Pu}}\right) = 0.984 \text{ g Pu}$$

The percentage of Pu in the sample

$$\frac{0.984 \text{ g Pu}}{2.00 \text{ g sample}} \times 100\% = 49.2\% \text{ Pu}$$

20.13 $$\text{Half life for Rn-222} = 3.82 \text{ days}\left(\frac{24 \text{ h}}{\text{day}}\right)\left(\frac{3600 \text{ s}}{1 \text{ h}}\right) = 330{,}048 \text{ s}$$

$$k = \frac{\ln 2}{t_{1/2}} = \frac{0.6931}{330{,}048 \text{ s}} = 2.10 \times 10^{-6} \text{ s}^{-1}$$

$$\text{Activity} = 4 \text{ pCi} = 4 \times 10^{-12} \text{ Ci}\left(\frac{3.7 \times 10^{10} \text{ dps}}{1 \text{ Ci}}\right)$$

Activity = 0.1482 Bq

Activity = 0.148 disintegrations per second

Activity = disintegrations/sec = kN

$0.148 = (2.10 \times 10^{-6} \text{ s}^{-1})N$

$N = 7.05 \times 10^{4}$ atoms Rn-222

20.14 We make use of the Inverse Square Law

$$\frac{I_1}{I_2} = \frac{d_2^2}{d_1^2}$$

$$\frac{4.8 \text{ units}}{0.30 \text{ units}} = \frac{(x \text{ m})^2}{(5.0 \text{ m})^2}$$

x m = 20 m

20.15 We make use of the Inverse Square Law

$$\frac{I_1}{I_2} = \frac{d_2^2}{d_1^2}$$

$$\frac{1.4 \text{ units}}{I_2} = \frac{(1.2 \text{ m})^2}{(10 \text{ m})^2} = 100 \text{ units (assuming 1 significant figure)}$$

Review Problems

20.51 Equation 20.1 becomes

$$m = \frac{1.00}{\sqrt{1 - \left(\frac{v}{c}\right)^2}}$$

On substituting the various velocities for the value of v in the above expression, and using the value

$c = 3.00 \times 10^8$ m s^{-1}, we get (a) m = 1.01 kg (b) 3.91 kg (c) 12.3 kg

20.53 Solve the Einstein equation for Δm

$$\Delta m = \frac{\Delta E}{c^2}$$

1 kJ = 1.00×10^3 J = 1.00×10^3 kg m^2 s^{-2}

$$\Delta m = \frac{1.00 \times 10^3 \text{ kg m}^2 \text{ s}^{-2}}{(3.00 \times 10^8 \text{ m s}^{-1})^2} = 1.11 \times 10^{-14} \text{ kg} = 1.11 \times 10^{-11} \text{ g}$$

20.55 The joule is equal to one kg m^2/s^2, and this is employed directly in the Einstein equation

$$\Delta m = \frac{\Delta E}{c^2}$$

where ΔE is the enthalpy of formation of liquid water, which is available in Table 6.2.

$H_2(g) + O_2(g) \longrightarrow H_2O(l)$, $\quad \Delta H = -285.9$ kJ/mol

$$\Delta m = \frac{285.9 \times 10^3 \text{ kg m}^2 \text{ s}^{-2}}{(3.00 \times 10^8 \text{ m s}^{-1})^2} = -3.18 \times 10^{-12} \text{ kg}$$

$$(-3.18 \times 10^{-12} \text{ kg})\left(\frac{1000 \text{ g}}{1 \text{ kg}}\right)\left(\frac{10^9 \text{ ng}}{\text{g}}\right) = -3.18 \text{ ng}$$

The negative value for the mass implies that mass is lost in the reaction.

The percent of mass that is lost is

$$\left(\frac{3.18 \times 10^{-12} \text{ kg}}{18.02 \times 10^{-3} \text{ kg}}\right) \times 100\% = 1.77 \times 10^{-8}\%$$

20.57 The mass of the deuterium nucleus is the mass of the proton (1.0072764668 u) plus that of a neutron (1.0086649160 u), or 2.0159413828 u. The difference between this calculated value and the observed value is equal to Δm

$\Delta m = (2.015941 - 2.0135) = 2.44 \times 10^{-3}$ u

$\Delta E = \Delta mc^2 = (2.44 \times 10^{-3} \text{ u})(1.6605 \times 10^{-27} \text{ kg/u})(3.00 \times 10^8 \text{ m/s})^2$

$\Delta E = 3.65 \times 10^{-13}$ kg m^2/s^2 = 3.65×10^{-13} J

Since there are two nucleons per deuterium nucleus, we have

$$\Delta E = \frac{3.65 \times 10^{-13}\ \text{J}}{2\ \text{nucleons}} = 1.8 \times 10^{-13}\ \text{J per nucleon}$$

20.59 $^{63}_{29}$Cu has 29 protons and 34 neutrons.

29 protons have a mass of	29×1.0072764669 u = 29.2110175401 u
34 neutrons have a mass of	34×1.0086649156 u = 34.02946071304 u
The total rest mass of the isotope is	63.50562477 u
The mass defect is	63.50562477 u – 62.9295975 u = 0.5760271 u
The binding energy is given by	$E = mc^2$

$$E = 0.5760271\ \text{u} \times \left(\frac{1.66065389 \times 10^{-27}\ \text{kg}}{\text{u}}\right)\left(2.9979 \times 10^{8}\ \text{m s}^{-1}\right)^2$$

$= 8.5966 \times 10^{-11}$ J

20.61 (a) $^{211}_{83}\text{Bi}$ (b) $^{177}_{72}\text{Hf}$ (c) $^{216}_{84}\text{Po}$ (d) $^{19}_{9}\text{F}$

20.63 (a) $^{242}_{94}\text{Pu} \longrightarrow {}^{4}_{2}\text{He} + {}^{238}_{92}\text{U}$

(b) $^{28}_{12}\text{Mg} \longrightarrow {}^{0}_{-1}e + \nu + {}^{28}_{13}\text{Al}$

(c) $^{26}_{14}\text{Si} \longrightarrow {}^{0}_{1}e + \bar{\nu} + {}^{26}_{13}\text{Al}$

(d) $^{37}_{18}\text{Ar} + {}^{0}_{-1}e \longrightarrow {}^{37}_{17}\text{Cl}$

20.65 (a) $^{261}_{102}\text{No}$ (b) $^{211}_{82}\text{Pb}$ (c) $^{141}_{61}\text{Pm}$ (d) $^{179}_{74}\text{W}$

20.67 $^{87}_{36}\text{Kr} \longrightarrow {}^{86}_{36}\text{Kr} + {}^{1}_{0}n$

20.69 The more likely process is positron emission, because this produces a product having a higher neutron–to–proton ratio $^{38}_{19}\text{K} \longrightarrow {}^{0}_{1}e + {}^{38}_{18}\text{Ar}$

20.71 The most probable decay is alpha emission.

$^{209}_{84}\text{Po} \longrightarrow {}^{205}_{82}\text{Pb} + {}^{4}_{2}\text{He}$

20.73 Six half-life periods correspond to the fraction 1/64 of the initial material. That is, one sixty-fourth of the initial material is left after 6 half lives 3.00 mg × 1/64 = 0.0469 mg remaining.

20.75 $^{53}_{24}\text{Cr}^*$ forms. The reactions are $^{51}_{23}\text{V} + {}^{2}_{1}\text{H} \longrightarrow {}^{53}_{24}\text{Cr}^* \longrightarrow {}^{1}_{1}p + {}^{52}_{23}\text{V}$

20.77 $^{80}_{35}\text{Br}$

20.79 $^{55}_{26}\text{Fe}$; $^{55}_{25}\text{Mn} + {}^{1}_{1}p \longrightarrow {}^{1}_{0}n + {}^{55}_{26}\text{Fe}$

20.81 $^{70}_{30}\text{Zn} + {}^{208}_{82}\text{Pb} \longrightarrow {}^{278}_{112}\text{Uub} \rightarrow {}^{1}_{0}n + {}^{277}_{112}\text{Uub}$

20.83 Radiation $\propto \dfrac{1}{d^2}$

$$\frac{I_1}{I_2} = \frac{d_2^{\,2}}{d_1^{\,2}}$$

$$d_2 = d_1\sqrt{\frac{I_1}{I_2}} = 2.0\text{ m}\sqrt{\frac{2.8}{0.28}} = 6.3\text{ m}$$

20.85 This calculation makes use of the Inverse Square Law

$$\frac{I_1}{I_2} = \frac{d_2^{\,2}}{d_1^{\,2}}$$

$$\frac{8.4\text{ rem}}{0.50\text{ rem}} = \frac{d_2^{\,2}}{(1.60\text{ m})^2}$$

$d_2 = 6.6$ m

20.87 This is one curie. It is also $1.0\text{ Ci} \times \left(\dfrac{3.7\times 10^{10}\text{ Bq}}{\text{Ci}}\right) = 3.7 \times 10^{10}$ Bq.

20.89 Activity = kN and $t_{1/2} = \dfrac{\ln 2}{k}$ or $k = \dfrac{\ln 2}{t_{1/2}}$

$$\text{Activity} = \frac{\ln 2}{t_{1/2}}N$$

$$N = 0.20\text{ mg}\left(\frac{1\text{ g }^{241}\text{Am}}{1000\text{ mg}}\right)\left(\frac{1\text{ mol }^{241}\text{Am}}{241\text{ g }^{241}\text{Am}}\right)\left(\frac{6.022\times 10^{23}\text{ atoms }^{241}\text{Am}}{1\text{ mol }^{241}\text{Am}}\right) = 5.0 \times 10^{17}\text{ atoms}$$
^{241}Am

$$\text{Activity} = \left(\frac{\ln 2}{1.70\times 10^{5}\text{ d}}\right) \times (5.0\times 10^{17}\text{ atoms }^{241}\text{Am})$$

$$= \left(\frac{2.0\times10^{12}\text{ Am decays}}{\text{d}}\right)\left(\frac{1\text{ d}}{24\text{ h}}\right)\left(\frac{1\text{ h}}{3600\text{ s}}\right) = 2.4\times 10^{7}\text{ s}^{-1}$$

$$2.4\times 10^{7}\text{ s}^{-1} = 2.4\times 10^{7}\text{ Bq} = 2.4\times 10^{7}\text{ Bq}\left(\frac{1\text{ Ci}}{3.7\times 10^{10}\text{ Bq}}\right)\left(\frac{1000\text{ mCi}}{1\text{ Ci}}\right) = 6.5\times 10^{-1}\text{ mCi}$$

$$= 6.5\times 10^{2}\ \mu\text{Ci}$$

20.91 Activity = kN

$$k = \frac{\text{activity}}{N}$$

$$N = \text{number of }^{131}\text{I atoms} = 1.00\text{ mg }^{131}\text{I}\left(\frac{1\text{ g }^{131}\text{I}}{1000\text{ mg }^{131}\text{I}}\right)\left(\frac{1\text{ mol }^{131}\text{I}}{131\text{ g }^{131}\text{I}}\right)\left(\frac{6.022\times 10^{23}\text{ g }^{131}\text{I}}{1\text{ mol }^{131}\text{I}}\right)$$

$$= 4.60\times 10^{18}\text{ atoms }^{131}\text{I}$$

$$k = \frac{4.6 \times 10^{12} \text{ Bq}}{4.60 \times 10^{18} \text{ atoms } ^{131}\text{I}}\left(\frac{1 \text{ s}^{-1}}{1 \text{ Bq}}\right) = 1.0 \times 10^{-6} \text{ s}^{-1}$$

$$t_{1/2} = \frac{\ln 2}{k} = \left(\frac{\ln 2}{1.0 \times 10^{-6} \text{ s}^{-1}}\right) = 6.9 \times 10^{5} \text{ s (This is about 8 days.)}$$

20.93 The chemical product is $BaCl_2$. Recall that for a first order process

$$k = \frac{0.693}{t_{1/2}}$$

$$\text{So } k = \frac{0.693}{30.1 \text{ yr}} = 2.30 \times 10^{-2}/\text{yr}.$$

Also,

$$\ln\frac{[A]_0}{[A]_t} = kt$$

$$[A]_t = [A]_0 \exp(-kt)$$

$$\frac{[A]_0}{[A]_t} = \exp[-(2.30 \times 10^{-2} \text{ yr}^{-1})(150 \text{ yr})]$$

$$\frac{[A]_0}{[A]_t} = 3.16 \times 10^{-2}$$

so 3.2% of the original sample remains.

20.95 This calculation makes use of the first order rate equation, where knowing $[A]_t$, we need to calculate $[A]_0$

$$\ln\frac{[A]_0}{[A]_t} = kt$$

$$k = \frac{0.693}{t_{1/2}} = \frac{0.693}{8.07 \text{ d}} = 8.59 \times 10^{-2} \text{ d}^{-1}$$

$$\ln\frac{[A]_0}{(25.6 \times 10^{-5} \text{ Ci g}^{-1})} = (8.59 \times 10^{-2} \text{ d}^{-1})(28.0 \text{ d})$$

Taking the exponential of both sides of the above equation gives

$$\frac{[A]_0}{(25.6 \times 10^{-5} \text{ Ci g}^{-1})} = e^{2.41} = 11.1$$

Solving for the value of $[A]_0$ gives $[A]_0 = 2.84 \times 10^{-3}$ Ci g^{-1}

20.97 Amount of $CaSO_4$ in the precipitate

$$65.6 \text{ mg CaSO}_4\left(\frac{1 \text{ g CaSO}_4}{1000 \text{ mg CaSO}_4}\right)\left(\frac{1 \text{ mol CaSO}_4}{136.14 \text{ g CaSO}_4}\right)\left(\frac{6.022 \times 10^{23} \text{ CaSO}_4}{1 \text{ mol CaSO}_4}\right)$$

$$= 2.90 \times 10^{20} \text{ formula units CaSO}_4$$

$$1.75 \times 10^{-6} \text{ Ci } ^{45}\text{Ca}\left(\frac{3.7 \times 10^{10} \text{ Bq}}{1 \text{ Ci}}\right)\left(\frac{1 \text{ dis s}^{-1}}{1 \text{ Bq}}\right) = 6.47 \times 10^{4} \text{ dis s}^{-1}$$

$$t_{1/2} = 163\ \text{d}\left(\frac{24\ \text{h}}{1\ \text{d}}\right)\left(\frac{3600\ \text{s}}{1\ \text{h}}\right) = 1.41 \times 10^{7}\ \text{s}$$

$$k = \frac{\ln 2}{t_{1/2}} = \frac{\ln 2}{1.41 \times 10^{7}\ \text{s}} = 4.92 \times 10^{-8}\ \text{s}^{-1}$$

$$\text{Activity} = \frac{\Delta N}{\Delta t} = kN$$

$$N = \frac{\text{Activity}}{k} = \frac{6.47 \times 10^{4}\ \text{dis s}^{-1}}{4.92 \times 10^{-8}\ \text{s}^{-1}} = 1.31 \times 10^{12}\ \text{nuclei Ca-45}$$

$$\%\ \text{Ca-45 in precipitate} = \frac{1.31 \times 10^{12}\ \text{nuclei Ca-45}}{2.90 \times 10^{20}\ \text{nuclei Ca}} \times 100\% = 4.52 \times 10^{-7}\%\ \text{Ca-45}$$

In the solution, the ratio of Ca-45 to Ca is the same as in the precipitate.

$$2.25 \times 10^{-6}\ \text{Ci Ca-45}\left(\frac{3.7 \times 10^{10}\ \text{Bq}}{1\ \text{Ci}}\right)\left(\frac{1\ \text{dis s}^{-1}}{1\ \text{Bq}}\right) = 8.32 \times 10^{4}\ \text{dis s}^{-1}\ \text{Ca-45}$$

$$N = \frac{\text{Activity}}{k} = \frac{8.32 \times 10^{4}\ \text{dis s}^{-1}}{4.92 \times 10^{-8}\ \text{s}^{-1}} = 1.68 \times 10^{12}\ \text{nuclei Ca-45}$$

$$4.52 \times 10^{-7}\%\ \text{Ca-45} = \frac{1.68 \times 10^{12}\ \text{nuclei Ca-45}}{x\ \text{nuclei Ca}} \times 100\%$$

$$x = 3.72 \times 10^{20}\ \text{nuclei Ca}$$

$$3.72 \times 10^{20}\ \text{Ca}\left(\frac{1\ \text{mol Ca}}{6.022 \times 10^{23}\ \text{Ca atoms}}\right) = 6.18 \times 10^{-4}\ \text{mol Ca}$$

$$[Ca^{2+}] = \frac{6.18 \times 10^{-4}\ \text{mol}\ Ca^{2+}}{0.100\ \text{L}\ H_2O} = 6.18 \times 10^{-3}\ M\ Ca^{2+}$$

20.99 In order to solve this problem, it must be assumed that all of the argon-40 that is found in the rock must have come from the potassium-40, i.e., that the rock contains no other source of argon-40. If the above assumption is valid, then any argon-40 that is found in the rock represents an equivalent amount of potassium-40, since the stoichiometry is 1:1. Since the ratio of K-40 to Ar-40 is 2:1, this indicates that originally, there was

$$2.45 \times 10^{-6} + 1.22 \times 10^{-6}\ \text{mol K-40} = 3.67 \times 10^{-6}\ \text{mol K-40}$$

We can calculate the rate constant

$$k = \frac{\ln 2}{1.3 \times 10^{9}} = 5.3 \times 10^{-10}\ \text{yr.}^{-1}$$

Using the integrated rate law for a first order reaction

$$\ln\frac{[\text{K-40}]_0}{[\text{K-40}]_t} = \ln\frac{3.67 \times 10^{-6}}{2.45 \times 10^{-6}} = kt = (5.3 \times 10^{-10}\ \text{yr.}^{-1})t$$

$$t = 7.6 \times 10^{8}\ \text{yr.}^{-1}$$

20.101 $\ln\left(\frac{^{14}\text{C}}{^{12}\text{C}}\right) = \left(1.2 \times 10^{-4}\right)t$

Taking the natural log we determine

$$\ln\left(\frac{1.2 \times 10^{-12}}{4.8 \times 10^{-14}}\right) = \left(1.2 \times 10^{-4}\right)t$$

$$t = \left(\frac{1}{1.2 \times 10^{-4}}\right)\ln\left(\frac{1.2 \times 10^{-12}}{4.8 \times 10^{-14}}\right) = 2.7 \times 10^{4} \text{ yr}$$

The tree died 2.7×10^4 years ago. This is when the volcanic eruption occurred.

20.103 $^{235}_{92}\text{U} + {}^{1}_{0}n \longrightarrow {}^{94}_{38}\text{Sr} + {}^{140}_{54}\text{Xe} + 2\,{}^{1}_{0}n$

Chapter Twenty-One
Metal Complexes

Practice Exercises

21.1 $[Ag(S_2O_3)_2]^{3-}$ $(NH_4)_3Ag(S_2O_3)_2$

21.2 $AlCl_3{\bullet}6H_2O$ $[Al(H_2O)_6]^{3+}$

21.3 $[CrCl_2(H_2O)_4]^+$ The counter ion would be a halide

21.4 (a) $[SnCl_6]^{2-}$
(b) $(NH_4)_2[Fe(CN)_4(H_2O)_2]$
(c) $OsBr_2(H_2NCH_2CH_2NH_2)_2$ or $OsBr_2(en)_2$

21.5 (a) potassium hexacyanoferrate(III)
(b) dichlorobis(ethylenediamine)chromium(III) sulfate
(c) hexaaquacobalt(II) hexafluorochromate(II)

21.6 Since there are three ligands and $C_2O_4^{2-}$ is a bidentate ligand, the coordination number is six.

21.7 (a) The coordination number is six. There are two bidentate ligands and two monodentate ligands.
(b) The coordination number is six. Both $C_2O_4^{2-}$ and ethylenediamine are bidentate ligands. Since there are three bidentate ligands, the coordination number must be six.
(c) EDTA is a hexadentate ligand so the coordination number is six.

21.8 (a) The coordination number is four. There are four monodentate ligands attached to the metal.
The coordination geometry is square planar.
(b) The coordination number is two. There are two monodentate ligands attached to the metal.
The coordination geometry is linear with the thiosulfate ions attached to the silver ion through the sulfur atoms.

21.9 (a) The two isomers have the same linkage, but are mirror images of each other. They are optical isomers, or enantiomers.
(b) The two isomers have different linkages, they are geometric isomers.

21.10 Six isomers can be drawn.

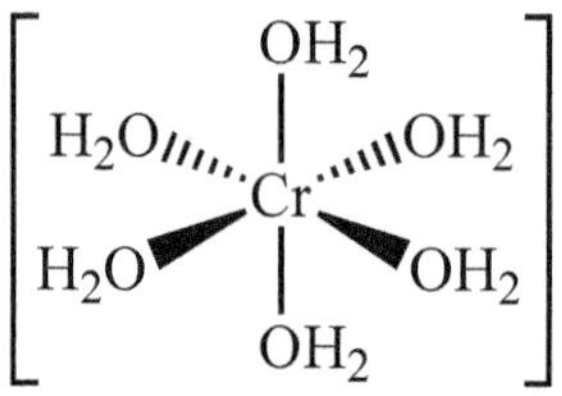

OH_2
H_2O OH_2
Cr
H_2O OH_2
OH_2
$3Br^-$

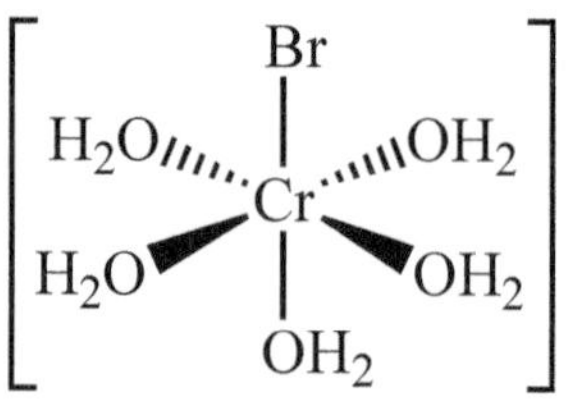

Br
H_2O OH_2
Cr
H_2O OH_2
OH_2
$2Br^-$ H_2O

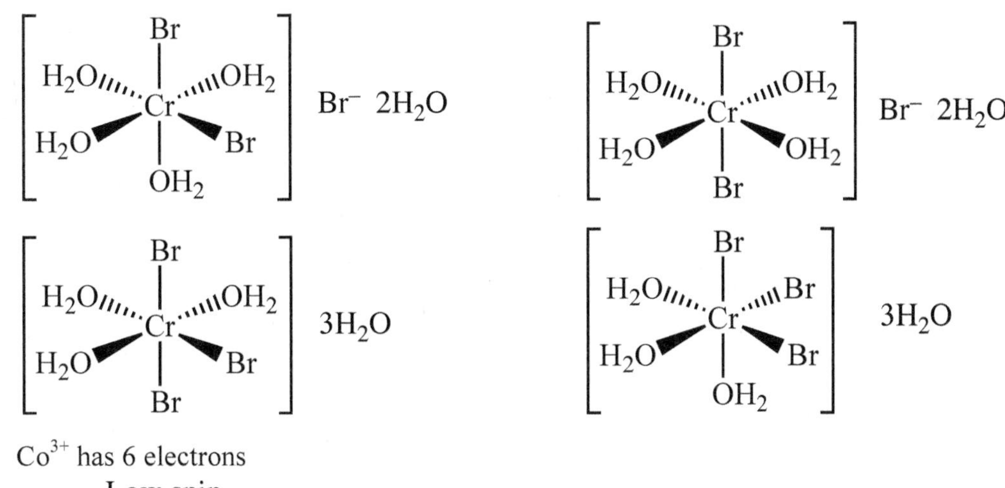

21.11 Co^{3+} has 6 electrons

Low spin

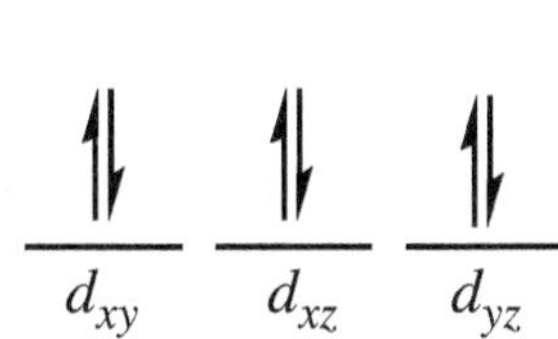

High spin

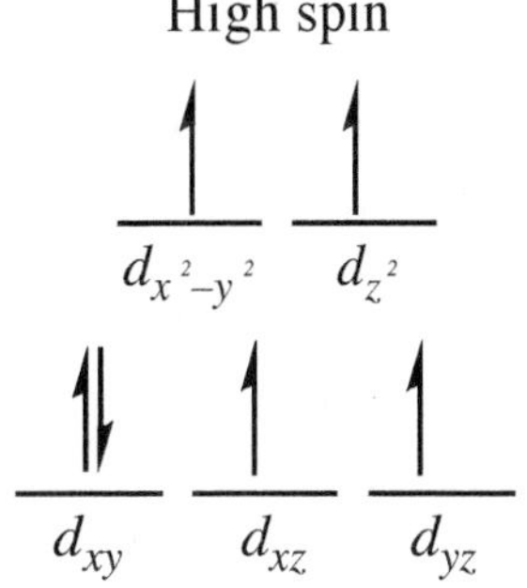

d_{xy} d_{xz} d_{yz}

d_{xy} d_{xz} d_{yz}

High spin has zero unpaired electrons and is diamagnetic.
Low spin has four unpaired electrons and is paramagnetic.
Diamagnetic compounds have no unpaired electrons.

21.12 The Ni(0) complex has no unpaired electrons.

d_{xy} d_{xz} d_{yz}

$d_{x^2-y^2}$ d_{z^2}

21.13 The ligands for the iron in hemoglobin are a porphyrin, imidazole and oxygen.

Review Problems

21.53 The net charge is –3, and the formula is $[Fe(CN)_6]^{3-}$. The IUPAC name for the complex is hexacyanoferrate(III) ion

21.55 $[Cr(NH_3)_2(NO_2)_4]^-$

21.57 (a) NH_3 ammine (b) N^{3-} azido
(c) SO_4^{2-} sulfato (d) $C_2H_3O_2^-$ acetato

21.59 (a) hexaamminenickel(II) chloride
(b) triamminetrichlorochromate(II) ion
(c) hexanitrocobaltate(III) ion
(d) diamminetetracyanomanganate(II) ion
(e) potassium trioxalatoferrate(III) or potassium trisoxalatoferrate(III)

21.61 (a) $[Fe(CN)_2(H_2O)_4]^+$
(b) $[Ni(C_2O_4)(NH_3)_4]$
(c) $[Al(H_2O)_5(OH)]Cl_2$
(d) $K_3[Mn(SCN)_6]$
(e) $[CuCl_4]^{2-}$

21.63 $C_2O_4^{2-}$ is bidentate × 2
NO_2^- is monodentate × 2
Coordination number = 6

21.65 (a)

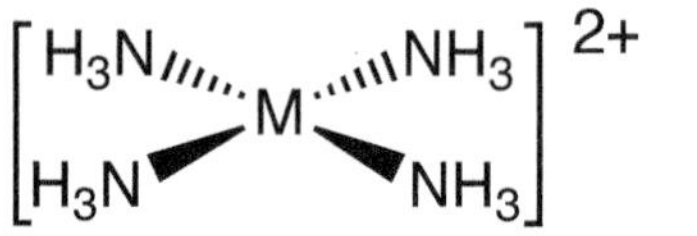

(b)

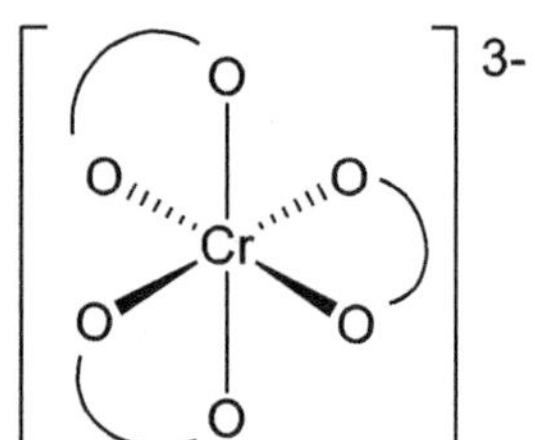

The curved lines represent the backbone of the oxalate ion.

21.67

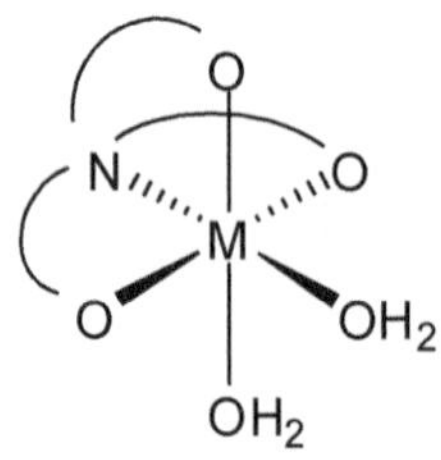

The curved lines represent $-CH_2-C(=O)-$ groups

21.69 Since both are the *cis* isomer, they are identical. One can be superimposed on the other after simple rotation.

21.71 cis:

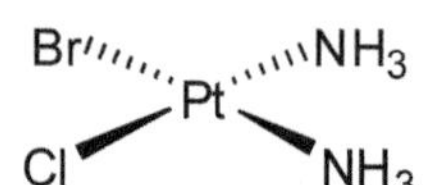

trans:

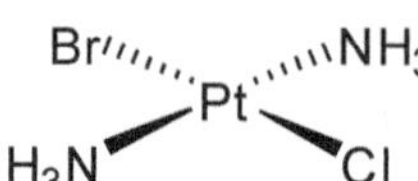

21.73 The *cis* isomer is chiral:

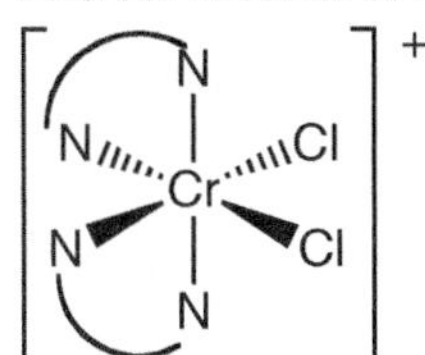

The *trans* isomer is nonchiral:

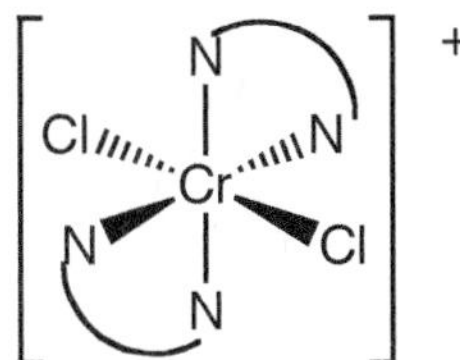

21.75 (a) $[Cr(H_2O)_6]^{3+}$ (b) $[Cr(en)_3]^{3+}$

21.77 $[Cr(CN)_6]^{3-}$

21.79 (a) The value of Δ increases down a group. Therefore, we choose: $[RuCl(NH_3)_5]^{2+}$
(b) The value of Δ increases with oxidation state of the metal. Therefore, we choose: $[Ru(NH_3)_6]^{3+}$

21.81 This is the one with the strongest field ligand, since Co^{2+} is a d^7 ion: CoA_6^{3+}. You are removing a higher energy electron in CoA_6^{2+} than in CoB_6^{2+}. *A* produces a larger ligand field splitting than does *B*.

21.83 This is a weak field complex of Co^{2+}, and it should be a high-spin d^7 case. It cannot be diamagnetic; even if it were low spin, we would still have one unpaired electron.

21.85 For $Fe(H_2O)_6^{3+}$, we expect a relatively small value for Δ, and we predict the high-spin case having five unpaired electrons:

$Fe(H_2O)_6^{3+}$

For $Fe(CN)_6^{3-}$, we expect a relatively large value for Δ, and we predict the low–spin case having one unpaired electron:

$Fe(CN)_6^{3-}$

Chapter Twenty-Two
Organic Compounds, Polymers, and Biochemicals

Practice Exercises

22.1 (a) C_6H_{14}, $CH_3CH_2CH_2CH_2CH_2CH_3$
(b) C_6H_{14}, $(CH_3)_2CHCH(CH_3)_2$
(c) CH_3Cl_3, $ClCH_2CHCl_2$

22.2

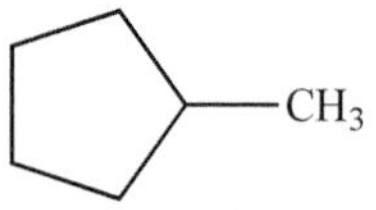

(cyclopentane ring)—CH_3

C_6H_{12}

22.3 2,2-dimethylpropane

22.4 (a) 3-methylhexane
(b) 4-ethyl-2,3-dimethylheptane
(c) 5-ethyl-2,4,6-trimethyloctane

22.5

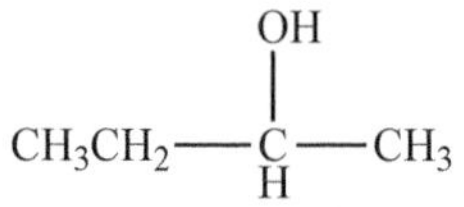

$$CH_3CH_2-\underset{H}{\overset{OH}{\overset{|}{C}}}-CH_3$$

22.6 (a)

$$CH_3\overset{O}{\overset{\|}{C}}H \quad \text{or} \quad CH_3\overset{O}{\overset{\|}{C}}OH$$

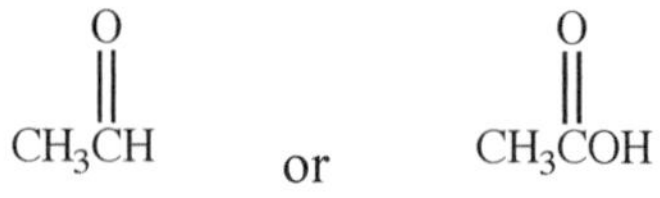

(depending on the strength of the oxidizing agent)

(b)

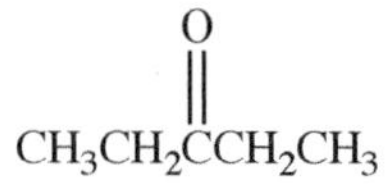

$$CH_3CH_2\overset{O}{\overset{\|}{C}}CH_2CH_3$$

(c) Tertiary alcohols do not undergo oxidation in the same manner as primary and secondary alcohols.

22.7 (a)

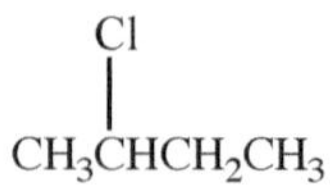

$$CH_3\overset{Cl}{\overset{|}{C}}HCH_2CH_3$$

(b)

$$CH_3CH=CH_2$$

22.8

$$CH_3CH_2\overset{O}{\overset{\|}{C}}OH + CH_3OH \xrightleftharpoons{H^+ \text{ catalyst}} CH_3CH_2\overset{O}{\overset{\|}{C}}OCH_3 + H_2O$$

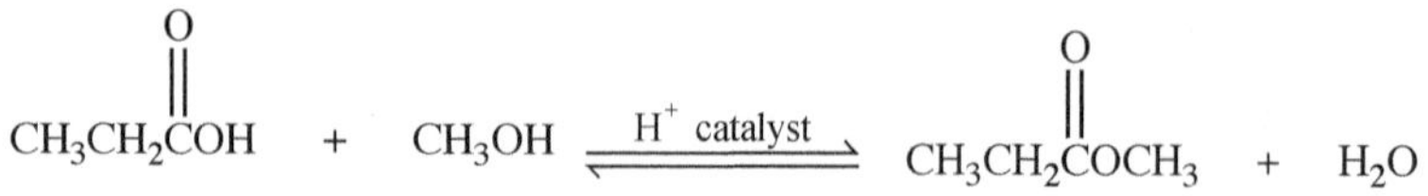

The product is methyl propanoate

22.9 (a)

$$CH_3\overset{CH_3}{\overset{|}{C}}HCH_2\overset{O}{\overset{\|}{C}}H$$

(b)

$$CH_3CH_2\overset{O}{\overset{\|}{C}}OCH_2CH_3$$

(c)

$$CH_3\overset{O}{\overset{\|}{C}}CH_2CH_3$$

22.10 The products are:

CH_3CH_2OH and $CH_3\overset{O}{\overset{\|}{C}}O^-$

22.11

$$H_2NCH_2CH_2CH_3 + CH_3CH_2\overset{O}{\overset{\|}{C}}OH \longrightarrow CH_3CH_2\overset{O}{\overset{\|}{C}}NHCH_2CH_2CH_3 + H_2O$$

22.12 (a)

$$CH_3CH_2\overset{O}{\overset{\|}{C}}OH + CH_3CH_2NH_2$$

(b)

$$CH_3CH_2\overset{O}{\overset{\|}{C}}O^- + HOCH(CH_3)_2$$

(c)

$$CH_3CH_2CH{=}CH_2$$

22.13

$$-\underset{CH_3}{\underset{|}{\overset{H}{\overset{|}{C}}}}-\underset{H}{\underset{|}{\overset{CH_3}{\overset{|}{C}}}}-\underset{CH_3}{\underset{|}{\overset{H}{\overset{|}{C}}}}-\underset{H}{\underset{|}{\overset{CH_3}{\overset{|}{C}}}}-\underset{CH_3}{\underset{|}{\overset{H}{\overset{|}{C}}}}-\underset{H}{\underset{|}{\overset{CH_3}{\overset{|}{C}}}}-$$

Repeat unit

22.14

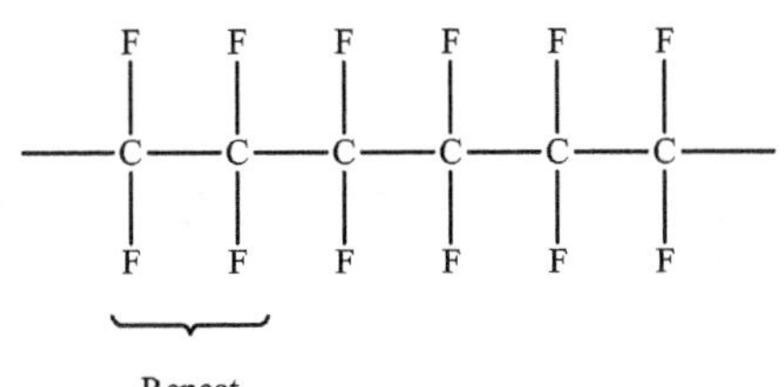

22.15 (a) lipid
(b) carbohydrate
(c) amino acid
(d) carbohydrate
(e) amino acid
(f) lipid

22.16

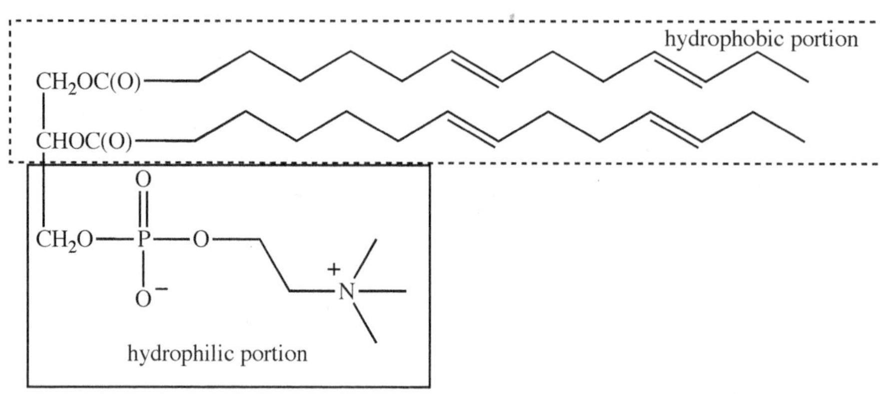

22.17 ribose

```
HO
  CH2              OH
        O
   H         H
H                  H
   OH        OH
```

deoxyribose

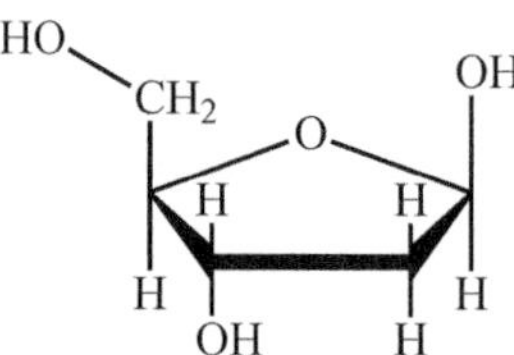

22.18 A hydrogen bonds to T using N–H–O and N–H–N hydrogen bonds.

G hydrogen bonds to C using N–H–O, N–H–N, and O–H–N hydrogen bonds

Review Problems

22.89

(a)

```
   H   H
   |   |
H—C—N—H
   |
   H
```

(b)

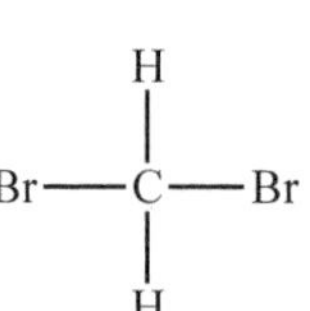

(c)

```
   Cl
   |
H—C—Cl
   |
   Cl
```

(d)

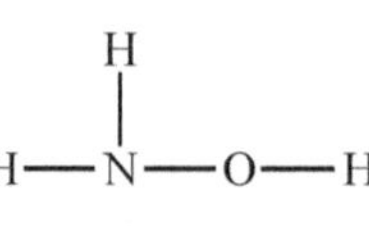

(e)

H—C≡C—H

(f)

```
   H   H
   |   |
H—N—N—H
```

22.91 (a) alkene (d) carboxylic acid
(b) alcohol (e) amine
c) ester (f) alcohol

22.93 The saturated compounds are b, c, d, e, and f.

22.95 (a) amine (b) amine (c) amide (d) amine, ketone

22.97 (a) These are identical, being oriented differently only.
(b) These are identical, being drawn differently only.
(c) These are unrelated, being alcohols with different numbers of carbon atoms.
(d) These are isomers, since they have the same molecular formula, but different structures.
(e) These are identical, being oriented differently only.
(f) These are identical, being drawn differently only.
(g) These are isomers, since they have the same molecular formula, but different structures.

22.99 (a) pentane
(b) 2-methylpentane
(c) 2,4-dimethylhexane

22.101 (a) No isomers
(b)

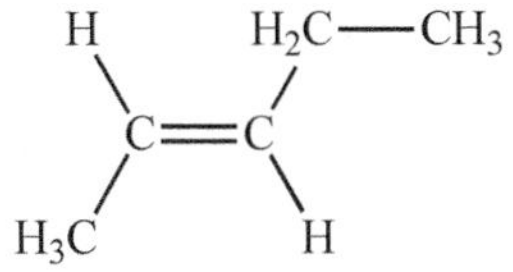

H, $H_2C—CH_3$ / C=C / H_3C, H
trans

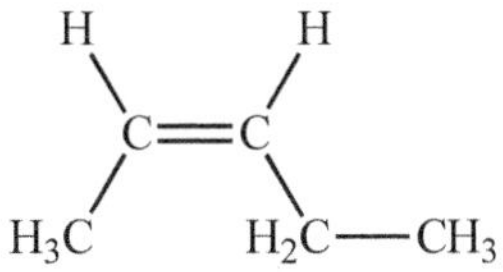

H, H / C=C / H_3C, $H_2C—CH_3$
cis

(c)

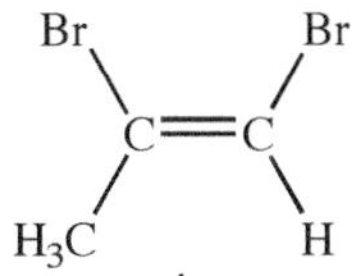

Br, Br / C=C / H_3C, H
cis

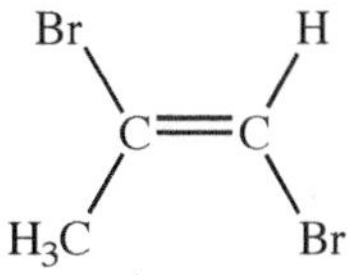

Br, H / C=C / H_3C, Br
trans

22.103 (a) CH_3CH_3
(b) $ClCH_2CH_2Cl$
(c) $BrCH_2CH_2Br$
(d) CH_3CH_2Cl
(e) CH_3CH_2Br
(f) CH_3CH_2OH

22.105 (a) $CH_3CH_2CH_2CH_3$
(b)

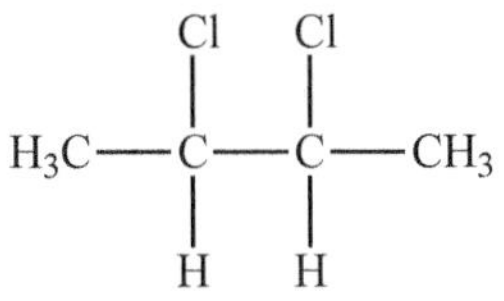

$H_3C—C(Cl)(H)—C(Cl)(H)—CH_3$

(c)

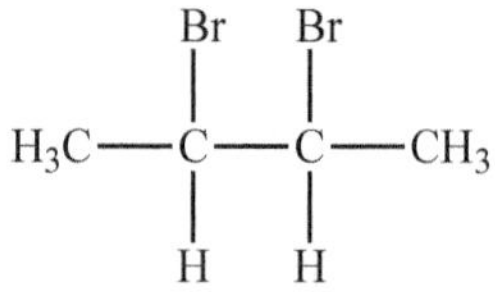

```
      Br   Br
      |    |
H3C—C—C—CH3
      |    |
      H    H
```

(d)

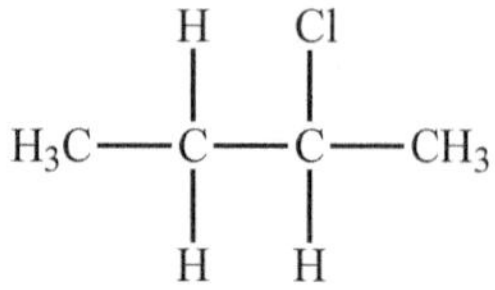

```
      H    Cl
      |    |
H3C—C—C—CH3
      |    |
      H    H
```

(e)

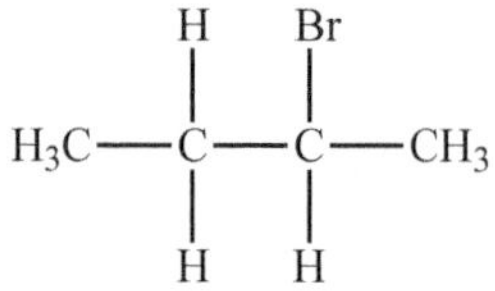

```
      H    Br
      |    |
H3C—C—C—CH3
      |    |
      H    H
```

(f)

```
      H    OH
      |    |
H3C—C—C—CH3
      |    |
      H    H
```

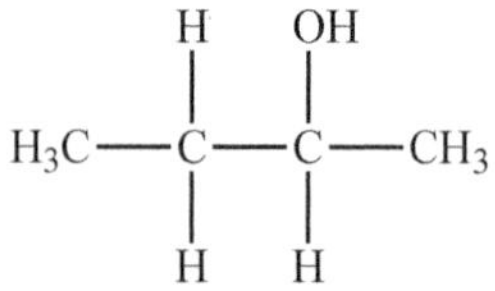

22.107 Benzene does not "add" Br_2, if it did, it would add across a double bond.

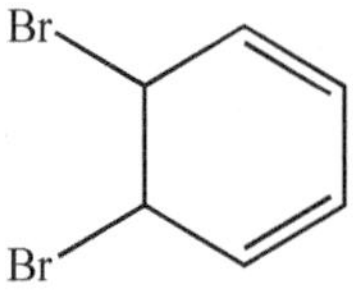

Br

Br

This does not occur because the loss of resonance energy would not be favorable. Rather, in the presence of a catalyst, it gives bromobenzene:

$C_6H_6 + Br_2 \longrightarrow C_6H_5Br + HBr$ ($FeBr_3$ catalyst)

22.109 CH_3OH IUPAC name = methanol; common name = methyl alcohol
CH_3CH_2OH IUPAC name = ethanol; common name = ethyl alcohol
$CH_3CH_2CH_2OH$ IUPAC name = 1-propanol; common name = propyl alcohol
$(CH_3)_2COH$ or as shown below

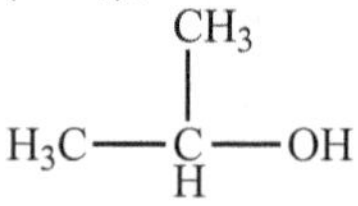

```
      CH3
      |
H3C—C—OH
      H
```

IUPAC name = 2-propanol; common name = isopropyl alcohol

22.111 $CH_3CH_2CH_2–O–CH_3$ methyl propyl ether
$CH_3CH_2–O–CH_2CH_3$ diethyl ether or just "ether"
$(CH_3)_2CH–O–CH_3$ methyl 2-propyl ether

22.113 (a)

(b)

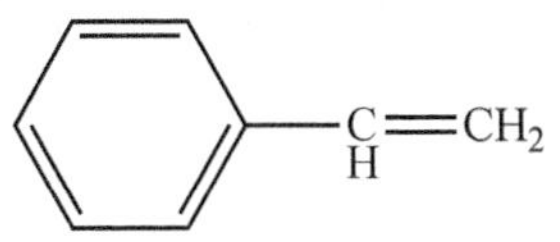

(c)

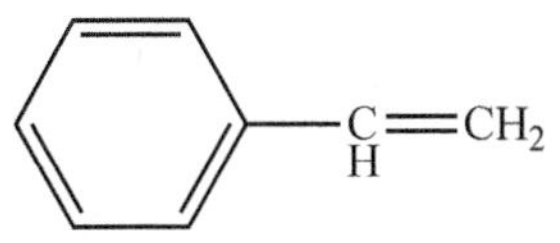

22.115 (a)

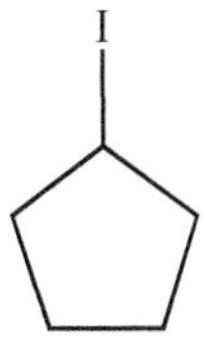

(b)

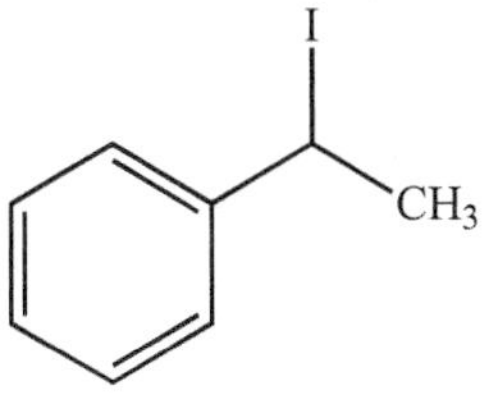

(c)

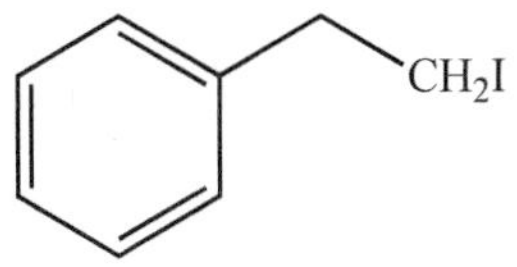

22.117 The elimination of water can result in a C=C double bond in two locations:

$CH_2{=}CHCH_2CH_3$	$CH_3CH{=}CHCH_3$
1-butene	2-butene

22.119 The aldehyde is more easily oxidized. The product is:

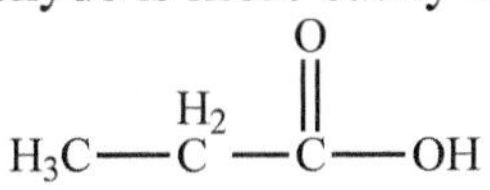

22.121 (a) $CH_3CH_2CO_2H$

(b) $CH_3CH_2CO_2H + CH_3OH$

(c) $Na^+ + CH_3CH_2CH_2CO_2^- + H_2O$

22.123 $CH_3CO_2H + CH_3CH_2NHCH_2CH_3$

22.125

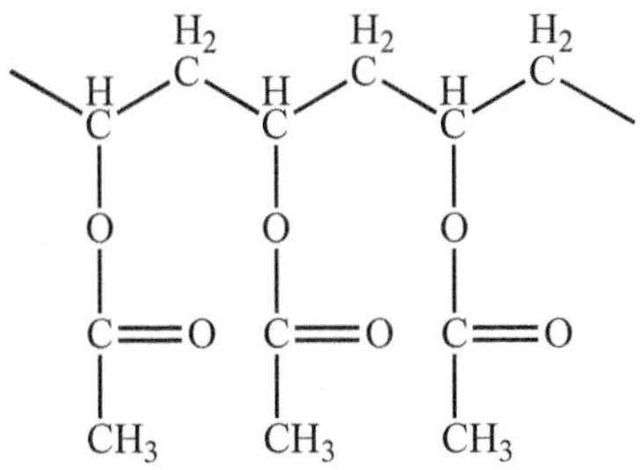

22.127

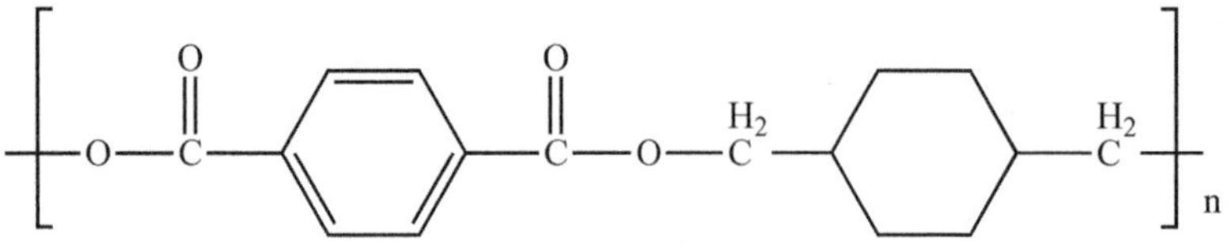

$$\left[-O-\overset{O}{\overset{\|}{C}}-C_6H_4-\overset{O}{\overset{\|}{C}}-O-\overset{H_2}{C}-C_6H_{10}-\overset{H_2}{C}- \right]_n$$

22.129

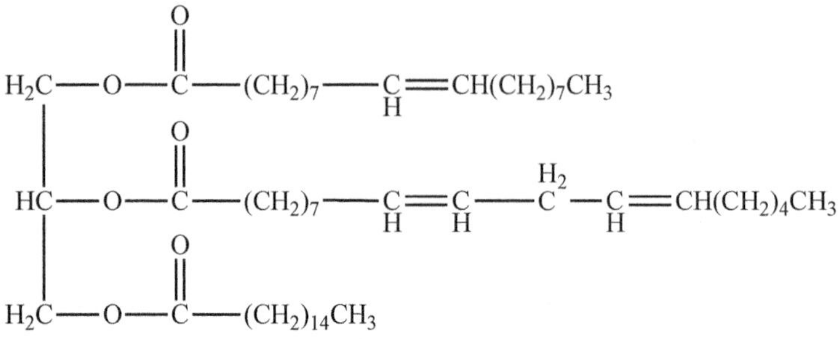

$$H_2C-O-\overset{O}{\overset{\|}{C}}-(CH_2)_7-\underset{H}{C}=CH(CH_2)_7CH_3$$
$$HC-O-\overset{O}{\overset{\|}{C}}-(CH_2)_7-\underset{H}{C}=\underset{H}{C}-\overset{H_2}{C}-\underset{H}{C}=CH(CH_2)_4CH_3$$
$$H_2C-O-\overset{O}{\overset{\|}{C}}-(CH_2)_{14}CH_3$$

oleic acid

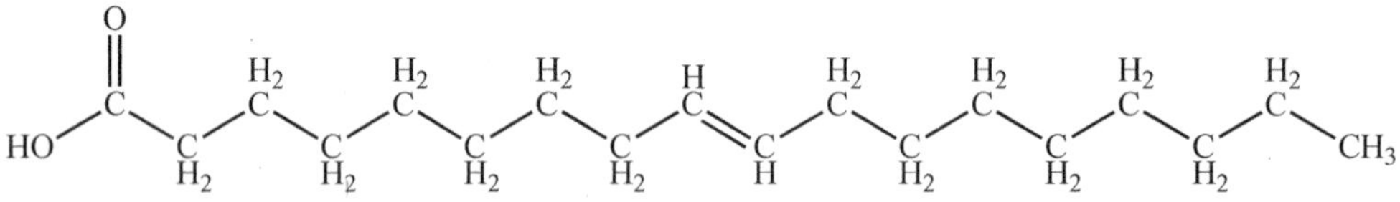

$$HO-\overset{O}{\overset{\|}{C}}-\underset{H_2}{C}-\overset{H_2}{C}-\underset{H_2}{C}-\overset{H_2}{C}-\underset{H_2}{C}-\overset{H_2}{C}-\underset{H_2}{C}-\overset{H}{C}=\underset{H}{C}-\overset{H_2}{C}-\underset{H_2}{C}-\overset{H_2}{C}-\underset{H_2}{C}-\overset{H_2}{C}-\underset{H_2}{C}-\overset{H_2}{C}-CH_3$$

myristic acid

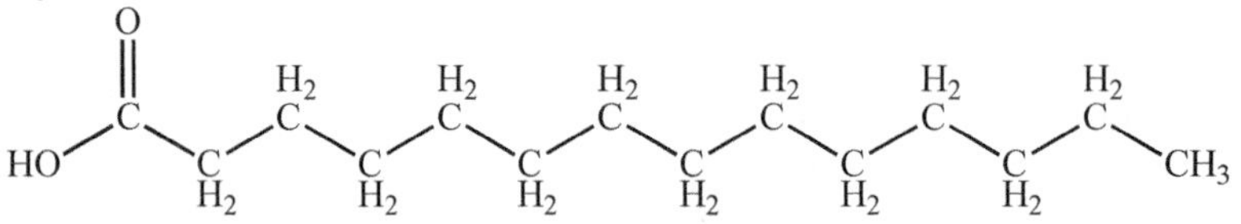

$$HO-\overset{O}{\overset{\|}{C}}-\underset{H_2}{C}-\overset{H_2}{C}-\underset{H_2}{C}-\overset{H_2}{C}-\underset{H_2}{C}-\overset{H_2}{C}-\underset{H_2}{C}-\overset{H_2}{C}-\underset{H_2}{C}-\overset{H_2}{C}-\underset{H_2}{C}-\overset{H_2}{C}-CH_3$$

22.131

$$H_2C-O-\overset{O}{\overset{\|}{C}}-(CH_2)_{16}CH_3$$
$$HC-O-\overset{O}{\overset{\|}{C}}-(CH_2)_{16}CH_3$$
$$H_2C-O-\overset{O}{\overset{\|}{C}}-(CH_2)_{14}CH_3$$

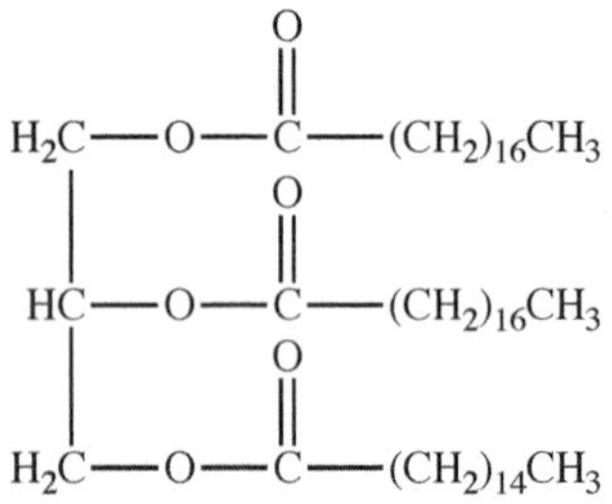

22.133 Hydrophobic sites are composed of fatty acid units. Hydrophilic sites are composed of charged units.

22.135

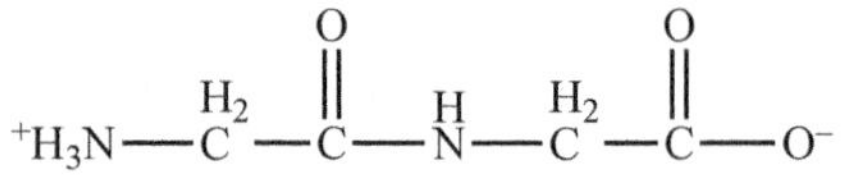

$$^{+}H_3N-\overset{H_2}{C}-\overset{O}{\overset{\|}{C}}-\overset{H}{N}-\overset{H_2}{C}-\overset{O}{\overset{\|}{C}}-O^{-}$$

22.137

$^{+}H_3N{-}CH_2{-}C(=O){-}NH{-}CH(CH_2C_6H_5){-}C(=O){-}O^{-}$

$^{+}H_3N{-}CH(CH_2C_6H_5){-}C(=O){-}NH{-}CH_2{-}C(=O){-}O^{-}$